Environmental Science and Engineering

The ultimate goal of this series is to contribute to the protection of our environment, which calls for both profound research and the ongoing development of solutions and measurements by experts in the field. Accordingly, the series promotes not only a deeper understanding of environmental processes and the evaluation of management strategies, but also design and technology aimed at improving environmental quality. Books focusing on the former are published in the subseries Environmental Science, those focusing on the latter in the subseries Environmental Engineering.

Sylvester Chibueze Izah · Matthew Chidozie Ogwu
Editors

Innovative Approaches in Environmental Health Management

Processes, Technologies, and Strategies for a Sustainable Future

Springer

Editors
Sylvester Chibueze Izah
Bayelsa Medical University
Yenagoa, Nigeria

Matthew Chidozie Ogwu
Goodnight Family Sustainable
Development Department
Appalachian State University
Boone, NC, USA

ISSN 1863-5520 ISSN 1863-5539 (electronic)
Environmental Science and Engineering
ISBN 978-3-031-81968-1 ISBN 978-3-031-81966-7 (eBook)
https://doi.org/10.1007/978-3-031-81966-7

This Springer imprint is published by the registered company Springer Nature Switzerland AG
The registered company address is: Gewerbestrasse 11, 6330 Cham, Switzerland

Preface

Welcome to *Innovative Approaches in Environmental Health Management: Processes, Technologies, and Strategies for a Sustainable Future.* This book guides the multifaceted relationship between our environment and public health, offering insights into fundamental processes and cutting-edge solutions that are transforming the field of environmental health.

In an era where environmental challenges increasingly affect human health, understanding the intricate web that connects ecological processes with well-being is crucial. Part I, *Understanding Environmental Processes and Their Health Impacts* establishes the foundation of this relationship by exploring how natural and human-influenced processes influence air, water, soil, and biodiversity—all essential to life and community resilience. From examining ecosystem dynamics to assessing pollution's impact on air quality and the importance of soil health for food security, this section brings forward the critical pathways that connect environmental health to human health. Chapters here also focus on how biodiversity loss and improper waste management contribute to broader public health challenges. In Part II, *Technological Innovations for Environmental Quality Improvement*, we focused on the transformative role of technology in enhancing environmental quality. Today's innovations, including renewable energy solutions, smart ecological monitoring, and green infrastructure, are paving the way for healthier ecosystems and communities. The chapters explore how advancements in clean energy and data-driven monitoring empower us to manage environmental health risks proactively. These tools provide better insights into environmental changes and offer sustainable ways to support urban health and improve food systems. Part III, *Strategic Management Approaches for Sustainable Environmental Health*, highlights the importance of strategic planning and community engagement in fostering environmental resilience. Effective environmental health management requires participatory approaches that encourage community involvement and ensure solutions are grounded in local knowledge and needs. This part addresses the strategic tools for evaluating health impacts, assessing risks, and implementing sustainable solutions. These chapters emphasize the importance of a collaborative approach to environmental health by integrating scientific data with community perspectives.

This book aims to provide a comprehensive look at the processes, technologies, and strategies essential for managing environmental health sustainably. Each chapter reflects a commitment to innovative approaches that respect and restore the balance between human health and ecological integrity. Whether you are a researcher, practitioner, or student, we hope this resource will inspire you to contribute to a future where environmental and human health are interwoven in sustainable harmony.

Yenagoa, Nigeria | Sylvester Chibueze Izah, Ph.D.
Boone, USA | Matthew Chidozie Ogwu, Ph.D.
2025

Contents

About the Editors

Dr. Sylvester Chibueze Izah holds a Ph.D. in Public Health, Applied Microbiology and Environmental Health, an M.Sc. in Applied Microbiology and a B.Sc. in Biological Sciences from Niger Delta University, Nigeria. He is a licensed Environmental Health Specialist in Nigeria. He brings over 10 years of experience from his work with an environmental consultancy firm specializing in soil, water, and atmospheric assessments. During the early stages of his career, Dr. Izah conducted studies on vegetation cover and biodiversity, focusing on species diversity, composition, and utilization. Presently, he serves as a lecturer at Bayelsa Medical University in Yenagoa, Nigeria, where he holds the position of Assistant Director of Academic Planning, Research, and Innovations. Dr. Izah's extensive research portfolio encompasses water, air, and soil quality, applied microbiology, biotechnology, public and environmental health, risk assessment, bioenergy, toxicology, medicinal plants, herbal medicine, and biodiversity. He has an impressive track record of over 250 peer-reviewed publications, including journal articles, book chapters, and edited books. Dr. Izah Google Scholar total citation, h-index, and i10-Index are >7740, 47, and 200, respectively. His current research interests are centered around Sustainable Human-Environmental Health Interactions, covering areas such as environmental sustainability trends, toxicology, hygiene, sanitation, food science, public environmental health assessment, medicinal plants, bioactive and natural compounds, vegetation, and wildlife. Dr. Izah has also collaborated on research projects with colleagues worldwide and actively contributes as an editorial and review board member for several esteemed journals. e-mail: sylvester.izah@bmu.edu.ng

Prof. Dr. Matthew Chidozie Ogwu is an Assistant Professor of Integrated Ecology and Sustainable Development in the Goodnight Family Sustainable Development Department at Appalachian State University, USA. He obtained his Ph.D. in Biological Sciences with a research focus in the Molecular Biology of Soil Microbial and Geographical Ecology from Seoul National University. He holds an M.Sc. (Distinction) and B.Sc. (First Class) in Plant Diversity and Conservation and Plant Biology and Biotechnology respectively from the University of Benin, Nigeria. After receiving his Ph.D., he worked as a Researcher at the Centre for Floristic Research of the Apennines, Italy where he contributed to numerous European Union-funded research including the famous NATURA 2000 project for sustainable ecosystem and biodiversity management. Before that, Dr. Ogwu was a Lecturer at the University of Benin, Nigeria, where he taught courses in plant biology and biotechnology, biometry, economic botany, cytogenetics, etc., and supervised and co-supervised honors and graduate students' theses respectively. Dr. Ogwu is an interdisciplinary academic with transdisciplinary skills pertinent to the assessment of coupled human and natural as well as socio-ecological systems and has numerous awards, research grants, and scholarships to his name. His research and teaching revolve around sustainable biodiversity and ecosystem management by assessing the realm of coupled systems for sustainable environmental development as well as addressing sustainable development issues arising from various scales of biogeographical interactions. Some of his research interests include soil microbial and geographical ecology, molecular and genomic characterization of economic plant species for breeding, conservation, and sustainable utilization as well as biodiversity issues and policies, climate change science, gut and fecal microbiota, bioinformatics, biosystems and bioeconomy, sustainable agriculture, and food production systems, ethnoscience (including traditional biotechnological practices), land use patterns, waste management, microbial diversity, environmental and social systems integrity, landscape preservation, and pollution ecology. He is spearheading some convergence works (One Health and Eco Health) in the Human Environmental and Agricultural Laboratory (HEAL lab) at Appalachian State University. Dr. Ogwu has published over 100 peer-reviewed works including several in high-impact journals and has presented his results at different international conferences. Dr. Ogwu has attended numerous professional courses and serves on the board of and as a reviewer for many peer-reviewed international journals. He continues to volunteer his time and skills to promote sustainable community development, especially in the Global South. Dr. Ogwu is fluent in several languages and loves to travel and meet new people.

Part I
Understanding Environmental Processes and Their Health Impacts

Chapter 1
Ecosystem Dynamics and Environmental Health Management: Linking Processes to Human Wellbeing

Glory Richard

Abstract The intricate relationship between ecosystem dynamics and human wellbeing emphasizes the importance of understanding environmental health. Healthy ecosystems, characterized by energy flow, nutrient cycling, biodiversity, and resilience, are essential for human survival, economic prosperity, and cultural richness. As human activities exert increasing pressure on natural systems, environmental stressors such as pollution, climate change, habitat degradation, invasive species, and overexploitation threaten the stability of ecosystems. These stressors interact in complex ways, creating challenges for traditional conservation strategies. Addressing these multifaceted challenges requires practical management approaches that enhance ecosystem resilience while considering the interconnectedness of ecological and human health outcomes. Therefore, there is a need for future research to elucidate the mechanisms driving these interactions and promote integrated strategies for conservation and sustainable development. Furthermore, engaging communities in environmental management and addressing socio-economic inequalities will foster resilience and ensure ecosystems and human populations thrive. Ultimately, protecting ecosystem health through conservation efforts, restoration initiatives, and incorporating ecosystem health into public health frameworks is crucial for sustaining the benefits of nature. Through prioritization of interdisciplinary collaboration and adaptive management, stakeholders can navigate the complexities of ecosystem dynamics, ensuring that the invaluable contributions of healthy ecosystems to our collective wellbeing endure for generations to come.

Keywords Biodiversity · Public health · Environmental health · Resilience · Ecosystem dynamics · Ecosystem services

G. Richard (✉)
School of Community Health and Policy, Morgan State University, 4530 Portage Avenue Campus Near East Cold Spring Lane, Baltimore, MD 21239, USA
e-mail: richieglo63@gmail.com

Department of Community Medicine, Faculty of Clinical Sciences, Niger Delta University, Wilberforce Island, Bayelsa State, Nigeria

S. Chibueze Izah and M. Chidozie Ogwu (eds.), *Innovative Approaches in Environmental Health Management*, Environmental Science and Engineering,
https://doi.org/10.1007/978-3-031-81966-7_1

1.1 Introduction

Ecosystem dynamics and environmental health are critical components of the broader discourse surrounding sustainability and human wellbeing. Ecosystems, defined by their intricate structures and processes, provide many services essential for human survival and quality of life. These services can be categorized into provisioning, regulating, cultural, and supporting services, each contributing uniquely to the fabric of human existence (Ma et al. 2021; Kalaba 2014). The Millennium Ecosystem Assessment (MEA) highlighted that approximately 60% of the World's ecosystem services are currently in decline, stressing the urgent need for effective management and restoration strategies (Liu et al. 2022). Understanding these dynamics is not merely an academic exercise but vital for informing policies promoting environmental health and human wellbeing.

The interdependence between ecosystem processes and human wellbeing is evident in various studies that elucidate how ecosystem services directly influence health, livelihoods, and overall quality of life (Morya and Punia 2022; Zhang et al. 2023). For instance, the degradation of ecosystems due to urbanization and industrial activities has been shown to adversely affect the availability of essential services such as clean water, air quality, and biodiversity (Morya and Punia 2022; Zhang et al. 2023; Izah and Seiyaboh 2018a, b; Izah 2018; Izah et al. 2018, 2023a, b). Such impacts are particularly pronounced in urban areas where rapid development often leads to losing green spaces and natural habitats, diminishing the ecological functions supporting human health (Abrego et al. 2020). Recognizing these relationships is crucial for developing integrated approaches that address environmental conservation and human health outcomes.

Moreover, the role of ecosystems in regulating climate and water quality cannot be overstated. Coastal and freshwater ecosystems, for example, play a significant role in mitigating climate change effects and maintaining hydrological cycles (Apostolaki et al. 2020). These ecosystem services are vital for ecological balance, sustaining agricultural productivity, and ensuring food security (Jiang et al. 2024). As highlighted by various authors, the economic valuation of these ecosystem services is essential for fostering a deeper appreciation of their contributions to human wellbeing and guiding sustainable management practices (Wood et al. 2018). Integrating ecosystem services into economic frameworks can facilitate better decision-making processes prioritizing environmental health and human development.

The importance of understanding ecosystem dynamics extends to the socio-cultural dimensions of human life. Ecosystems are not merely resources to be exploited but integral to cultural identity and community wellbeing (Rani and Sangwan 2022; Bunch 2016). The concept of biocultural approaches emphasizes the interconnectedness of human societies and natural environments, advocating for policies that recognize and incorporate cultural values in ecosystem management (Sterling et al. 2017). This perspective is particularly relevant in indigenous and local communities, where traditional ecological knowledge is crucial in sustaining biodiversity and ecosystem health (Bunch 2016; Sterling et al. 2017). By fostering a holistic understanding of these relationships, policymakers can develop strategies that enhance ecological integrity and social equity.

This chapter explores the fundamental components of ecosystem dynamics, including energy flow, nutrient cycling, biodiversity, and ecosystem resilience. It examines the impact of environmental stressors, such as pollution and climate change, on ecosystem health and human dependence on ecosystem services like food, water, and climate regulation. The chapter also highlights strategies for protecting ecosystems through conservation, sustainable development, and integrating ecosystem health into public health frameworks.

1.2 Critical Components of Ecosystem Dynamics

Ecosystem dynamics involve a complex interplay of various components that contribute to the functioning and stability of ecological systems (Table 1.1). Understanding these dynamics requires comprehensively examining energy flow and nutrient cycling, biodiversity and species interactions, and ecosystem resilience and adaptability to environmental changes. Each component plays a critical role in maintaining the balance and health of ecosystems, and their interconnections are vital for sustaining ecological integrity.

Energy flow within ecosystems is primarily governed by energy transfer through trophic levels, where producers convert solar energy into biomass, which is then consumed by herbivores and predators. This energy flow is not linear; various factors influence it, including species composition, biomass differences, and food web structures. For instance, in integrated aquaculture systems, different species' energy efficiency (EE) values can vary significantly due to harvesting pressures and species interactions, which ultimately affect the overall energy flow within the ecosystem (Dong et al. 2021). Furthermore, the role of detritivores in nutrient cycling is crucial, as they facilitate the breakdown of organic matter, thus contributing to energy transfer within food webs (Fukumori et al. 2016). The complexity of these interactions highlights the necessity for a nuanced understanding of how energy flows through ecosystems, particularly in the context of human impacts and climate change (Sales et al. 2020).

Nutrient cycling is another fundamental aspect of ecosystem dynamics, involving the movement of nutrients through biotic and abiotic components. Nutrient inputs can occur through atmospheric deposition, nitrogen fixation, and weathering of rocks, while losses may result from leaching and disturbances such as fires (Gravel et al. 2010). The cycling of nutrients is essential for maintaining soil fertility and supporting plant growth, which sustains herbivores and higher trophic levels. Studies have shown that the diversity of detritivores can significantly enhance the contribution of detritus to predators' diets, thereby influencing the overall nutrient dynamics within aquatic ecosystems (Fukumori et al. 2016). Moreover, the spatial distribution of nutrients across ecosystems can lead to variations in productivity and biodiversity,

Table 1.1 Key components of ecosystem dynamics

Component	Ecological processes involved	Benefits and examples
Energy flow	Movement of energy through an ecosystem from sunlight to producers and consumers via food chains and webs	Powers all living organisms. Example: photosynthesis in plants captures and transfers solar energy
Nutrient cycling	The essential nutrients, like carbon and nitrogen, are recycled through biogeochemical cycles that sustain ecosystems	Maintains soil fertility and ecosystem productivity. Example: the nitrogen cycle is vital for plant growth
Biodiversity	A variety of life within an ecosystem, involving species richness and genetic diversity, is essential for ecosystem function	Enhances resilience and stability. Example: coral reefs support a wide range of marine species
Species interactions	Relationships among species, such as predation, competition, and symbiosis, shape ecosystem dynamics	Promotes biodiversity and balance. Example: bee pollination supports plant reproduction
Ecosystem resilience	The ability of an ecosystem to recover from disturbances is influenced by biodiversity and nutrient availability	Aids recovery from environmental impacts. Example: forests regrowing after wildfires
Adaptability to changes	Ecosystems' capacity to adjust to environmental changes, including species migration and community shifts	Ensures survival amid changes. Example: bird migration patterns shift in response to climate variations

underscoring the interconnectedness of nutrient cycling and energy flow (Ellis-Soto et al. 2021).

Biodiversity is a critical component of ecosystem dynamics, as it influences species interactions and the resilience of ecosystems to disturbances. Species' functional roles within ecosystems extend beyond mere biomass production; they include interactions such as predation, competition, and mutualism, which collectively shape community structure and ecosystem functioning (Dehling and Stouffer 2018). For example, diverse plant species can enhance decomposition rates and herbivory by affecting aboveground arthropod communities (Ebeling et al. 2014). Additionally, the extinction of ecological interactions, rather than just species loss, poses a significant threat to biodiversity and ecosystem stability (Valiente-Banuet et al. 2015). The intricate web of interactions among species contributes to the resilience of ecosystems, enabling them to withstand and adapt to environmental changes. Ecosystem resilience refers to the capacity of an ecosystem to absorb disturbances while maintaining its essential functions and processes. Resilience is influenced by biodiversity, as diverse ecosystems exhibit more excellent stability and adaptability to changes such as climate fluctuations and anthropogenic pressures (Zamora-Gutiérrez et al. 2021). For instance, removing a single pollinator species can disrupt the entire pollination network, leading to cascading effects on plant reproduction and community composition (Zamora-Gutiérrez et al. 2021). Furthermore, meta-ecosystems emphasize the importance of spatial heterogeneity and connectivity in enhancing resilience, as interconnected ecosystems can share resources and mitigate the impacts of local disturbances (Massol et al. 2011). This interconnectedness is crucial for maintaining energy and nutrient flows across landscapes, supporting overall ecosystem health.

Adaptability to environmental changes is another crucial aspect of ecosystem dynamics, as ecosystems must continuously adjust to shifting conditions. The ability of ecosystems to adapt is often linked to their structural complexity and the diversity of species present. For example, ecosystems with a high degree of spatial heterogeneity can better accommodate changes in environmental conditions, as different species may respond variably to stressors (Gravel et al. 2016). Additionally, keystone species can play an essential role in maintaining ecosystem functions during change, as their interactions can stabilize food webs and nutrient cycling processes (Klemmer et al. 2020). Understanding the mechanisms underlying adaptability is essential for effective conservation and management strategies to enhance ecosystem resilience.

The interplay between energy flow, nutrient cycling, biodiversity, and ecosystem resilience underscores the complexity of ecosystem dynamics (Fig. 1.1). These components are not isolated; instead, they are intricately linked, with changes in one aspect often reverberating throughout the system. For instance, alterations in energy flow due to species invasions can disrupt nutrient cycling and lead to declines in biodiversity, ultimately compromising ecosystem resilience (Abernethy et al. 2014). Therefore, a holistic approach that considers these interconnections is essential for understanding and managing ecosystems in the face of ongoing environmental changes.

1.3 Environmental Stressors and Ecosystem Health

Environmental stressors, including pollution, climate change, and habitat degradation, profoundly impact ecosystem health (Izah et al. 2023c). Pollution, particularly from industrial and agricultural sources, introduces harmful substances into ecosystems, leading to a decline in biodiversity and the disruption of ecological processes (Aigberua et al. 2023, 2021). For instance, nutrient pollution from fertilizers can lead to eutrophication, which depletes oxygen in water bodies and results in dead zones where aquatic life cannot survive (Staudt et al. 2013). This phenomenon directly impacts the species and alters food webs and nutrient cycling, leading to broader ecological consequences (Gallardo et al. 2016). Furthermore, pollutants such as heavy metals and plastics can accumulate in the food chain, posing risks to wildlife and human health (Staudt et al. 2013; Lopez et al. 2022). The synergistic effects of pollution and climate change exacerbate these issues, as stressed ecosystems are less resilient to additional environmental pressures (Staudt et al. 2013). Table 1.2 provides an overview of environmental stressors and ecosystem health.

Climate change is another critical stressor that influences ecosystem health. Rising temperatures, altered precipitation patterns, and increased frequency of extreme

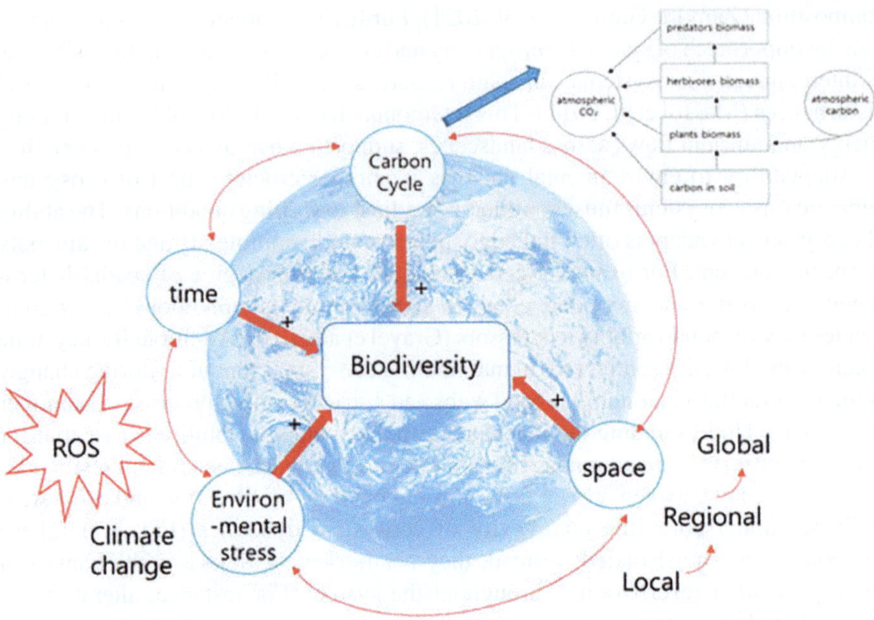

Fig. 1.1 Dynamic interplay between energy flow, nutrient cycling, and biodiversity for building ecosystem resilience. *Source* Kim et al. (2024). https://creativecommons.org/licenses/by/4.0/

weather events disrupt the delicate balance of ecosystems (Staudt et al. 2013; Pecl et al. 2017). For example, shifts in temperature can lead to mismatches in the timing of biological events, such as flowering and pollination, which can have cascading effects on food availability for various species (Pecl et al. 2017). Additionally, climate change can facilitate the spread of invasive species, which often thrive in altered conditions, further threatening native biodiversity (Stephens et al. 2019). The interaction between climate change and other stressors, such as habitat degradation, can lead to rapid and irreversible changes in ecosystem structure and function (Staudt et al. 2013; Pecl et al. 2017). Habitat degradation, often driven by human activities such as deforestation, urbanization, and agriculture, poses significant threats to ecosystem health. The loss of habitat reduces the available space for species to thrive and fragments populations, making it difficult for them to reproduce and maintain genetic diversity (Staudt et al. 2013; Lopez et al. 2022). This fragmentation can lead to local extinctions and decreased overall biodiversity, which is crucial for ecosystem resilience (Staudt et al. 2013; Pecl et al. 2017). Moreover, habitat degradation often alters ecosystems' physical and chemical properties, affecting nutrient cycling and energy flow (Staudt et al. 2013; Gallardo et al. 2016). For instance, the destruction of wetlands can lead to increased flooding and loss of water quality, further stressing the organisms that depend on these habitats (Staudt et al. 2013; Gallardo et al. 2016).

Invasive species represent a significant challenge to ecosystem stability, as they can outcompete native species for resources, disrupt food webs, and alter habitat

Table 1.2 Environmental stressors and ecosystem health

Environmental stressor	Description	Impacts on ecosystem health	Examples	Mitigation strategies	Refs.
Pollution	Contamination of air, water, or soil by harmful substances	Reduces biodiversity, degrades habitats, and harms species	Industrial runoff in rivers and airborne pollutants	Implement stricter regulations and promote green technologies	Münzel et al. (2023)
Climate change	Long-term changes in temperature and weather patterns due to human activities	Altered species distribution increases extreme weather events and leads to habitat loss	Coral bleaching and shifts in bird migratory patterns	Reduce greenhouse gas emissions and promote renewable energy	Shivanna (2022)
Deforestation	Large-scale removal of forests for agriculture or development	Causes habitat loss, soil erosion, and decreased carbon storage	Amazon rainforest clearing and urban expansion	Support reforestation and sustainable land management	Leite-Filho et al. (2021)
Invasive species	Non-native species that disrupt local ecosystems	Decreases native biodiversity and alters ecosystem functions	Zebra mussels in the Great Lakes and Burmese pythons in Florida	Implement early detection and public awareness campaigns	Simberloff (2015)
Overexploitation	Unsustainable use of natural resources	This leads to a decline in species populations and disrupted food webs	Overfishing of cod stocks and unsustainable logging	Enforce quotas and support conservation initiatives	Durant et al. (2021)
Habitat fragmentation	Division of habitats into smaller, isolated patches	Reduces genetic diversity and hinders species movement	Urban development creates isolated parks and roads, dividing habitats	Create wildlife corridors and minimize fragmentation in planning	Haddad et al. (2015)
Soil degradation	Decline in soil quality from erosion, compaction, or contamination	Lowers agricultural productivity and disrupts soil organisms	Desertification and contaminated farmland	Adopt sustainable agriculture and soil restoration practices	Gregory et al. (2015)

(continued)

Table 1.2 (continued)

Environmental stressor	Description	Impacts on ecosystem health	Examples	Mitigation strategies	Refs.
Water scarcity	Insufficient freshwater resources due to overuse or climate change	Reduces aquatic biodiversity and increases resource competition	Droughts affecting rivers and lakes	Promote water conservation and invest in management technologies	Ingrao et al. (2023)
Natural disasters	Events like hurricanes, floods, and wildfires that impact ecosystems	Causes immediate habitat destruction and long-term recovery challenges	Forest fires in California and hurricanes affecting coasts	Implement disaster preparedness plans and restore habitats post-disaster	Shivanna (2022)
Agricultural runoff	Nutrient and pesticide runoff from farms into water bodies	This leads to eutrophication and harms aquatic life	Algal blooms in lakes and contaminated drinking water sources	Implement best management practices and reduce fertilizer use	Bashir et al. (2020)

structure (Raghunathan et al. 2021; Russell et al. 2014). Introducing non-native species often leads to declining native biodiversity, as these invaders can exploit resources more efficiently or lack natural predators in their new environments (Raghunathan et al. 2021; Russell et al. 2014). For example, invasive aquatic species such as the zebra mussel have been shown to dramatically alter nutrient cycling and energy flow in freshwater ecosystems, leading to shifts in community composition and function (Higgins and Zanden 2010; Gallardo et al. 2016). The impacts of invasive species are not limited to direct competition; they can also have indirect effects by altering the physical environment, such as changing sediment dynamics or water clarity (Gallardo et al. 2016; Russell et al. 2014). The effects of invasive species can be exacerbated by overexploitation of native species, further destabilizing ecosystems. Overfishing, for instance, can lead to the collapse of fish populations, disrupting the entire marine food web and affecting species that rely on these fish for survival (Staudt et al. 2013; Lopez et al. 2022).

Similarly, the overharvesting of plants can lead to soil erosion and habitat loss for various organisms (Staudt et al. 2013; Lopez et al. 2022). The combination of invasive species and overexploitation creates a feedback loop that can lead to significant ecological degradation as native species struggle to compete and recover from the compounded pressures (Staudt et al. 2013; Lopez et al. 2022). The interaction between invasive species and environmental stressors such as climate change and pollution can lead to complex ecological dynamics. For example, invasive species may thrive in polluted environments where native species are already stressed, allowing them to establish and spread more rapidly (Gutiérrez et al. 2014; Lopez et al. 2022). This can result in a scenario where the combined effects of multiple stressors lead to more significant ecological impacts than any single stressor alone (Lopez et al. 2022). Understanding these interactions is crucial for effective management and conservation strategies, as it highlights the need for integrated approaches that consider the multifaceted nature of ecosystem health (Gutiérrez et al. 2014; Lopez et al. 2022).

Ecosystem engineers, a subset of invasive species, can particularly affect ecosystem structure and function. These species modify their environment in ways that can facilitate or hinder the establishment of other organisms (Emery-Butcher et al. 2020; Rilov et al. 2012). For instance, invasive plants may alter soil chemistry and nutrient availability, impacting the growth of native flora and fauna (Chen et al. 2015; Kacheche and Mzuza 2021). The presence of invasive ecosystem engineers can lead to shifts in community composition, as they may create conditions that favor their proliferation while disadvantaging native species (Emery-Butcher et al. 2020; Rilov et al. 2012). These dynamics underscore the importance of recognizing invasive species' role as competitors and modifiers of their environments. The impacts of invasive species on ecosystem health are not uniform; they can vary based on the specific characteristics of the invaders and the ecosystems they invade. For example, some invasive species may benefit certain ecosystem functions, such as increasing nutrient availability or providing habitat for other species (Emery-Butcher et al. 2020; Gutiérrez et al. 2014). However, these potential benefits are often outweighed by the negative consequences of reduced biodiversity and altered ecosystem processes

(Emery-Butcher et al. 2020; Gutiérrez et al. 2014). The complexity of these interactions necessitates a nuanced understanding of the ecological roles of invasive species and their potential impacts on ecosystem health.

1.4 Human Dependence on Ecosystem Services

Human dependence on ecosystem services is a critical aspect of sustainable development, as these services underpin the wellbeing of societies globally. Ecosystem services can be categorized into four main types: provisioning, regulating, cultural, and supporting services (Table 1.3). Provisioning services include essential resources such as food, water, and raw materials, vital for human survival and economic activities. Regulating services involve the natural processes that maintain environmental balance, such as climate regulation and water purification. Cultural services provide non-material benefits, including recreation and spiritual enrichment, while supporting services are foundational processes that sustain ecosystems, such as nutrient cycling and habitat provision for biodiversity.

The provisioning services provided by ecosystems are indispensable for human sustenance. Agriculture, for instance, relies heavily on ecosystem services such as pollination, soil fertility, and water supply. Studies have shown that biodiversity is crucial in enhancing crop production through pollination and pest control, which are vital for food security. The interdependence of biodiversity and provisioning services highlights the need for sustainable agricultural practices that protect and enhance ecosystem functions. Moreover, biodiversity loss can lead to decreased agricultural productivity, as diverse ecosystems are more resilient and capable of withstanding environmental changes. This relationship underscores the importance of integrating biodiversity conservation into agricultural policies to ensure long-term food security. Regulating services are equally important, as they maintain the environmental conditions necessary for human life. Ecosystems regulate climate by

Table 1.3 Human dependence on ecosystem services

Ecosystem service	Examples	Benefits
Provisioning services	Food (crops, livestock), freshwater, raw materials (timber, fibers)	Supports nutrition and health; provides economic resources
Regulating services	Climate regulation (carbon storage), water purification, pollination	Mitigates climate change improves water quality, and supports agriculture
Cultural services	Recreation, tourism, spiritual and historical significance, aesthetic enjoyment	Enhances mental health and fosters cultural identity and community ties
Supporting services	Nutrient cycling, soil formation, habitat provision	It is essential for food production and maintains biodiversity

sequestering carbon, thus mitigating climate change impacts. For example, forests play a significant role in carbon storage and climate regulation, which is critical for maintaining global temperatures within safe limits.

Additionally, ecosystems contribute to water purification, which is essential for clean drinking water. Wetlands and riparian zones act as natural filters, removing pollutants from water before they reach human consumption. The degradation of these regulating services due to human activities, such as deforestation and urbanization, poses significant risks to human health and environmental stability. While often overlooked, cultural services provide essential benefits that enhance human wellbeing. These services involve recreational opportunities, aesthetic enjoyment, and spiritual connections to nature. Urban green spaces, for instance, improve air quality and offer recreational areas that promote physical and mental health. The cultural value of ecosystems is increasingly recognized in urban planning, where integrating nature into city landscapes can enhance community wellbeing and foster a sense of belonging. Furthermore, the spiritual significance of natural landscapes can influence conservation efforts, as communities often rally to protect areas that hold cultural importance. This highlights the need for policies recognizing and incorporating cultural values in ecosystem management.

Though less visible, supporting services are fundamental to the functioning of ecosystems and the provision of other services. These include soil formation, nutrient cycling, and habitat provision for various species. The health of ecosystems is directly linked to the diversity of species they support, as diverse ecosystems are generally more productive and resilient (Abdelkhalek et al. 2023). The loss of biodiversity can disrupt these supporting services, leading to diminished capacity for ecosystems to provide provisioning and regulating services. Therefore, conservation efforts must prioritize biodiversity protection to ensure the continued provision of ecosystem services vital for human survival. Trade-offs between different ecosystem services often arise due to competing human interests. For instance, increasing agricultural production (a provisioning service) can lead to the degradation of regulating and cultural services, such as water quality and biodiversity. Understanding these trade-offs is essential for effective land management and policy-making. Decision-makers must balance the need for food production with the preservation of ecosystem health to achieve sustainable outcomes. This requires a comprehensive approach that considers the interconnectedness of various ecosystem services and the potential long-term impacts of management decisions.

The integration of ecosystem services into policy frameworks is crucial for promoting sustainable development. Policymakers need to recognize the value of ecosystem services in economic planning and environmental management. Governments can better allocate resources and prioritize conservation efforts by incorporating ecosystem service assessments into decision-making processes. For example, protected area networks can enhance the provision of ecosystem services, such as groundwater recharge and biodiversity conservation, which are essential for maintaining ecological balance. This approach benefits the environment and supports human wellbeing by ensuring the availability of vital resources.

Public awareness and engagement are also critical components in fostering a culture of sustainability. Educating communities about the importance of ecosystem services can enhance support for conservation initiatives and encourage sustainable practices. Community involvement in ecosystem management can lead to more effective conservation outcomes as local knowledge and values are integrated into decision-making processes. Furthermore, recognizing the intrinsic value of nature can inspire individuals to advocate for policies that protect ecosystems and the services they provide. This collective effort is essential for addressing the challenges posed by environmental degradation and climate change.

1.5 Linking Ecosystem Health to Human Wellbeing

The intricate relationship between ecosystem health and human wellbeing has garnered increasing attention in recent years, particularly as environmental changes escalate (Table 1.4). Ecosystems provide essential services that underpin human health, livelihoods, and community stability. These services include clean air and water, food production, and disease regulation, all compromised by environmental degradation. For instance, the degradation of natural habitats can lead to increased exposure to pathogens and pollutants, exacerbating local health issues (Singh 2010). Furthermore, as ecosystems deteriorate, the economic stability of communities that rely on these resources for their livelihoods is threatened, leading to a cycle of poverty and health disparities (Givens et al. 2019).

Environmental changes, such as climate change, habitat destruction, and pollution, significantly impact human health. For example, rising temperatures and altered precipitation patterns can influence the distribution of vector-borne diseases, such as malaria and dengue fever, thereby increasing the risk of outbreaks in previously unaffected areas (Singh 2010). Additionally, the loss of biodiversity can disrupt the ecological balance, leading to the proliferation of disease vectors and pathogens (Frishkoff et al. 2018). The interconnectedness of these factors highlights the necessity for a holistic approach to understanding how environmental changes affect human health and wellbeing (Orr et al. 2020). Studies have illustrated the profound impacts of ecosystem degradation on public health and economic stability. In the Greater Mara Region of Kenya, for instance, overexploitation of natural resources has led to significant declines in wildlife populations, which are crucial for local tourism and subsistence livelihoods (Ongare et al. 2013). This decline threatens the economic stability of communities that depend on tourism and affects food security and health outcomes, as local populations may struggle to access nutritious food sources (Fig. 1.2; Givens et al. 2019). Similarly, in peatland ecosystems, changes in plant community composition due to climate change have been shown to alter methane dynamics, which can exacerbate climate change and its associated health impacts (Robroek et al. 2015; Dieleman et al. 2015).

The role of community engagement in addressing environmental changes is critical. Research indicates that communities actively participating in natural resource

Table 1.4 Linking ecosystem health to human wellbeing

Environmental change	Impacts on human health	Impacts on livelihoods	Impacts on communities	Refs.
Climate change	Increases respiratory diseases and heat-related illnesses	It affects agriculture, leading to food insecurity and income loss	Disrupts community cohesion and increases migration pressures	Myers et al. (2022)
Biodiversity loss	Reduces access to medicinal resources and increases disease risk	Limits opportunities in agriculture, fishing, and tourism	Weakens cultural identity and reduces recreational opportunities	Ogwu et al.
Pollution	Causes health problems like asthma, cancer, and neurological disorders	Impairs local economies reliant on clean resources (e.g., fisheries)	This leads to community discontent and increased healthcare costs	Rentschler and Leonova (2023)
Deforestation	It affects air quality and contributes to climate change-related health issues	Disrupts livelihoods dependent on forest resources and ecotourism	Diminishes cultural practices linked to forest environments	Wigand et al. (2022)
Water scarcity	This leads to waterborne diseases and malnutrition	Threatens agriculture and livelihoods reliant on water access	Creates conflicts over water resources and affects social stability	Ingrao et al.

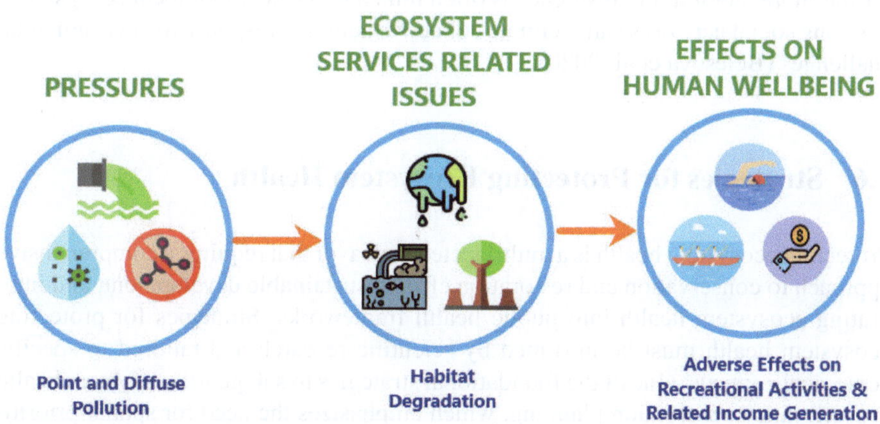

Fig. 1.2 Benefits of ecosystem services on human health. *Source* Akinsete et al. (2019). https://creativecommons.org/licenses/by/4.0/

management are more likely to adopt sustainable practices that benefit the environment and their health (Ongare et al. 2013). For instance, community-based initiatives in the Greater Mara Region have demonstrated that when local populations are involved in decision-making, there is a greater likelihood of successful conservation outcomes, supporting public health and economic stability (Ongare et al. 2013). This participatory approach fosters a sense of ownership and responsibility towards environmental stewardship, ultimately leading to healthier ecosystems and communities. Moreover, the effects of environmental changes are often compounded by socioeconomic factors, leading to "ecologically unequal exchange." This concept refers to marginalized communities' disproportionate burden of environmental degradation, which can exacerbate health disparities (Givens et al. 2019). For example, communities with limited resources may lack access to healthcare and clean water, making them more vulnerable to the health impacts of environmental changes (Givens et al. 2019). Multiple ecological stressors further complicate the interplay between ecosystem health and human wellbeing. Research has shown that habitat loss, climate change, and pollution often interact in complex ways, leading to unexpected consequences for both ecosystems and human health (Orr et al. 2020). For instance, increased nutrient loading and habitat alteration can lead to harmful algal blooms, which pose significant risks to water quality and public health (Orr et al. 2020). Understanding these interactions is crucial for developing effective strategies to mitigate the impacts of environmental changes on human health. In addition to direct health impacts, environmental changes can also influence mental health and community cohesion. The loss of natural spaces and biodiversity can lead to increased stress and anxiety among community members and a sense of loss and disconnection from the environment (Belesova et al. 2020). This is particularly evident in urban areas where green spaces are diminished, highlighting the importance of preserving and restoring ecosystems for the mental wellbeing of communities (Belesova et al. 2020). Furthermore, community resilience is often linked to the health of local ecosystems, as strong social networks can facilitate collective action in response to environmental challenges (Belesova et al. 2020).

1.6 Strategies for Protecting Ecosystem Health

Protecting ecosystem health is a multifaceted endeavor that requires a comprehensive approach to conservation and restoration efforts, sustainable development, and integrating ecosystem health into public health frameworks. Strategies for protecting ecosystem health must be informed by scientific research and tailored to specific ecological contexts. One of the foundational strategies to safeguard ecosystem health is systematic conservation planning, which emphasizes the need for spatial prioritization in conservation efforts. Kukkala and Moilanen (2013) highlighted that effective conservation management must incorporate biodiversity objectives, ensuring that habitat reserves are sufficient in number and size and appropriately distributed across landscapes to maintain biological diversity. This approach is critical in the face of climate change, as species and ecosystems are increasingly vulnerable to shifting environmental conditions. Schloss et al. (2011) argued that conservation strategies must adapt to the dynamic nature of ecosystems, recognizing that areas that protect current biodiversity may not suffice for future conservation needs. Thus, a proactive approach to anticipating ecological changes is essential for long-term ecosystem health. Also, community participation is another vital aspect of successful conservation and restoration efforts. Sacandé and Berrahmouni (2016) emphasized the importance of involving local communities in selecting species for restoration projects, particularly in regions like the Sahel, where ecological conditions are rapidly changing. Engaging communities fosters a sense of ownership and enhances the environmental relevance of restoration efforts by ensuring that local ecological knowledge is integrated into decision-making processes. This participatory approach is echoed in the work of Guetz et al. (2022), who advocate for prioritizing dam removals in the western United States to restore riverine ecosystems, emphasizing the need for community involvement in identifying restoration priorities.

Furthermore, integrating ecosystem services frameworks into conservation strategies can enhance the effectiveness of restoration efforts. Ford et al. (2015) discussed how recognizing the interdependence of human and ecosystem health through ecosystem services can facilitate the development of flexible frameworks that accommodate both positive and negative impacts of ecosystems on human wellbeing. This holistic perspective is crucial for addressing the complex challenges of environmental degradation and public health concerns. Policymakers can better articulate the benefits of preserving natural systems by framing conservation efforts within the context of ecosystem services, thereby garnering broader support for conservation initiatives.

Sustainable development policies must also align with conservation goals to ensure the long-term health of ecosystems. The Convention on Biological Diversity's Aichi targets underscores the need for nations to develop recovery programs for threatened species and restore degraded ecosystems (Fernández and Morales 2016). However, as highlighted by Zhang et al. (2018), the gap between policy intentions and actual implementation remains a significant barrier to achieving these targets. Effective policy development requires a commitment to integrating ecological considerations into economic planning and urban development, as demonstrated by White et al., who advocate for cross-sectoral policies that simultaneously address public health and ecosystem health (White et al. 2020).

Integrating ecosystem health into public health frameworks is essential for recognizing the interconnectedness of human and environmental wellbeing. Keune et al. (2013) argued that ecosystem approaches to health must consider the intricate relationships between biodiversity, public health, and urbanization. By acknowledging that human health is intrinsically linked to the health of ecosystems, public health initiatives can be designed to promote both ecological sustainability and community wellbeing. This integration is particularly relevant in urban contexts, where environmental degradation can exacerbate health disparities and reduce the quality of life for residents (Su et al. 2013).

The restoration of coastal wetlands serves as a pertinent example of how conservation efforts can be effectively implemented to protect ecosystem health. Moore's review highlights the historical reliance on top-down management approaches in coastal restoration, often limiting local community engagement. However, recent trends indicate a shift towards more inclusive practices that recognize the value of local knowledge and participation in restoration efforts. Renzi et al. (2019) further emphasized the importance of harnessing positive species interactions to enhance coastal wetland restoration, demonstrating that collaborative approaches can yield significant ecological benefits.

In addition to community engagement, prioritizing restoration sites based on ecological criteria is crucial for maximizing the effectiveness of conservation efforts. The spatial multi-criteria decision analysis approach discussed illustrated how prioritizing sites based on biodiversity value can lead to more targeted and impactful restoration initiatives. This method systematically evaluates potential restoration sites, ensuring that efforts are focused on areas with the highest ecological significance. Moreover, the role of dispersal in restoration success cannot be overlooked. Brederveld et al. (2011) highlighted that dispersal limitations can significantly affect the colonization of restored habitats by plants and macroinvertebrates, underscoring the need for strategic planning in restoration projects. By considering the dispersal abilities of target species, restoration practitioners can enhance the likelihood of successfully re-establishing biodiversity in restored areas.

Integrating genetic diversity considerations into restoration efforts is also vital for ensuring the resilience of restored ecosystems. Ort et al. (2014) demonstrated the importance of conserving genetic diversity in eelgrass restoration projects, advocating for incorporating genetic considerations into aquatic plant restoration programs. This focus on genetic diversity is essential for maintaining the adaptive capacity of species in the face of environmental change. As addressing climate change intensifies, the need for ambitious restoration initiatives becomes increasingly apparent. Suding (2011) emphasized that restoration ecology must evaluate its effectiveness in informing management efforts, particularly in global challenges such as climate change and biodiversity loss. The Bonn Challenge, which aims to restore 350 million hectares of degraded landscapes by 2030, exemplifies the international commitment to large-scale restoration efforts (Beirne et al. 2021). Such initiatives require coordinated action across multiple sectors and disciplines to achieve meaningful outcomes.

1.7 Conclusion

Ecosystem dynamics are intricately linked to human wellbeing, highlighting the profound connections between the health of the natural environments and the quality of life for people. Healthy ecosystems provide essential services such as clean air and water, fertile soil, and biodiversity, vital for human survival, economic prosperity, and cultural identity. Critical components like energy flow, nutrient cycling, biodiversity, and resilience influence the functioning of these ecosystems. As these components interact, they contribute to ecosystems' overall stability and health. However, increasing environmental stressors threaten these natural systems, including pollution, climate change, habitat degradation, invasive species, and overexploitation. The degradation of ecosystems jeopardizes their ability to provide crucial services. It has significant implications for public health and economic stability, underscoring the need for integrated approaches considering ecological and human health outcomes. Collaborative action among stakeholders is essential to ensure ecosystem sustainability for future generations. Engaging communities in environmental management, addressing socio-economic inequalities, and fostering resilience are critical steps towards preserving biodiversity and maintaining ecosystems' essential services. Future research and policy initiatives must focus on understanding the complex interactions within ecosystems and the mechanisms that enhance their resilience in the face of global change. This calls for integrating ecosystem health into public health frameworks and sustainable development policies, prioritizing community engagement, and utilizing ecosystem services frameworks. By adopting evidence-based strategies for conservation and restoration, we can enhance the resilience of ecosystems and promote human wellbeing. Ultimately, interdisciplinary collaboration and adaptive management will be vital in addressing our ecosystems' multifaceted challenges today, ensuring that nature and humanity can thrive together in a changing world.

References

Abernethy EF, Turner KL, Beasley JC, DeVault TL, Pitt WC, Rhodes O (2014) Impacts of invasive species on ecosystem energy flow on the Big Island of Hawai'i: excuse me, but are you going to eat that cane toad? In: Proceedings of the vertebrate pest conference, vol 26, No 26. https://doi.org/10.5070/v426110551

Abdelkhalek ST, Moussa MA, Gomaa SI, Qiu CL, Wang MQ (2023) Agrochemicals: safety evaluation and characterization for humans and biodiversity. In: One health implications of agrochemicals and their sustainable alternatives. Springer Nature Singapore, Singapore, pp 3–51. https://doi.org/10.1007/978-981-99-3439-3_1

Abrego N, Crosier B, Somervuo P, Ivanova N, Abrahamyan A, Abdi A, Hämäläinen K, Junninen K, Maunula M, Purhonen J et al (2020) Fungal communities decline with urbanization—more in air than in soil. ISME J 14(11):2806–2815. https://doi.org/10.1038/s41396-020-0732-1

Aigberua AO, Izah SC, Aigberua AA (2023) Occurrence, source delineation, and health hazard of polycyclic aromatic hydrocarbons in tissues of *Sarotherodon melanotheron* and *Chrysichthys nigrodigitatus* from Okulu River, Nigeria. Environ Monit Assess 195:364. https://doi.org/10.1007/s10661-023-10970-y

Aigberua AO, Izah SC, Richard G (2021) Hazard analysis of trace metals in muscle of *Sarotherodon melanotheron* and *Chrysichthys nigrodigitatus* from Okulu River, Rivers State, Nigeria. J Environ Health Sustain Dev 6(3):1340–1356. https://doi.org/10.18502/jehsd.v6i3.7242

Akinsete E, Apostolaki S, Chatzistamoulou N, Koundouri P, Tsani S (2019) The link between ecosystem services and human wellbeing in the implementation of the European water framework directive: assessing four river basins in Europe. Water 11(3):508. https://doi.org/10.3390/w11030508

Apostolaki S, Akinsete E, Koundouri P, Samartzis P (2020) Freshwater: the importance of freshwater for providing ecosystem services. In: Encyclopedia of the world's biomes, pp 71–79

Bashir I, Lone FA, Bhat RA, Mir SA, Dar ZA, Dar SA (2020) Concerns and threats of contamination on aquatic ecosystems. In: Hakeem K, Bhat R, Qadri H (eds) Bioremediation and biotechnology: sustainable approaches to pollution degradation. Springer, Cham, pp 1–26. https://doi.org/10.1007/978-3-030-35691-0_11-26

Beirne C, Sun C, Tattersall ER, Burgar JM, Fisher JT, Burton AC (2021) Multispecies modelling reveals potential for habitat restoration to re-establish boreal vertebrate community dynamics. J Appl Ecol 58(12):2821–2832. https://doi.org/10.1111/1365-2664.14020

Belesova K, Haines A, Ranganathan J, Seddon J, Wilkinson P (2020) Monitoring environmental change and human health: planetary health watch. Lancet 395(10218):96–98. https://doi.org/10.1016/s0140-6736(19)33042-9

Brederveld RJ, Jähnig SC, Lorenz AW, Brunzel S, Soons MB (2011) Dispersal as a limiting factor in the colonization of restored mountain streams by plants and macroinvertebrates. J Appl Ecol 48(5):1241–1250. https://doi.org/10.1111/j.1365-2664.2011.02026.x

Bunch M (2016) Ecosystem approaches to health and wellbeing: navigating complexity, promoting health in social–ecological systems. Syst Res Behav Sci 33(5):614–632. https://doi.org/10.1002/sres.2429

Chen X, Liu Y, Liu H, Wang H, Yang D, Huangfu C (2015) Impacts of four invasive *Asteraceae* on soil physico-chemical properties and AM fungi community. Am J Plant Sci 6(17):2734–2743. https://doi.org/10.4236/ajps.2015.617274

Dehling DM, Stouffer DB (2018) Bringing the Eltonian niche into functional diversity. Oikos 127(12):1711–1723. https://doi.org/10.1111/oik.05415

Dieleman CM, Branfireun BA, McLaughlin JW, Lindo Z (2015) Climate change drives a shift in peatland ecosystem plant community: implications for ecosystem function and stability. Glob Chang Biol 21(1):388–395. https://doi.org/10.1111/gcb.12643

Dong SH, Wang FL, Zhang DS, Yu LH, Pu WQ, Xu XF, Xie Y (2021) Assessment of the carrying capacity of integrated pond aquaculture of *Portunus trituberculatus* at the ecosystem level. Front Mar Sci 8:747891. https://doi.org/10.3389/fmars.2021.747891

Durant JM, Aarvold L, Langangen Ø (2021) Stock collapse and its effect on species interactions: cod and herring in the Norwegian-Barents Seas system as an example. Ecol Evol 11(23):16993–17004. https://doi.org/10.1002/ece3.8336

Ebeling A, Meyer ST, Abbas M, Eisenhauer N, Hillebrand H, Lange M, Scherber C, Vogel A, Weigelt A, Weisser WW (2014) Plant diversity impacts decomposition and herbivory via changes in aboveground arthropods. PLoS ONE 9(9):e106529. https://doi.org/10.1371/journal.pone.0106529

Ellis-Soto D, Ferraro KM, Rizzuto M, Briggs E, Monk JD, Schmitz OJ (2021) A methodological roadmap to quantify animal-vectored spatial ecosystem subsidies. J Anim Ecol 90(7):1605–1622. https://doi.org/10.1111/1365-2656.13538

Emery-Butcher HE, Beatty SJ, Robson BJ (2020) The impacts of invasive ecosystem engineers in freshwaters: a review. Freshw Biol 65(5):999–1015. https://doi.org/10.1111/fwb.13479

Fernández IC, Morales NS (2016) A spatial multi-criteria decision analysis for selecting priority sites for plant species restoration: a case study from the Chilean biodiversity hotspot. Restor Ecol 24(5):599–608. https://doi.org/10.1111/rec.12354

Ford AE, Graham H, White PC (2015) Integrating human and ecosystem health through ecosystem services frameworks. EcoHealth 12(4):660–671. https://doi.org/10.1007/s10393-015-1041-4

Frishkoff LO, Echeverri A, Chan KM, Karp DS (2018) Do correlated responses to multiple environmental changes exacerbate or mitigate species loss? Oikos 127(12):1724–1734. https://doi.org/10.1111/oik.05288

Fukumori K, Yoshizaki E, Takamura N, Kadoya T (2016) Detritivore diversity promotes a relative contribution rate of detritus to the diet of predators in ponds. Ecosphere 7(3):e01211. https://doi.org/10.1002/ecs2.1211

Gallardo B, Clavero M, Sánchez M, Vilà M (2016) Global ecological impacts of invasive species in aquatic ecosystems. Glob Chang Biol 22(1):151–163. https://doi.org/10.1111/gcb.13004

Givens JE, Huang X, Jorgenson AK (2019) Ecologically unequal exchange: a theory of global environmental injustice. Sociol Compass 13(5):e12693. https://doi.org/10.1111/soc4.12693

Gravel D, Guichard F, Loreau M, Mouquet N (2010) Source and sink dynamics in meta-ecosystems. Ecology 91(7):2172–2184. https://doi.org/10.1890/09-0843.1

Gravel D, Massol F, Leibold MA (2016) Stability and complexity in model meta-ecosystems. Nat Commun 7(1):12457. https://doi.org/10.1038/ncomms12457

Gregory AS, Ritz K, McGrath SP, Quinton JN, Goulding KWT, Jones RJA, Harris JA, Bol R, Wallace P, Pilgrim ES et al (2015) A review of the impacts of degradation threats on soil properties in the UK. Soil Use Manag 31:1–15. https://doi.org/10.1111/sum.12212

Guetz KM, Joyal T, Dickson B, Perry D (2022) Prioritizing dams for removal to advance restoration and conservation efforts in the western United States. Restor Ecol 30(5):e13583. https://doi.org/10.1111/rec.13583

Gutiérrez JL, Jones CG, Sousa R (2014) Toward an integrated ecosystem perspective of invasive species impacts. Acta Oecologica 54:131–138. https://doi.org/10.1016/j.actao.2013.10.003

Haddad NM, Brudvig LA, Clobert J, Davies KF, Gonzalez A, Holt RD, Lovejoy TE, Sexton JO, Austin MP, Collins CD et al (2015) Habitat fragmentation and its lasting impact on earth's ecosystems. Sci Adv 1(2):e1500052. https://doi.org/10.1126/sciadv.1500052

Higgins SN, Vander Zanden MJ (2010) What a difference a species makes: a meta-analysis of dreissenid mussel impacts on freshwater ecosystems. Ecol Monogr 80(2):179–196. https://doi.org/10.1890/09-1249.1

Ingrao C, Strippoli R, Lagioia G, Huisingh D (2023) Water scarcity in agriculture: an overview of causes, impacts and approaches for reducing the risks. Heliyon 9(8):e18507. https://doi.org/10.1016/j.heliyon.2023.e18507

Izah SC (2018) Ecosystem of the Niger Delta region of Nigeria: potentials and threats. Biodivers Int J 2(4):338–345

Izah SC, Seiyaboh EI (2018a) Challenges of wildlife with therapeutic properties in Nigeria: a conservation perspective. Int J Avian Wildlife Biol 3(4):252–257. https://doi.org/10.15406/ijawb.2018.03.00096

Izah SC, Seiyaboh EI (2018b) Changes in the protected areas of Bayelsa State, Nigeria. Int J Mol Evol Biodivers 8(1):1–11. https://doi.org/10.5376/ijmeb.2018.08.0001

Izah SC, Aigberua AO, Nduka JO (2018) Factors affecting the population trend of biodiversity in the Niger Delta region of Nigeria. Int J Avian Wildlife Biol 3(3):199–207. https://doi.org/10.15406/ijawb.2018.03.00085

Izah SC, Iyiola AO, Richard G (2023a) Impacts of pollution on the hydrogeochemical and microbial community of aquatic ecosystems in Bayelsa State, Southern Nigeria. In: Madhav S, Singh VB, Kumar M, Singh S (eds) Hydrogeochemistry of aquatic ecosystems, pp 283–305. https://doi.org/10.1002/9781119870562.ch13

Izah SC, Iyiola AO, Yarkwan B, Richard G (2023b) Impact of air quality as a component of climate change on biodiversity-based ecosystem services. In: Srivastav AL, Dubey AK, Kumar A, Narang SK, Khan MA (eds) Visualization techniques for climate change with machine learning and artificial intelligence. Elsevier, pp 123–148. ISBN: 978-0-323-99714-0. https://doi.org/10.1016/B978-0-323-99714-0.00005-4

Izah SC, Richard G, Stanley HO, Sawyer WE, Ogwu MC, Uwaeme OR (2023c) Integrating the One Health approach and statistical analysis for sustainable aquatic ecosystem management and trace metal contamination mitigation. ES Food Agrofor 14(2):1012. https://doi.org/10.30919/esfaf1012

Jiang L, Hu S, Li X, Han B (2024) Preliminary evaluation and spatio-temporal evolution characteristics of the ecosystem service value in Jiangsu Province, East China. Pol J Environ Stud 33(1):697–710. https://doi.org/10.15244/pjoes/172031

Kacheche RD, Mzuza MK (2021) Environmental impacts of invasive alien plant species on the biodiversity of the Nyika National Park, Rumphi District, Malawi. Am J Plant Sci 12(10):1503–1514. https://doi.org/10.4236/ajps.2021.1210106

Kalaba FK (2014) A conceptual framework for understanding forest socio-ecological systems. Biodivers Conserv 23(14):3391–3403. https://doi.org/10.1007/s10531-014-0792-5

Keune H, Kretsch C, De Blust G, Gilbert M, Flandroy L, Van den Berge K, Versteirt V, Hartig T, De Keersmaecker L, Eggermont H (2013) Science–policy challenges for biodiversity, public health and urbanization: Examples from Belgium. Environ Res Lett 8(2):025015. https://doi.org/10.1088/1748-9326/8/2/025015

Kim AY, Lee WS, Son Y (2024) The interaction between climate change and biodiversity can be assessed from a material cycle perspective. Diversity 16(8):506. https://doi.org/10.3390/d16080506

Klemmer AJ, Galatowitsch ML, McIntosh AR (2020) Cross-ecosystem bottlenecks alter reciprocal subsidies within meta-ecosystems. Proc R Soc B 287(1929):20200550. https://doi.org/10.1098/rspb.2020.0550

Kukkala AS, Moilanen A (2013) Core concepts of spatial prioritisation in systematic conservation planning. Biol Rev 88(2):443–464. https://doi.org/10.1111/brv.12008

Leite-Filho AT, Soares-Filho BS, Davis JL, Abrahão GM, Börner J (2021) Deforestation reduces rainfall and agricultural revenues in the Brazilian Amazon. Nat Commun 12(1):2591. https://doi.org/10.1038/s41467-021-22840-7

Liu M, Wei H, Dong X, Wang X, Zhao B, Zhang Y (2022) Integrating land use, ecosystem service, and human wellbeing: a systematic review. Sustainability 14(11):6926. https://doi.org/10.3390/su14116926

Lopez BE, Allen JM, Dukes JS, Lenoir J, Vilà M, Blumenthal DM, Beaury EM, Fusco EJ, Laginhas BB, Morelli TL et al (2022) Global environmental changes more frequently offset than intensify detrimental effects of biological invasions. Proc Natl Acad Sci 119(22):e2117389119. https://doi.org/10.1073/pnas.2117389119

Ma L, Qin Y, Zhang H, Zheng J, Hou Y, Wen Y (2021) Improving wellbeing of farmers using ecological awareness around protected areas: evidence from Qinling region, China. Int J Environ Res Public Health 18(18):9792. https://doi.org/10.3390/ijerph18189792

Massol F, Gravel D, Mouquet N, Cadotte MM, Fukami T, Leibold MA (2011) Linking community and ecosystem dynamics through spatial ecology. Ecol Lett 14(3):313–323. https://doi.org/10.1111/j.1461-0248.2011.01588.x

Morya CP, Punia M (2022) Impact of urbanization processes on availability of ecosystem services in National Capital Region of Delhi (1992–2010). Environ Dev Sustain 24(5):7324–7348. https://doi.org/10.1007/s10668-021-01748-8

Münzel T, Hahad O, Daiber A, Landrigan PJ (2023) Soil and water pollution and human health: what should cardiologists worry about? Cardiovasc Res 119(2):440–449. https://doi.org/10.1093/cvr/cvac082

Myers S, Fanzo J, Wiebe K, Huybers P, Smith M (2022) Current guidance underestimates risk of global environmental change to food security. BMJ (Clinical Research Ed.) 378:e071533. https://doi.org/10.1136/bmj-2022-071533

Ongare D, Macharia A, Mwakaje A, Muchai M, Warui C, Mugoya C, Masiga C, Nikundiwe A, Muiti A, Wakibara J (2013) Environmental communication: a review of information sources and communication channels for enhanced community-based natural resource management in the Greater Mara region of Kenya. J Educ Sustain Dev 7(1):65–74. https://doi.org/10.1177/0973408213495608

Orr JA, Vinebrooke RD, Jackson MC, Kroeker KJ, Kordas RL, Mantyka-Pringle C, Van den Brink PJ, De Laender F, Stoks R, Holmstrup M et al (2020) Towards a unified study of

multiple stressors: divisions and common goals across research disciplines. Proc R Soc B 287(1926):20200421. https://doi.org/10.1098/rspb.2020.0421
Ort BS, Cohen CS, Boyer KE, Reynolds LK, Tam SM, Wyllie-Echeverria S (2014) Conservation of eelgrass (Zostera marina) genetic diversity in a mesocosm-based restoration experiment. PLoS ONE 9(2):e89316. https://doi.org/10.1371/journal.pone.0089316
Pecl GT, Araújo MB, Bell JD, Blanchard JL, Bonebrake TC, Chen IC, Clark TD, Colwell RK, Danielsen F, Evengård B et al (2017) Biodiversity redistribution under climate change: impacts on ecosystems and human wellbeing. Science 355(6332):eaai9214. https://doi.org/10.1126/science.aai9214
Raghunathan C, Mondal T, Chandra K (2021) Invasion and potential risks of introduced exotic aquatic species in Indian islands. Aquat Ecosyst Health Manag 24(2):76–85. https://doi.org/10.14321/aehm.024.02.11
Rani N, Sangwan S (2022) Advantages of ecosystem services to human beings. Curr J Appl Sci Technol 41(17):19–24. https://doi.org/10.9734/cjast/2022/v41i1731730
Rentschler J, Leonova N (2023) Global air pollution exposure and poverty. Nat Commun 14(1):4432. https://doi.org/10.1038/s41467-023-39797-4
Renzi JJ, He Q, Silliman BR (2019) Harnessing positive species interactions to enhance coastal wetland restoration. Front Ecol Evol 7:131. https://doi.org/10.3389/fevo.2019.00131
Rilov G, Mant R, Lyons D, Bulleri F, Benedetti-Cecchi L, Kotta J, Queirós AM, Chatzinikolaou E, Crowe T, Guy-Him T (2012) How strong is the effect of invasive ecosystem engineers on the distribution patterns of local species, the local and regional biodiversity, and ecosystem functions? Environ Evid 1(1):10. https://doi.org/10.1186/2047-2382-1-10
Robroek BJM, Jassey VEJ, Kox MAR, Berendsen RL, Mills RTE, Cécillon L, Puissant J, Meima-Franke M, Bakker PA, Bodelier PL et al (2015) Peatland vascular plant functional types affect methane dynamics by altering microbial community structure. J Ecol 103(4):925–934. https://doi.org/10.1111/1365-2745.12413
Russell JC, Sataruddin NS, Heard AD (2014) Over-invasion by functionally equivalent invasive species. Ecology 95(8):2268–2276. https://doi.org/10.1890/13-1672.1
Sacandé M, Berrahmouni N (2016) Community participation and ecological criteria for selecting species and restoring natural capital with native species in the Sahel. Restor Ecol 24(4):479–488. https://doi.org/10.1111/rec.12337
Sales LP, Galetti M, Pires MM (2020) Climate and land-use change will lead to a faunal "savannization" on tropical rainforests. Glob Chang Biol 26(12):7036–7044. https://doi.org/10.1111/gcb.15374
Schloss CA, Lawler JJ, Larson ER, Papendick HL, Case MJ, Evans DM, DeLap JH, Langdon JG, Hall SA, McRae BH (2011) Systematic conservation planning in the face of climate change: bet-hedging on the Columbia Plateau. PLoS ONE 6(12):e28788. https://doi.org/10.1371/journal.pone.0028788
Shivanna KR (2022) Climate change and its impact on biodiversity and human welfare. Proc Indian Natl Sci Acad 88(2):160–171. https://doi.org/10.1007/s43538-022-00073-6
Simberloff D (2015) Non-native invasive species and novel ecosystems. F1000Prime Rep 7:47. https://doi.org/10.12703/P7-47
Singh BK (2010) Climate change and human health: an environmental perspective. Climate Res 41(1):41–44. https://doi.org/10.3354/cr00839
Staudt A, Leidner AK, Howard J, Brauman KA, Dukes JS, Hansen LJ, Paukert C, Sabo J, Solórzano LA (2013) The added complications of climate change: understanding and managing biodiversity and ecosystems. Front Ecol Environ 11(9):494–501. https://doi.org/10.1890/120275
Stephens KL, Dantzler-Kyer ME, Patten MA, Souza L (2019) Differential responses to global change of aquatic and terrestrial invasive species: evidence from a meta-analysis. Ecosphere 10(4):e02680. https://doi.org/10.1002/ecs2.2680

Sterling EJ, Filardi C, Toomey A, Sigouin A, Betley E, Gazit N, Newell J, Albert S, Alvira D, Bergamini N et al (2017) Biocultural approaches to wellbeing and sustainability indicators across scales. Nat Ecol Evol 1(12):1798–1806. https://doi.org/10.1038/s41559-017-0349-6

Su M, Zhang Y, Liu G, Xu L, Zhang L, Yang Z (2013) Urban ecosystem health assessment: perspectives and Chinese practice. Int J Environ Res Public Health 10(11):5874–5885. https://doi.org/10.3390/ijerph10115874

Suding KN (2011) Toward an era of restoration in ecology: successes, failures, and opportunities ahead. Annu Rev Ecol Evol Syst 42(1):465–487. https://doi.org/10.1146/annurev-ecolsys-102710-145115

Valiente-Banuet A, Aizen MA, Alcántara JM, Arroyo J, Cocucci AA, Galetti M, García MB, García D, Gómez JM, Jordano P et al (2015) Beyond species loss: the extinction of ecological interactions in a changing world. Funct Ecol 29(3):299–307. https://doi.org/10.1111/1365-2435.12356

White PC, Guégan JF, Keune H, Bell S, Geijzendorffer IR, Hermans T, Prieur-Richard AH, Iroegbu C, Stone D, Vanwambeke S et al (2020) Integrative policy development for healthier people and ecosystems: a European case analysis. Area 52(3):495–504. https://doi.org/10.1111/area.12618

Wigand ME, Timmermann C, Scherp A, Becker T, Steger F (2022) Climate change, pollution, deforestation, and mental health: research trends, gaps, and ethical considerations. GeoHealth 6(11):e2022GH000632. https://doi.org/10.1029/2022GH000632

Wood SL, Jones SK, Johnson JA, Brauman KA, Chaplin-Kramer R, Fremier AK, Girvetz E, Gordon LJ, Kappel CV, Mandle L et al (2018) Distilling the role of ecosystem services in the sustainable development goals. Ecosyst Serv 29:70–82. https://doi.org/10.1016/j.ecoser.2017.10.010

Zamora-Gutiérrez V, Rivera-Villanueva AN, Martinez Balvanera S, Castro-Castro A, Aguirre-Gutiérrez J (2021) Vulnerability of bat–plant pollination interactions due to environmental change. Glob Chang Biol 27(14):3367–3382. https://doi.org/10.1111/gcb.15611

Zhang YS, Cioffi WR, Cope R, Daleo P, Heywood E, Hoyt C, Smith CS, Silliman BR (2018) A global synthesis reveals gaps in coastal habitat restoration research. Sustainability 10(4):1040. https://doi.org/10.3390/su10041040

Zhang S, Wang H, Fu X, Tang M, Wu D, Li S, Wu G (2023) Analysis of the effect of ecosystem services and urbanization on human wellbeing in inner Mongolia province. Sustainability 15(22):16021. https://doi.org/10.3390/su152216021

Chapter 2
Air Quality and Public Health: Assessing Pollution Processes and Their Mitigation

Ayobami O. Aigberua and Kurotimipa Frank Ovuru

Abstract Air pollution poses a significant threat to public health, contributing to respiratory and cardiovascular diseases, especially among vulnerable populations such as children and the elderly. Key pollutants, including particulate matter, nitrogen oxides, sulfur dioxide, and volatile organic compounds, largely stem from industrial activities, transportation, agriculture, and urbanization. Understanding these pollutants' formation, dispersion, and health impacts is crucial for developing effective mitigation strategies. This paper explores the processes underlying air pollution, the health consequences of exposure, and the latest innovations in pollution mitigation technologies. Strategies such as advanced air filtration systems, low-emission transportation technologies, and green energy solutions offer promising avenues to improve air quality. Furthermore, effective environmental health management requires comprehensive policies integrating urban planning, regulatory frameworks, and public engagement. Collaboration among governments, industries, and communities is essential to address the global burden of air pollution. With urbanization and industrialization continuing to challenge air quality, adopting sustainable environmental management practices and innovative technologies is critical to safeguarding public health. Ongoing research and a collective commitment to air quality improvement will reduce pollution-related health impacts and foster a healthier environment for future generations.

Keywords Air pollution · Public health · Particulate matter · Renewable energy · Respiratory diseases · Pollution mitigation · Environmental health management · Sustainable urban planning

---

A. O. Aigberua (✉)
Quality Assurance Department, Infra Pipe Solutions Ltd, Saskatoon, Canada
e-mail: ozedee101@gmail.com

K. F. Ovuru
Department of Disease Control and Immunization, Bayelsa State Primary Health Care Board, Yenagoa, Bayelsa State, Nigeria

S. Chibueze Izah and M. Chidozie Ogwu (eds.), *Innovative Approaches in Environmental Health Management*, Environmental Science and Engineering,
https://doi.org/10.1007/978-3-031-81966-7_2

2.1 Introduction

Air pollution is a critical global issue that poses significant threats to public health, particularly in urban environments where industrial activities and vehicular emissions are prevalent (Izah et al. 2024a, b, c, d). The World Health Organization (WHO) has identified air pollution as the most significant environmental health risk, responsible for approximately 4.2 million premature deaths annually (Yan 2023). The detrimental health impacts of air pollution are multifaceted, affecting respiratory, cardiovascular, and neurological systems, leading to increased morbidity and mortality rates (Hassan Bhat et al. 2021; Izah et al. 2024e, f). In countries like China and India, where air quality is often severely compromised, the public health implications are particularly acute, necessitating urgent and effective policy interventions (Huang et al. 2018; Majumdar 2021). The growing body of evidence linking air quality to health outcomes underscores the need for comprehensive strategies to mitigate air pollution and protect public health.

The relationship between air pollution and public health is well-documented, with numerous studies highlighting the adverse effects of particulate matter (PM) and other pollutants on human health. For instance, research indicates that long-term exposure to ambient PM pollution is associated with increased premature mortality rates and years of life lost (Matkovic et al. 2020). Moreover, the health burden of air pollution is not uniformly distributed; disadvantaged populations, including low-income communities, often bear a disproportionate share of the health risks associated with poor air quality (Ma et al. 2020). This inequity necessitates targeted interventions that address both the environmental and social determinants of health, ensuring that vulnerable populations receive adequate protection from air pollution (McCarron et al. 2023). Sustainable ecological health management is essential for addressing the public health crisis posed by air pollution (Ogwu and Izah 2024; Ogwu et al. 2024a, b). Effective management strategies must integrate scientific research, public policy, and community engagement to create a holistic approach to air quality improvement (Leal Filho et al. 2022). For example, implementing air quality monitoring systems and public awareness campaigns can empower communities to take action against pollution (Liu et al. 2017). Furthermore, investments in clean technologies and infrastructure are crucial for achieving sustainable development goals (SDGs) related to health and well-being (Leal Filho et al. 2022). Sustainable environmental health management can improve air quality and public health by fostering collaboration between government agencies, private sectors, and civil society.

The need for sustainable environmental health management is underscored by the increasing urbanization and industrialization that characterize many developing countries. Rapid urban growth often leads to increased emissions from transportation and industrial activities, exacerbating air pollution levels (Rumambi 2023). In response, countries like China have initiated comprehensive air pollution control plans, which have shown promising results in improving air quality and reducing health impacts during specific events, such as the 2008 Beijing Olympics (Huang et al. 2018). However, these efforts must be sustained and expanded to address urban air pollution challenges (Majumdar 2021). Policymakers must prioritize long-term strategies for immediate pollution reduction and promoting sustainable urban planning and development.

Public perception and awareness of air quality issues are crucial in driving policy change and fostering community engagement in environmental health initiatives. Studies have shown that increased public awareness of air pollution and its health impacts can lead to more significant support for environmental policies and individual behavior changes (Liu et al. 2017; Gurajala et al. 2019). For instance, community-led initiatives to monitor air quality and advocate for cleaner air can significantly influence local policies and practices (McCarron et al. 2023). Effective communication strategies that convey the health risks of air pollution can motivate individuals to adopt protective behaviors, such as reducing outdoor activities during high-pollution days (Ramondt and Ramírez 2020). Thus, enhancing public understanding of air quality issues is vital to sustainable environmental health management. Integrating air quality and public health benefits into broader ecological policies is essential for achieving meaningful improvements in air quality and health outcomes. Research has demonstrated that decarbonization strategies, which aim to reduce greenhouse gas emissions, can also yield significant co-benefits for air quality and public health (Gallagher and Holloway 2020; Farzaneh et al. 2019). For example, transitioning to cleaner energy sources and promoting sustainable transportation can reduce emissions of harmful pollutants, improve air quality, and reduce health risks (Farzaneh et al. 2019). Policymakers must recognize the interconnectedness of climate change, air quality, and public health and develop integrated strategies that address these issues simultaneously (Yamineva and Romppanen 2017). By aligning environmental and health policies, governments can create a more sustainable and healthier future for their populations. The role of technology in improving air quality and public health cannot be overstated. Innovations in air quality monitoring, data analysis, and predictive modeling can enhance understanding of pollution sources and health impacts (Relvas et al. 2018). For instance, artificial intelligence and machine learning can facilitate the development of more accurate air quality prediction models, enabling timely interventions to protect public health (Relvas et al. 2018). Additionally, technological advancements can support innovative city initiatives that prioritize air quality management and promote sustainable urban living (Relvas et al. 2018). By leveraging technology, policymakers can make informed decisions that lead to more effective air quality management strategies.

The chapter identifies vital air pollutants, particulate matter, nitrogen oxides, volatile organic compounds, and their primary sources, including industry, transportation, and agriculture. It explores the health impacts of air pollution, particularly respiratory and cardiovascular diseases, emphasizing vulnerable populations and the global disease burden. The chapter also discusses innovative mitigation technologies, pollution monitoring techniques, and environmental health management strategies, highlighting the challenges and future directions for sustainable air quality improvement and global collaboration.

2.2 Key Air Pollutants and Sources

Air pollution is a critical environmental issue that significantly impacts public health and ecosystems (Izah et al. 2021; Uzoekwe et al. 2021). Table 2.1 provides an overview of key air pollutants and their sources. Among the various pollutants, particulate matter (PM), nitrogen oxides (NOx), sulfur dioxide (SO_2), and volatile organic compounds (VOCs) are of particular concern due to their widespread sources and adverse health effects. Particulate matter, especially PM2.5, and PM10, is a complex mixture of tiny particles that can penetrate deep into the lungs and enter the bloodstream, leading to severe respiratory and cardiovascular diseases (Tunno et al. 2012; Gyasi et al. 2022; Noh et al. 2016). Nitrogen oxides, primarily from vehicle emissions and industrial processes, contribute to ground-level ozone and smog formation, exacerbating respiratory conditions and cardiovascular diseases (Noh et al. 2016; El-Sheekh and Hassan 2021). Sulfur dioxide, mainly emitted from fossil fuel combustion, can lead to acid rain and respiratory problems (Nosek et al. 2016; Huang et al. 2021). VOCs, released from industrial processes, vehicle emissions, and household products, can contribute to the formation of ground-level ozone and have been linked to various health issues, including respiratory diseases (Singh 2023; Acciai et al. 2017).

The primary sources of these pollutants can be categorized into several sectors, including industry, transportation, agriculture, and urbanization. Industrial activities significantly contribute to air pollution, releasing various pollutants, including PM, NOx, SO_2, and VOCs. For instance, studies have shown that industrial emissions are a significant source of PM2.5 and PM10 in urban areas, with emissions from factories contributing to the deterioration of air quality (Singh 2023; Acciai et al. 2017; Azid et al. 2014). Transportation is another critical source, with vehicles emitting large quantities of NOx and PM. The increasing number of vehicles on the road, particularly in urban areas, has increased these pollutants' heightened levels, resulting in profound health implications for residents (Noh et al. 2016; El-Sheekh and Hassan 2021; Huang et al. 2021). Furthermore, agricultural practices, including using fertilizers and pesticides, contribute to releasing ammonia and other pollutants, which can react with other compounds in the atmosphere to form secondary pollutants (Bui et al. 2022).

Urbanization plays a significant role in exacerbating air pollution. As cities expand, the demand for transportation and industrial activities increases, leading to higher emissions of air pollutants. The concentration of people and vehicles in urban areas creates a unique environment where pollutants accumulate, resulting in poor air quality (El-Sheekh and Hassan 2021; Azid et al. 2014). Moreover, urban heat islands can intensify the effects of air pollution, as higher temperatures can increase the formation of ground-level ozone, further impacting public health (Huang et al. 2021). The interplay between urbanization and air pollution highlights the need for effective urban planning and pollution control measures to mitigate the adverse effects on air quality and public health.

Table 2.1 Key air pollutants and sources

Air pollutant	Description	Primary sources	Health impacts
Particulate matter (PM)	Tiny particles (solid or liquid) that are suspended in the air. Includes PM2.5 and PM10	Combustion processes (vehicles, power plants), agriculture, industry	Respiratory issues, cardiovascular diseases, aggravation of asthma, premature death
Nitrogen oxides (NOx)	Gases that include nitrogen dioxide (NO_2) and nitrogen monoxide (NO), primarily from combustion	Vehicle emissions, industrial facilities, power plants	Respiratory inflammation, increased susceptibility to respiratory infections, lung function impairment
Sulfur dioxide (SO_2)	A gas produced by burning fossil fuels, particularly coal and oil	Power plants, industrial processes, volcanic eruptions	Irritation of eyes, nose, and throat, respiratory issues, worsening of asthma and lung diseases
Volatile organic compounds (VOCs)	Organic chemicals that quickly evaporate into the air, contributing to ozone formation	Paints, solvents, gasoline, industrial processes	Eye, nose, and throat irritation, headaches, damage to liver, kidney, and central nervous system
Carbon monoxide (CO)	A colorless, odorless gas resulting from incomplete combustion of carbon-containing fuels	Vehicle emissions, industrial activities, residential heating	Dizziness, headaches, impaired vision, at high levels: poisoning and death
Ozone (O_3)	A secondary pollutant formed by the reaction of NOx and VOCs in sunlight	Formed in the atmosphere from reactions involving NOx and VOCs	Lung irritation, reduced lung function, aggravation of asthma, respiratory diseases
Ammonia (NH_3)	A gas primarily from agricultural activities, especially livestock waste and fertilizer use	Agriculture, industrial processes	Irritation of the respiratory system, lung inflammation, and long-term exposure may lead to chronic respiratory problems

Particulate matter, particularly PM2.5, has been extensively studied for its health impacts. Research indicates that long-term exposure to PM2.5 is associated with increased mortality rates and the progression of cardiovascular diseases (Tonne and Wilkinson 2013; Raaschou-Nielsen et al. 2013). Epidemiological studies have shown a strong correlation between PM exposure and the incidence of lung cancer, with individuals living in areas with high PM levels at greater risk (Raaschou-Nielsen et al. 2013; Lee et al. 2022). Additionally, PM exposure has been linked to respiratory allergies and asthma exacerbations, particularly in vulnerable populations such as children and the elderly (Sompornrattanaphan et al. 2020; Noh et al. 2016). The health effects of PM are compounded by the presence of other pollutants, such as NOx and VOCs, which can interact with and exacerbate respiratory conditions (Noh et al. 2016; El-Sheekh and Hassan 2021).

The chemical composition of particulate matter varies depending on its source. For instance, studies have identified heavy metals and organic compounds in PM emitted from industrial sources, while traffic-related PM is often rich in nitrates and sulfates (Jandačka 2015; Ďurčanská and Jandačka 2015). Understanding the sources and composition of PM is crucial for developing targeted pollution control strategies. For example, research has shown that PM from biomass burning differs in composition from PM generated by vehicle emissions, necessitating different approaches to mitigation (Nosek et al. 2016; Ďurčanská and Jandačka 2015). Furthermore, identifying specific pollutants and their sources can aid public health interventions and inform regulatory policies to reduce exposure to harmful air pollutants (Singh 2023; Acciai et al. 2017).

In addition to health impacts, air pollution has significant environmental consequences. For example, sulfur dioxide and nitrogen oxides contribute to acid rain, harming aquatic ecosystems and damaging vegetation (Nosek et al. 2016; Huang et al. 2021). Furthermore, particulate matter can affect climate by influencing cloud formation and altering the Earth's radiation balance (Wu et al. 2018). The interaction between air pollution and climate change is complex, as pollutants can contribute to and be affected by changing climatic conditions (Wu et al. 2018). Addressing air pollution, therefore, requires a multifaceted approach that considers both health and environmental impacts. Mitigating air pollution necessitates regulatory measures, technological advancements, and public awareness initiatives. Governments play a crucial role in establishing air quality standards and regulations to limit emissions from industrial and transportation sources. For instance, implementing stricter vehicle emission standards and promoting cleaner technologies can significantly reduce NOx and PM emissions (Noh et al. 2016; El-Sheekh and Hassan 2021; Huang et al. 2021). Additionally, public awareness campaigns can educate communities about the sources and health effects of air pollution, encouraging individuals to adopt practices that reduce exposure, such as using public transportation or reducing energy consumption (Singh 2023; Acciai et al. 2017; Azid et al. 2014).

2.3 Health Impacts of Air Pollution

Air pollution is a significant public health concern worldwide, with profound implications for respiratory and cardiovascular health. Numerous studies have established a strong correlation between exposure to air pollutants and the incidence of respiratory diseases such as asthma, chronic obstructive pulmonary disease (COPD), and other chronic lung conditions. For instance, Liu and Grigg (2018) highlighted that air pollution is linked to a wide range of adverse health effects, particularly respiratory diseases, which account for approximately 40,000 excess deaths annually in the UK alone due to long-term exposure and exacerbation of pre-existing conditions. Furthermore, research by Syuhada et al. (2023) indicated that air pollution contributes to substantial mortality rates, with estimates of 15,300 premature deaths in Canada and 50,000 in Thailand attributed to poor air quality, underscoring the economic burden of air pollution on healthcare systems. Table 2.2 provides an overview of the health impacts of air pollution.

The impact of air pollution is particularly pronounced among vulnerable populations, including children, the elderly, and individuals with pre-existing health conditions. Children are especially susceptible due to their developing respiratory systems and higher rates of outdoor activity, which increase their exposure to air pollutants. Lian et al. (2023) systematic review emphasized that children and adolescents are more sensitive to air pollution, which can lead to long-term health consequences such as impaired lung function and increased risk of respiratory diseases. Moreover, studies have shown that the elderly are at a heightened risk due to age-related physiological changes and comorbidities. For example, research by Honda et al. (2017) indicated that long-term exposure to air pollution is associated with anemia and increased mortality risk in older adults, highlighting the need for targeted interventions in this demographic. The global burden of disease related to poor air quality is staggering, with air pollution being a leading cause of morbidity and mortality. Lelieveld et al. (2018) conducted a modeling analysis that revealed significant health risks associated with ambient air pollution, particularly in middle- and low-income countries with higher exposure levels. This study underscores the urgent need for comprehensive strategies to mitigate air pollution and its health impacts, particularly in regions with limited healthcare resources.

The World Health Organization (WHO) has also identified air pollution as a critical public health issue, calling for effective policies to reduce exposure and protect vulnerable populations (Tsocheva et al. 2023). The interplay between air pollution and health outcomes is complex and multifaceted. For instance, the relationship between air quality and respiratory health is influenced by various factors, including socioeconomic status, geographic location, and access to healthcare. Research by Costin and Clark (2022) highlighted the negative correlations between air pollution and social determinants of health, suggesting that disadvantaged communities often bear a disproportionate burden of air pollution-related health issues. Furthermore, studies have shown that indoor air quality can significantly impact health, particularly among the elderly, who may spend more time indoors. Bentayeb et al. (2015) found that indoor pollutants, such as carbon monoxide and nitrogen dioxide, are linked to respiratory diseases in elderly populations living in nursing homes.

The economic implications of air pollution are also noteworthy, as the costs associated with healthcare and lost productivity due to air pollution-related illnesses can be substantial. Syuhada et al.(2023) reported that Thailand's economic burden of air pollution amounted to approximately USD 60.9 billion, representing nearly 15% of the country's GDP. This financial strain is compounded by the long-term health effects of air pollution, which can lead to chronic diseases requiring ongoing medical

Table 2.2 Health impacts of air pollution

Health impact	Description	Associated pollutants	Refs.
Respiratory diseases	Air pollution exacerbates asthma, bronchitis, and other respiratory issues	Particulate matter (PM), ozone (O_3), NOx	Lee et al. (2021)
Cardiovascular diseases	Long-term exposure increases the risk of heart attacks, strokes, and other heart-related issues	Particulate matter (PM), NOx, SO_2	Palacio et al. (2023)
Premature mortality	Air pollution is linked to premature death, especially in individuals with pre-existing conditions	Particulate matter (PM), NOx, ozone (O_3)	Waidyatillake et al. (2021)
Lung cancer	Increased pollutant exposure, mainly from combustion sources, is associated with lung cancer risk	Particulate matter (PM), VOCs	Holme et al. (2023)
Impaired lung development in children	Children exposed to polluted air may experience reduced lung growth and function	Particulate matter (PM), ozone (O_3), NOx	Garcia et al. (2021)
Cognitive decline and neurological effects	Long-term exposure to air pollution has been linked to cognitive decline and neurodegenerative diseases	Particulate matter (PM), NOx	Roy and D'Anguiulli (2024)
Pregnancy complications	Air pollution exposure during pregnancy can lead to low birth weight, preterm births, and developmental issues	Particulate matter (PM), NOx, SO_2	Soesanti et al. (2023)
Increased risk of diabetes	Emerging evidence suggests that air pollution may increase the risk of developing type 2 diabetes	Particulate matter (PM), ozone (O_3), NOx	Li et al. (2019)

care and management. Thus, addressing air pollution is not only a public health imperative but also an economic necessity. In addition to respiratory and cardiovascular diseases, air pollution has been linked to a range of non-respiratory health hazards, particularly in children. Kelishadi and Poursafa (2010) emphasized that air pollution is a potential risk factor for developing conditions such as obesity and diabetes, which are increasingly prevalent in developing countries. This highlights the need for a holistic approach to public health that considers the multifaceted impacts of air pollution on various health outcomes.

Furthermore, the long-term effects of air pollution exposure during childhood can have lasting implications for health into adulthood, making early intervention critical. The role of air pollution in exacerbating existing health conditions cannot be overlooked. For instance, individuals with pre-existing respiratory or cardiovascular diseases are particularly vulnerable to the adverse effects of air pollution. Studies have shown that exposure to air pollutants can trigger acute asthma exacerbations and other respiratory conditions, leading to increased hospital admissions and healthcare utilization (Xiao et al. 2016). This underscores the importance of monitoring air quality and implementing effective public health interventions to protect at-risk populations.

2.4 Assessing Pollution Processes

Assessing pollution processes involves a multifaceted approach involving pollutant formation and dispersion mechanisms, monitoring techniques, and the influence of meteorological and geographical factors (Table 2.3). Understanding these components is crucial for developing effective strategies to mitigate air pollution and protect public health. Pollutant formation is a complex process influenced by various chemical and physical atmospheric interactions. For instance, Zhu et al. (2011) highlighted the significance of heterogeneous chemical processes in forming air pollution complexes, particularly in the context of regional atmospheric oxidizing capacity and the kinetics of photochemical oxidants. This complexity is further compounded by the thermal formation of persistent organic pollutants, as discussed by Liu et al. (2021), who emphasized the role of temperature in the formation of polychlorinated dibenzo-p-dioxins and dibenzofurans. Understanding these mechanisms is essential for predicting the behavior of pollutants in the atmosphere and their potential impacts on air quality.

Dispersion modeling is critical for assessing how pollutants spread in the atmosphere. Leelőssy et al. (2014) reviewed dispersion modeling techniques comprehensively, emphasizing the importance of incorporating stochastic elements to account for the inherent variability in weather processes. This approach allows for a more accurate representation of pollutant dispersion patterns. Wright et al. also contribute to this discussion by advocating for time-varying advection components in dispersion models, which better reflect the physical processes governing pollutant movement (Wright et al. 2022). Such advancements in modeling techniques are vital for understanding the spatial distribution of pollutants and informing regulatory decisions. Another essential tool is the Gaussian Plume Model, an important tool in environmental health monitoring, particularly for assessing air quality near pollutant sources such as industrial stacks, power plants, and traffic emissions (Johnson 2022). By modeling the atmospheric pollutants' dispersion, it estimates ground-level

Table 2.3 Assessing pollution processes

Aspect	Description	Examples	Refs.
Mechanisms of pollutant formation and dispersion	The processes by which pollutants are created (e.g., combustion, industrial processes) and how they disperse into the atmosphere	Pollutants like NOx and VOCs form ozone (O_3) in the presence of sunlight and disperse depending on wind patterns	Chang et al. (2023)
Monitoring techniques for air quality assessment	Monitoring air quality techniques include satellite-based, ground-based, and sensor network technologies	Satellite: NASA's MODIS measures aerosol optical depth (AOD) to estimate pollution Ground-based: Aair quality stations measure real-time PM, NOx, and SO_2 levels Sensor networks: IoT sensors placed in cities for continuous monitoring	Dutta et al. (2021)
Data analysis for air auality	Methods used to analyze data from monitoring systems, including temporal and spatial trends in air pollutant levels	Geographic information systems (GIS) map pollution hotspots, and time-series analysis determines seasonal pollutant variation	Yadav et al. (2022)
Role of meteorology in pollutant spread	Weather conditions (e.g., wind, temperature, humidity) influence pollutants' dispersion, transformation, and deposition	Wind speed and direction impact the spread of pollutants, while temperature inversions trap pollutants near the ground	Pérez et al. (2020)
Role of geography in pollutant spread	Geographical features such as mountains, valleys, and proximity to water bodies affect pollutant dispersion	Mountain ranges can block airflow, causing pollutants to accumulate in valleys and forming smog	Wu et al. (2022)

concentrations based on emission rate, wind speed, and atmospheric stability. This information is crucial for understanding exposure risks, as pollutants like sulfur dioxide, nitrogen oxides, and particulate matter can significantly impact respiratory health and exacerbate conditions such as asthma and heart disease. The model's predictions help regulatory agencies and health officials evaluate compliance with air quality standards, design interventions, and plan public health responses. While it assumes steady-state conditions and flat terrain, which can limit accuracy in complex environments, the Gaussian Plume Model remains widely used for its practicality in providing actionable data for environmental health protection.

Monitoring air quality is essential for assessing pollution levels and understanding the effectiveness of mitigation strategies. Various techniques, including satellite observations, ground-based monitoring stations, and sensor networks, are employed to gather data on air quality. Integrating satellite data with ground-based measurements enhances air quality assessments' spatial coverage and temporal resolution (Yudego et al. 2018). Additionally, using sensor networks allows for real-time monitoring of pollutants, providing valuable data for immediate response efforts. These monitoring techniques are crucial for identifying pollution sources and trends and guiding policy interventions. Meteorological factors play a significant role in the dispersion of pollutants. As Pérez et al. (2020) highlighted, various meteorological phenomena, such as land-sea breezes and urban heat islands, can influence the transport and concentration of pollutants in urban areas. The interaction between meteorology and geography is particularly evident in complex terrains, where orographic features can enhance or inhibit pollutant dispersion (Pérez et al. 2020). Understanding these interactions is essential for developing accurate dispersion models and effective air quality management strategies.

Geographical factors, including urbanization and land use, also significantly impact pollution processes. Yazid et al. (2014) discussed how urban configurations, such as street canyons, can trap pollutants and exacerbate local air quality issues. Studying pollutant dispersion in urban environments is critical, as rapid urbanization often increases emissions and deteriorates air quality. Furthermore, the geographical distribution of pollution sources, such as industrial facilities and transportation networks, must be considered when assessing overall air quality.

Integrating advanced computational tools and modeling techniques is transforming the field of air quality assessment. For example, using virtual geographic environments (VGE) allows for the simulation of air pollution dispersion in a geographically referenced context, facilitating a better understanding of spatial dynamics (Xu et al. 2011). Additionally, novel analytical solutions for three-dimensional dispersion modeling provide deeper insights into atmospheric processes, enabling more effective pollution management strategies (Farhane and Souhar 2023). These advancements underscore the importance of interdisciplinary approaches in addressing air quality challenges.

Data analysis is a crucial component of air quality assessment, as it enables researchers and policymakers to interpret monitoring data effectively. Applying statistical methods and machine learning techniques can enhance the analysis of air quality data, allowing for the identification of trends and correlations between pollution levels and meteorological conditions (Lysova et al. 2018). Moreover, as Martinėnas et al. (2011) demonstrated, developing semi-empirical models for traffic pollution dispersion illustrated the potential of combining empirical data with theoretical frameworks to improve air quality assessments. Such approaches are essential for informing evidence-based policy decisions.

2.5 Innovative Mitigation Technologies in Assessing Air Pollution Processes and Their Mitigation

Innovative mitigation technologies have emerged as critical components in the global effort to combat climate change and reduce greenhouse gas emissions. These technologies involve a broad spectrum of advancements, including air filtration and purification systems, low-emission technologies in transportation and industry, and green technologies aimed at reducing emissions through renewable energy and carbon capture (Table 2.4). Integrating these technologies is essential for achieving sustainable development and environmental sustainability, particularly in rising global temperatures and increasing pollution levels. Air filtration and purification systems are significant innovations in mitigating air pollution. These systems utilize various technologies, including high-efficiency particulate air (HEPA) filters, electrostatic precipitators, and ultraviolet (UV) light disinfection, to remove harmful pollutants from the air. Research indicates that effective air purification can significantly reduce the concentration of particulate matter and other toxic substances, thereby improving public health and environmental quality (Li et al. 2022; Imasiku et al. 2019). Moreover, intelligent air quality monitoring systems, which utilize sensors and data analytics, have enhanced the ability to track and respond to air quality issues in real-time, further contributing to pollution reduction efforts (Imasiku et al. 2019).

In transportation, low-emission technologies transform how goods and people are moved. Developing electric vehicles (EVs), hydrogen fuel cell technologies, and biofuels is pivotal in reducing carbon emissions associated with traditional fossil fuel-powered vehicles. Studies have shown that transitioning to EVs can substantially decrease greenhouse gas emissions, particularly when the electricity used to charge these vehicles comes from renewable sources (Shi et al. 2021; Davis et al. 2018a, b). Furthermore, the integration of intelligent transportation systems (ITS) can optimize traffic flow and reduce congestion, leading to lower emissions from vehicles on the road (Liu et al. 2022). These technologies address emissions and enhance energy efficiency within the transportation sector. Green technologies, particularly those focused on renewable energy production, are crucial in mitigating climate change. The shift towards solar, wind, and hydroelectric power has been shown to significantly reduce reliance on fossil fuels, thereby decreasing carbon emissions (Wang and Zhang 2021; Wang et al. 2021). For instance, research indicates that a 1% increase in renewable energy consumption correlates with a 0.1568% decrease in global carbon emissions (Wang and Zhang 2021).

Carbon capture and storage (CCS) technologies are also emerging as vital tools for reducing emissions from industrial processes. These technologies capture carbon dioxide emissions at their source and store them underground, preventing them from entering the atmosphere (Wang et al. 2021). The successful implementation of CCS could allow for continued use of fossil fuels while minimizing their environmental impact.

The role of green technology innovation in reducing carbon emissions cannot be overstated. Studies have demonstrated that green technological advancements

Table 2.4 Innovative mitigation technologies

Technology	Description	Examples	Refs.
Advances in air filtration and purification systems	Technologies designed to remove harmful particles, gases, and pollutants from the air in indoor and outdoor environments	HEPA filters: used in buildings and vehicles to remove particulate matter (PM2.5) Activated carbon filters: effective in removing VOCs and odors from industrial emissions	Chen et al. (2022a, b)
Low-emission technologies in transportation	Innovations that reduce or eliminate harmful emissions from vehicles and transportation systems	Electric vehicles (EVs): EVs produce no tailpipe emissions, significantly reducing NOx and CO_2 Hybrid vehicles: lower emissions by using internal combustion engines and electric motors	Carey (2023)
Low-emission technologies in industry	Technologies and practices designed to reduce industrial emissions of CO_2, NOx, SOx, and other pollutants	Scrubbers: used in power plants to remove sulfur dioxide from emissions Electrostatic precipitators: capture particulate matter in industrial processes	Asif et al. (2022)
Green technologies for reducing emissions	Eco-friendly technologies aimed at lowering greenhouse gas emissions and promoting sustainability	Renewable energy: solar and wind energy produce electricity without emitting greenhouse gases Carbon capture and storage (CCS): this process captures CO_2 from industrial sources and stores it underground to prevent atmospheric release	Oyebanji and Kirikkaleli (2023)
Energy-efficient building technologies	Innovations in building design and construction that reduce energy consumption and emissions	Passive house design: buildings designed for energy efficiency through insulation, ventilation, and solar energy use Smart thermostats: devices that optimize heating and cooling to save energy	Anand et al. (2023)

improve energy efficiency and reduce waste, which is essential for achieving sustainability goals (Meng et al. 2023; Li et al. 2022). For example, adopting energy-efficient appliances and industrial processes has been linked to significant reductions in energy consumption and associated emissions (Cai et al. 2021; Wu et al. 2023). Furthermore, the spillover effects of green technology innovation can enhance overall economic productivity while simultaneously addressing environmental concerns (Chen et al. 2022a, b; Wang et al. 2022). This dual benefit underscores the importance of investing in research and development of green technologies to foster sustainable economic growth. In addition to technological advancements, policy frameworks, and regulations play a critical role in promoting the adoption of innovative mitigation technologies. Governments worldwide are implementing policies incentivizing developing and deploying green technologies, such as tax credits for renewable energy investments and emissions trading systems (Cui et al. 2022; Cui et al. 2023). These policies encourage private sector investment in green technologies and create a conducive environment for innovation. The interplay between regulatory frameworks and technological advancements is essential for driving the transition towards a low-carbon economy.

The digital economy is also influencing the landscape of green technology innovation. Integrating digital tools and platforms facilitates the development of smart technologies that enhance energy efficiency and reduce emissions. For instance, big data analytics and artificial intelligence can optimize energy consumption in buildings and industrial processes, reducing emissions significantly (Hou et al. 2023; Cui et al. 2023). Additionally, digital platforms can improve the accessibility and affordability of green technologies, making them more widely available to consumers and businesses alike. Green technology innovation's spatial and temporal dimensions are essential in understanding its impact on carbon emissions. Research has shown that the effectiveness of green technologies can vary significantly across different regions and economic contexts (Gao et al. 2022; Liu et al. 2022). For instance, regions with higher industrialization and technological advancement levels tend to experience more significant benefits from green technology adoption than less developed areas. This highlights the need for tailored approaches considering local conditions and capacities when implementing green technologies.

Furthermore, the relationship between green technology innovation and carbon emissions is complex and multifaceted. While green technologies can lead to direct reductions in emissions through improved efficiency and reduced waste, they can also have indirect effects that may increase emissions if not managed properly (Cai et al. 2021; Wu et al. 2023). For example, expanding economic activities driven by green technology innovation may lead to increased energy consumption, necessitating careful consideration of the overall impact on emissions. Policymakers must navigate these complexities to ensure that the deployment of green technologies aligns with broader climate goals.

2.6 Environmental Health Management Strategies for Assessing Air Pollution Processes and Their Mitigation

Environmental health management strategies are critical in addressing the multifaceted challenges of urban air pollution and its associated health impacts. Practical policy approaches are essential for mitigating these challenges and encompass regulatory frameworks, urban planning and design, and public awareness campaigns. Each strategy is pivotal in enhancing air quality and promoting public health. Regulatory frameworks, such as the World Health Organization (WHO) air quality guidelines, provide a foundational basis for national policies to improve air quality. These guidelines set permissible limits for various air pollutants, including particulate matter (PM2.5 and PM10), nitrogen oxides, and ozone, which are crucial for protecting public health (Meo et al. 2021). Countries that adopt these guidelines often see a significant reduction in air pollution levels, which correlates with improved health outcomes. For instance, studies have shown that adherence to WHO guidelines can lead to decreased incidence of respiratory diseases and other health issues linked to poor air quality (Meo et al. 2021; Margaryan 2021). Moreover, national policies aligning with these guidelines can foster a coordinated approach to air quality management, ensuring local governments implement effective measures tailored to their contexts (Quarmby et al. 2019) (Table 2.5).

Urban planning and design are also integral to environmental health management strategies and drive demographic, urbanization, and climatic trends (Fig. 2.1; Grafakos et al. 2016). Incorporating green spaces in urban environments has been shown to significantly improve air quality by absorbing pollutants and providing a habitat for biodiversity (Meo et al. 2021; Huang et al. 2017). Green areas mitigate the urban heat island effect and enhance neighborhoods' aesthetic and recreational value, promoting physical activity and mental well-being (Huang et al. 2017; Zhang et al. 2017). For example, research indicates that urban green spaces contribute to lower levels of PM2.5 and PM10, which are critical for reducing the burden of respiratory diseases in urban populations (Huang et al. 2017; Darettamarlan et al. 2022). Furthermore, the establishment of low-emission zones (LEZ) in cities has proven effective in reducing traffic-related air pollution, thereby improving overall air quality and public health (Margaryan 2021; Gehrsitz 2017). Implementing low-emission zones represents a proactive approach to urban air quality management. By restricting access to high-emission vehicles in designated areas, cities can significantly reduce traffic-related air pollution (Margaryan 2021; Börjesson et al. 2021). Studies have demonstrated that LEZs lead to measurable improvements in air quality, positively impacting public health, particularly among vulnerable populations such as infants and the elderly (Gehrsitz 2017; Margaryan 2021). For instance, introducing LEZs in cities like London and Milan has substantially reduced nitrogen dioxide levels and other harmful pollutants (Mądziel 2023; Börjesson et al. 2021). Additionally, the economic implications of LEZs are noteworthy, as they can stimulate the adoption

Table 2.5 Environmental health management strategies

Strategy	Description	Examples	Refs.
Policy approaches: regulatory frameworks	We are implementing laws, regulations, and standards to control air pollution and protect public health	WHO air quality guidelines: set global standards for PM, NOx, SO_2, and O_3 concentrations National clean air act: legislation in various countries (e.g., U.S., Germany) to reduce air pollution through strict industrial and vehicular emission standards	WHO (2021)
Urban planning and design for air quality improvement	Urban strategies that integrate green infrastructure and zoning to minimize air pollution and enhance quality of life	Green spaces: parks and vegetation help filter air pollutants and reduce urban heat Low-emission zones (LEZs): cities like London and Paris restrict vehicle access to reduce emissions in high-traffic areas	Venter et al. (2024)
Public awareness campaigns and community engagement	Initiatives aimed at educating the public about air quality and empowering communities to take action	Public health campaigns: awareness programs on the risks of air pollution and how individuals can reduce their exposure Citizen science projects: community involvement in monitoring local air quality and advocating for policy changes	Ramírez et al. (2019)

of cleaner transportation options and promote the use of alternative fuels (Peters et al. 2021; Holnicki et al. 2023).

Public awareness campaigns and community engagement are essential to effective environmental health management strategies. Educating the public about the health impacts of air pollution and the importance of sustainable practices can foster community support for regulatory measures and urban planning initiatives (Quarmby et al. 2019; Wolch et al. 2014). Campaigns promoting the benefits of green spaces and low-emission zones can encourage residents to advocate for policies prioritizing air quality improvement (Rizki et al. 2022; Huang et al. 2017). Furthermore, community engagement in the planning process ensures that the needs and preferences of residents are considered, leading to more effective and accepted interventions (Sakina and Setiawan 2023; Wolch et al. 2014). Research has shown that communities with higher public engagement in environmental health initiatives experience better health outcomes and greater satisfaction with their living environments (Zhang et al. 2017; Sakina and Setiawan 2023). Integrating technology in monitoring air quality

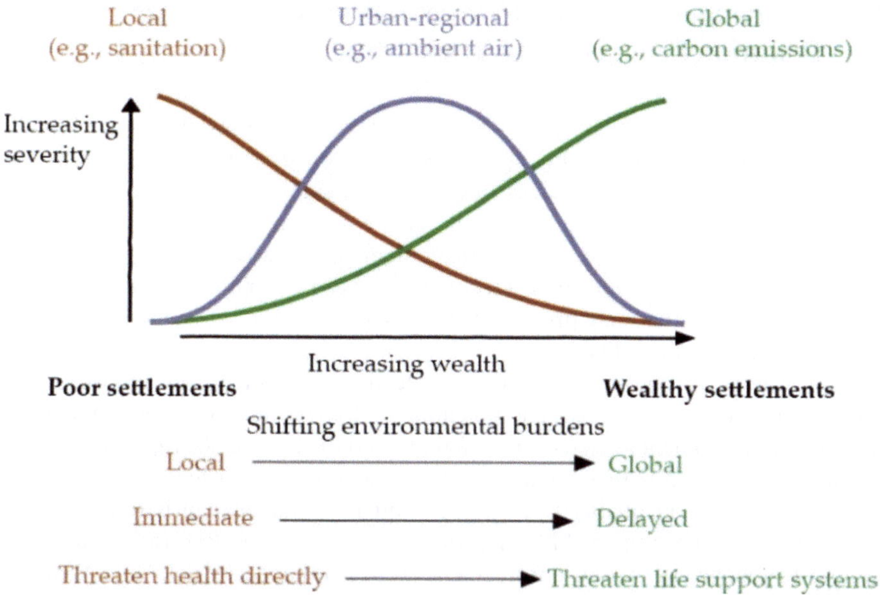

Fig. 2.1 Urban growth intervention strategies. *Source* Grafakos et al. (2016)

and community engagement can enhance the effectiveness of environmental health management strategies. Advanced monitoring systems that utilize real-time data can provide valuable insights into air quality trends and pollution sources, enabling policymakers to make informed decisions (Huang et al. 2017; Yu et al. 2017). Additionally, mobile applications that allow residents to report air quality issues and access information about local pollution levels can empower communities to take action (Quarmby et al. 2019; Yu et al. 2017). This technological integration improves transparency and fosters a sense of collective responsibility among residents, encouraging them to participate in air quality improvement efforts (Sakina and Setiawan 2023; Yu et al. 2017).

Moreover, intersecting public health and environmental justice is critical in developing ecological health management strategies. Disadvantaged communities often bear a disproportionate burden of air pollution and its health impacts (Wolch et al. 2014; Holnicki et al. 2023). Therefore, policies to improve air quality must also address social inequities to ensure that all populations benefit from cleaner air (Wolch et al. 2014; Holnicki et al. 2023). Strategies that prioritize the needs of vulnerable groups, such as low-income families and communities of color, can help mitigate health disparities and promote equitable access to clean air and green spaces (Wolch et al. 2014; Holnicki et al. 2023).

2.7 Challenges and Future Directions in Assessing Air Pollution Processes and Their Mitigation

Implementing sustainable air quality strategies faces numerous challenges broadly categorized into economic, political, and social factors. Economically, the transition towards more sustainable practices often requires significant upfront investments in technology and infrastructure, which can be a barrier for many governments and businesses. For instance, the cost of implementing advanced air quality monitoring systems and pollution control technologies can be prohibitive, particularly for developing nations that may prioritize immediate economic growth over long-term environmental sustainability (Coccia 2021a, b). Furthermore, the financial implications of stricter air quality regulations can lead to resistance from industries that fear increased operational costs and potential job losses, creating a conflict between economic development and environmental protection (Coccia 2021a, b). Politically, the lack of cohesive policies and regulatory frameworks can hinder the effective implementation of air quality strategies. Air quality management is fragmented in many regions, with responsibilities divided among various governmental levels, leading to inconsistencies in enforcement and compliance (Monks and Williams 2020).

Moreover, political will is often lacking, as policymakers may prioritize short-term gains over long-term environmental health, especially in the face of competing interests from powerful industrial lobbies (Woods and Potoski 2010). The need for international cooperation further compounds the challenge, as air pollution is a transboundary issue that requires coordinated efforts across borders. Yet, geopolitical tensions can impede collaborative initiatives.

Social factors also play a crucial role in implementing sustainable air quality strategies. Public awareness and engagement are vital for the success of any environmental policy. Yet, many communities remain uninformed about the health impacts of air pollution and the importance of sustainable practices (Ferreira et al. 2013). Additionally, socioeconomic disparities can exacerbate the effects of air pollution, as marginalized communities often bear the brunt of poor air quality while having limited resources to advocate for change (Rappold et al. 2012). This inequity can lead to a lack of public support for air quality initiatives, as those most affected may feel powerless to influence policy decisions (Eastham et al. 2023). Emerging technologies and research are pivotal in shaping the future of air quality management. Innovations such as low-cost air quality sensors, satellite monitoring, and artificial intelligence for predictive modeling are transforming how to monitor and respond to air pollution (Aderoba et al. 2023). These technologies enable real-time data collection and analysis, allowing for more responsive and targeted interventions. For example, machine learning algorithms can analyze vast datasets to identify pollution sources and predict air quality trends, facilitating proactive measures to mitigate health risks (Klausbruckner et al. 2016). Furthermore, advancements in renewable energy technologies can significantly reduce emissions from traditional fossil fuel sources, improving air quality (Ji et al. 2023).

Research into the health impacts of air pollution continues to evolve, providing critical insights that inform policy decisions. Studies have demonstrated a direct correlation between air quality and public health outcomes, highlighting the urgent need for effective air quality management strategies (Westervelt et al. 2019). For instance, research has shown that reducing particulate matter (PM2.5) levels can significantly decrease respiratory and cardiovascular diseases, thereby reducing healthcare costs and improving overall community health (Cui et al. 2020). This evidence underscores the importance of integrating health considerations into air quality policies, ensuring that public health is prioritized in decision-making. Global collaboration is essential for reducing air pollution and protecting public health. International agreements, such as the Paris Agreement, emphasize the need for collective action to address climate change and its associated air quality challenges. Collaborative efforts can take various forms, including knowledge sharing, joint research initiatives, and coordinated policy frameworks that align national efforts with global goals. For example, the World Health Organization (WHO) has been instrumental in promoting global air quality standards and providing countries with guidance on effective pollution control measures (Skirienė and Stasiškienė 2021). Such collaborations can enhance the capacity of nations to implement sustainable air quality strategies and share best practices.

Furthermore, partnerships between governments, non-governmental organizations (NGOs), and the private sector can foster innovation and resource sharing, leading to more effective air quality management solutions. Engaging local communities in these efforts is also crucial, as grassroots movements can drive change from the bottom up, advocating for policies that reflect the needs and concerns of those most affected by air pollution. By fostering a culture of collaboration and inclusivity, stakeholders can work together to overcome barriers and achieve sustainable air quality goals.

2.8 Conclusion

Mitigating the public health impacts of air pollution demands a holistic approach that integrates innovative technologies, effective policies, and active public engagement. As urbanization, industrialization, and transportation continue to drive air pollution, technological advancements—such as air filtration systems, low-emission transport, and renewable energy solutions—are crucial for reducing emissions and improving air quality. However, technology alone cannot solve the issue; robust regulatory frameworks and government policies aligned with international air quality standards are equally vital for achieving sustainable results. Furthermore, public awareness and community involvement play an essential role in addressing air pollution at the local level, fostering behavior change, and encouraging sustainable practices. Collaborative efforts across sectors involving policymakers, industries, and civil society are necessary to create environments that support public health and environmental sustainability. Looking forward, the future of mitigating air pollution's impact on global health rests on continued innovation and coordinated international action. Emerging technologies such as advanced pollution monitoring, carbon capture, and clean energy solutions offer promising avenues for reducing pollution at the source.

Additionally, interdisciplinary research and data analysis are critical for understanding the complex relationships between pollutants, climate change, and health outcomes. Governments and organizations must also ensure equitable access to clean air, with particular attention to vulnerable populations—children, the elderly, and those with pre-existing conditions—who bear the most significant health risks from poor air quality. Creating a healthier, sustainable future requires a comprehensive, global response. By aligning technological innovations with solid policies and engaging the public in solutions, the health burden of air pollution can be significantly reduced, thus safeguarding the well-being of both present and future generations.

References

Acciai C, Zhang Z, Wang F, Zhong Z, Lonati G (2017) Characteristics and source analysis of trace elements in PM2.5 in the urban atmosphere of Wuhan in spring. Aerosol Air Qual Res 17:2224–2234. https://doi.org/10.4209/aaqr.2017.06.0207

Aderoba AK, Kabra R, Kiarie JN (2023) Implementation strategies, facilitators, and barriers to scaling up and sustaining task-sharing in family planning: a protocol for a mixed-methods systematic review. Syst Rev 12(1):190

Anand V, Kadiri VL, Putcha C (2023) Passive buildings: a state-of-the-art review. J Infrastruct Preserv Resil 4(1):3. https://doi.org/10.1186/s43065-022-00068-z

Asif Z, Chen Z, Wang H, Zhu Y (2022) Update on air pollution control strategies for coal-fired power plants. Clean Technol Environ Policy 24(8):2329–2347. https://doi.org/10.1007/s10098-022-02328-8

Azid A, Juahir H, Aris AZ, Toriman ME, Latif MT, Zain SM, Yusof KMKK, Saudi ASM (2014) Spatial analysis of the air pollutant index in the Southern Region of Peninsular Malaysia using environmetric techniques. In: From sources to solution: proceedings of the international conference on environmental forensics 2013. Springer Singapore, pp 307–312. https://doi.org/10.1007/978-981-4560-70-2_56

Bentayeb M, Norback D, Bednarek M, Bernard A, Cai G, Cerrai S, Eleftheriou KK, Gratziou C, Holst GJ, Lavaud F, Nasilowski J (2015) Indoor air quality, ventilation and respiratory health in elderly residents living in nursing homes in Europe. Eur Respir J 45(5):1228–1238. https://doi.org/10.1183/09031936.00082414

Börjesson M, Bastian A, Eliasson J (2021) The economics of low emission zones. Transp Res Part A: Policy Pract 153:99–114. https://doi.org/10.1016/j.tra.2021.08.016

Bui HT, Odsuren U, Kim SY, Park BJ (2022) Seasonal variations in the particulate matter accumulation and leaf traits of 24 plant species in urban green space. Land 11(11):1981. https://doi.org/10.3390/land11111981

Cai A, Zheng S, Cai L, Yang H, Comite U (2021) How does green technology innovation affect carbon emissions? A spatial econometric analysis of China's provincial panel data. Front Environ Sci 9:813811. https://doi.org/10.3389/fenvs.2021.813811

Carey J (2023) The other benefit of electric vehicles. Proc Natl Acad Sci 120(3):e2220923120. https://doi.org/10.1073/pnas.2220923120

Chang JH, Lee YL, Chang LT, Chang TY, Hsiao TC, Chung KF, Ho KF, Kuo HP, Lee KY, Chuang KJ, Chuang HC (2023) Climate change, air quality, and respiratory health: a focus on particle deposition in the lungs. Ann Med 55(2):2264881. https://doi.org/10.1080/07853890.2023.2264881

Chen J, Long X, Lin S (2022b) Special economic zone, carbon emissions and the mechanism role of green technology vertical spillover: evidence from Chinese cities. Int J Environ Res Public Health 19(18):11535. https://doi.org/10.3390/ijerph191811535

Chen CF, Hsu CH, Chang YJ, Lee CH, Lee DL (2022a) Efficacy of HEPA air cleaner on improving indoor particulate matter 2.5 concentration. Int J Environ Res Public Health 19(18):11517. https://doi.org/10.3390/ijerph191811517

Coccia M (2021a) The effects of atmospheric stability with low wind speed and of air pollution on the accelerated transmission dynamics of COVID-19. Int J Environ Stud 78(1):1–27. https://doi.org/10.1080/00207233.2020.1802937

Coccia M (2021b) Effects of the spread of COVID-19 on public health of polluted cities: results of the first wave for explaining the dejà vu in the second wave of COVID-19 pandemic and epidemics of future vital agents. Environ Sci Pollut Res 28(15):19147–19154. https://doi.org/10.1007/s11356-020-11662-7

Costin A, Clark A (2022) Air pollution and social determinants of health throughout US counties. https://doi.org/10.21203/rs.3.rs-2351598/v1

Cui L, Zhou J, Peng X, Ruan S, Zhang Y (2020) Analyses of air pollution control measures and co-benefits in the heavily air-polluted Jinan city of China, 2013–2017. Sci Rep 10(1):5423. https://doi.org/10.1038/s41598-020-62475-0

Cui B, Shui Z, Yang S, Lei T (2022) Evolutionary game analysis of green technology innovation under the carbon emission trading mechanism. Front Environ Sci 10:997724. https://doi.org/10.3389/fenvs.2022.997724

Cui B, Liu Y, Wu J (2023) Digital economy, green technology innovation and carbon reduction effect. In: Proceedings of the 2023 2nd international conference on artificial intelligence, internet and digital economy (ICAID 2023), vol 9. Springer Nature, p 215. https://doi.org/10.2991/978-94-6463-222-4_21

Darettamarlan RR, Sanjaya H, Cahyono AB, Handayani HH, Darminto MR (2022) Changes in land requirements analysis for green space needed in Bogor city. Int J Adv Sci, Eng Inf Technol 12(5):2053–2059. https://doi.org/10.18517/ijaseit.12.5.15072

Davis SJ, Lewis NS, Shaner M, Aggarwal S, Arent D, Azevedo IL, Benson SM, Bradley T, Brouwer J, Chiang YM, Clack CT (2018a) Net-zero emissions energy systems. Science 360(6396):eaas9793. https://doi.org/10.1126/science.aas9793

Davis SJ, Lewis NS, Shaner M, Aggarwal S, Arent D, Azevedo IL, Benson SM, Bradley T, Brouwer J, Chiang YM, Clack CT (2018b) Net-zero emissions energy systems. Science 360(6396):eaas9793. https://doi.org/10.1126/science.aas9793

Dutta V, Kumar S, Dubey D (2021) Recent advances in satellite mapping of global air quality: evidences during COVID-19 pandemic. Environ Sustain 4(3):469–487. https://doi.org/10.1007/s42398-021-00166-w

Eastham SD, Monier E, Rothenberg D, Paltsev S, Selin NE (2023) Rapid estimation of climate—air quality interactions in integrated assessment using a response surface model. ACS Environ Au 3(3):153–163. https://doi.org/10.1021/acsenvironau.2c00054

El-Sheekh MM, Hassan IA (2021) Lockdowns and reduction of economic activities during the COVID-19 pandemic improved air quality in Alexandria, Egypt. Environ Monit Assess 193(1):11. https://doi.org/10.1007/s10661-020-08780-7

Farhane M, Souhar O (2023) A novel development in three-dimensional analytical solutions for air pollution dispersion modeling. https://doi.org/10.5772/intechopen.112225

Farzaneh H, de Oliveira JAP, McLellan B, Ohgaki H (2019) Towards a low emission transport system: evaluating the public health and environmental benefits. Energies 12(19):3747. https://doi.org/10.3390/en12193747

Ferreira S, Akay A, Brereton F, Cuñado J, Martinsson P, Moro M, Ningal TF (2013) Life satisfaction and air quality in Europe. Ecol Econ 88:1–10. https://doi.org/10.1016/j.ecolecon.2012.12.027

Gallagher CL, Holloway T (2020) Integrating air quality and public health benefits in US decarbonization strategies. Front Public Health 8:563358. https://doi.org/10.3389/fpubh.2020.563358

Gao P, Wang Y, Zou Y, Su X, Che X, Yang X (2022) Green technology innovation and carbon emissions nexus in China: does industrial structure upgrading matter? Front Psychol 13:951172. https://doi.org/10.3389/fpsyg.2022.951172

Garcia E, Rice MB, Gold DR (2021) Air pollution and lung function in children. J Allergy Clin Immunol 148(1):1–14. https://doi.org/10.1016/j.jaci.2021.05.006

Gehrsitz M (2017) The effect of low emission zones on air pollution and infant health. J Environ Econ Manag 83:121–144. https://doi.org/10.1016/j.jeem.2017.02.003

Grafakos S, Gianoli A, Tsatsou A (2016) Towards the development of an integrated sustainability and resilience benefits assessment framework of urban green growth interventions. Sustainability 8(5):461. https://doi.org/10.3390/su805046

Gurajala S, Dhaniyala S, Matthews JN (2019) Understanding public response to air quality using tweet analysis. Soc Media+ Soc 5(3):2056305119867656. https://doi.org/10.1177/2056305119867656

Gyasi PA, Amu-Darko JNO, Issaka E, Amarteifio ENA (2022) Air pollution and health: particulate matter PM2.5 measurements in cantonments, Ghana, Epra Int J Multidiscip Res (IJMR) 131–138. https://doi.org/10.36713/epra11460

Hassan Bhat T, Jiawen G, Farzaneh H (2021) Air pollution health risk assessment (AP-HRA), principles and applications. Int J Environ Res Public Health 18(4):1935. https://doi.org/10.3390/ijerph18041935

Holme JA, Vondráček J, Machala M, Lagadic-Gossmann D, Vogel CF, Le Ferrec E, Sparfel L, Øvrevik J (2023) Lung cancer associated with combustion particles and fine particulate matter (PM2.5)—the roles of polycyclic aromatic hydrocarbons (PAHs) and the aryl hydrocarbon receptor (AhR). Biochem Pharmacol 115801. https://doi.org/10.1016/j.bcp.2023.115801

Holnicki P, Kałuszko A, Nahorski Z (2023) A projection of environmental impact of a low emission zone planned in Warsaw, Poland. Sustainability 15(23):16260. https://doi.org/10.3390/su152316260

Honda T, Pun VC, Manjourides J, Suh H (2017) Anemia prevalence and hemoglobin levels are associated with long-term exposure to air pollution in an older population. Environ Int 101:125–132. https://doi.org/10.1016/j.envint.2017.01.017

Hou J, Bai W, Sha D (2023) Does the digital economy successfully facilitate carbon emission reduction in China? Green technology innovation perspective. Sci, Technol Soc 28(4):535–560. https://doi.org/10.1177/09717218231161235

Huang C, Yang J, Lu H, Huang H, Yu L (2017) Green spaces as an indicator of urban health: Evaluating its changes in 28 mega-cities. Remote Sens 9(12):1266. https://doi.org/10.3390/rs9121266

Huang J, Pan X, Guo X, Li G (2018) Health impact of China's air pollution prevention and control action plan: an analysis of national air quality monitoring and mortality data. Lancet Planet Health 2(7):e313–e323. https://doi.org/10.1016/s2542-5196(18)30141-4

Huang Y, Lei C, Liu CH, Perez P, Forehead H, Kong S, Zhou JL (2021) A review of strategies for mitigating roadside air pollution in urban street canyons. Environ Pollut 280:116971. https://doi.org/10.1016/j.envpol.2021.116971

Imasiku K, Thomas V, Ntagwirumugara E (2019) Unraveling green information technology systems as a global greenhouse gas emission game-changer. Adm Sci 9(2):43. https://doi.org/10.3390/admsci9020043

Izah SC, Uzoekwe S, Aigberua A (2021) Source, geochemical spreading and risks of trace metals in particulate matter 2.5 within a gas flaring area in Bayelsa State, Nigeria. Adv Environ Technol 7(2):101–118. https://doi.org/10.22104/aet.2021.5053.1368

Izah SC, Amabie T, Aigberua AO (2024a) Technological advancements for sustainable clean air. In: Ogwu MC, Izah SC (eds) Sustainable strategies for air pollution mitigation: development, economics and technologies. Part of the book series: the handbook in environmental chemistry. Springer, Berlin, Heidelberg. https://doi.org/10.1007/698_2024_1120

Izah SC, Ogwu MC, Etim NG, Shahsavani A, Namvar Z (2024b) Short-term health effects of air pollution. In: Izah SC, Ogwu MC, Shahsavani A (eds) Air pollutants in the context of one health: fundamentals, sources and impacts. The handbook in environmental chemistry. Springer, Berlin, Heidelberg

Izah SC, Ogwu MC, Shahsavani A, Kermani M, Hasanzadeh V (2024c) Understanding the one health implications of air pollution. In: Izah SC, Ogwu MC, Shahsavani A (eds) Air pollutants in the context of one health: fundamentals, sources and impacts. The handbook in environmental chemistry. Springer, Berlin, Heidelberg. https://doi.org/10.1007/698_2024_1129

Izah SC, Onwudiegwu CA, Sylva L, Etim NG (2024d) Health and environmental justice strategies for mitigating air pollution. In: Ogwu MC, Izah SC (eds) Sustainable strategies for air pollution mitigation: development, economics and technologies. Part of the book series: the handbook in environmental chemistry. Springer

Izah SC, Sawyer WE, Ogwu MC, Shahsavani A, Kermani M, Hasanzadeh V (2024e) Air pollution principles, research, and innovation. In: Izah SC, Ogwu MC, Shahsavani A (eds) Air pollutants in the context of one health: fundamentals, sources and impacts. The handbook in environmental chemistry. Springer. https://doi.org/10.1007/698_2024_1131

Izah SC, Sylva L, Ogwu MC, Shahsavani A, Bazzazpour S, Rahmatinia M (2024f) Modeling and statistical approaches for air pollution analysis. In: Izah SC, Ogwu MC, Shahsavani A (eds) Air pollutants in the context of one health: fundamentals, sources and impacts. The handbook in environmental chemistry. Springer. https://doi.org/10.1007/698_2024_1138

Jandačka D (2015) Sources of PM air pollution in rural area in the vicinity of a highway in Žilina self-governing region, Slovakia. Civ Environ Eng 11(1):58–68. https://doi.org/10.1515/cee-2015-0008

Ji D, Liu Z, Wang L, Peng Y (2023, October) Research on PM2.5 influencing factors of LSTM based on attention mechanism. In: Fifth international conference on artificial intelligence and computer science (AICS 2023), vol 12803. SPIE, pp 979–984. https://doi.org/10.1117/12.3009515

Johnson JB (2022) An introduction to atmospheric pollutant dispersion modelling. Environ Sci Proc 19(1):18. https://doi.org/10.3390/ecas2022-12826

Kelishadi R, Poursafa P (2010) Air pollution and non-respiratory health hazards for children. Arch Med Sci 6(4):483–495. https://doi.org/10.5114/aoms.2010.14458

Klausbruckner C, Annegarn H, Henneman LR, Rafaj P (2016) A policy review of synergies and trade-offs in South African climate change mitigation and air pollution control strategies. Environ Sci Policy 57:70–78. https://doi.org/10.1016/j.envsci.2015.12.001

Leal Filho W, Vidal DG, Chen C, Petrova M, Dinis MAP, Yang P, Rogers S, Álvarez-Castañón L, Djekic I, Sharifi A, Neiva S (2022) An assessment of requirements in investments, new technologies, and infrastructures to achieve the SDGs. Environ Sci Eur 34(1):1–17. https://doi.org/10.1186/s12302-022-00629-9

Lee YG, Lee PH, Choi SM, An MH, Jang AS (2021) Effects of air pollutants on airway diseases. Int J Environ Res Public Health 18(18):9905. https://doi.org/10.3390/ijerph18189905

Lee EH, Ryu D, Hong NS, Kim JY, Park KD, Lee WJ, Lee SJ, Kim SH, Do Y, Jang YH (2022) Defining the relationship between daily exposure to particulate matter and hospital visits by psoriasis patients. Ann Dermatol 34(1):40. https://doi.org/10.5021/ad.2022.34.1.40

Leelőssy Á, Molnár F, Izsák F, Havasi Á, Lagzi I, Mészáros R (2014) Dispersion modeling of air pollutants in the atmosphere: a review. Open Geosci 6(3):257–278. https://doi.org/10.2478/s13533-012-0188-6

Lelieveld J, Haines A, Pozzer A (2018) Age-dependent health risk from ambient air pollution: a modelling and data analysis of childhood mortality in middle-income and low-income countries. Lancet Planet Health 2(7):e292–e300. https://doi.org/10.1016/s2542-5196(18)30147-5

Li Y, Xu L, Shan Z, Teng W, Han C (2019) Association between air pollution and type 2 diabetes: an updated review of the literature. Ther Adv Endocrinol Metab 10:2042018819897046. https://doi.org/10.1177/2042018819897046

Li Y, Zhang C, Li S, Usman A (2022) Energy efficiency and green innovation and its asymmetric impact on CO_2 emission in China: a new perspective. Environ Sci Pollut Res 29(31):47810–47817. https://doi.org/10.1007/s11356-022-19161-7

Lian X, Wan X, Si F, Guo J, Tang Z, Wang Y, Xie S, Li J (2023) The association between air pollutants and depression in children and adolescents: a systematic review. Environ Res Commun 5(10):102002. https://doi.org/10.1088/2515-7620/ad00a6

Liu X, Yang L, Liu G, Zheng M (2021) Formation of environmentally persistent free radicals during thermochemical processes and their correlations with unintentional persistent organic pollutants. Environ Sci Technol 55(10):6529–6541. https://doi.org/10.1021/acs.est.0c08762

Liu Y, Chen L, Huang C (2022) Study on the carbon emission spillover effects of transportation under technological advancements. Sustainability 14(17):10608. https://doi.org/10.3390/su141710608

Liu NM, Grigg J (2018) Diesel, children and respiratory disease. BMJ Paediatr Open 2(1). https://doi.org/10.1136/bmjpo-2017-000210

Liu HaiYing LH, Kobernus M, Liu Hong LH (2017) Public perception survey study on air quality issues in Wuhan, China. J Environ Prot 08(10):1194–1218. https://doi.org/10.4236/jep.2017.810075

Lysova E, Paramonova O, Samarskaya N, Gyrova O, Tsarevskaya I (2018) Application of physical and energetic approach to estimation and selection of atmospheric protection systems for energetic devices. In: MATEC web of conferences, vol 170. EDP Sciences, p 04013. https://doi.org/10.1051/matecconf/201817004013

Ma X, Li X, Kwan MP, Chai Y (2020) Who could not avoid exposure to high levels of residence-based pollution by daily mobility? Evidence of air pollution exposure from the perspective of the neighborhood effect averaging problem (NEAP). Int J Environ Res Public Health 17(4):1223. https://doi.org/10.3390/ijerph17041223

Mądziel M (2023) Future cities carbon emission models: hybrid vehicle emission modelling for low-emission zones. Energies 16(19):6928. https://doi.org/10.3390/en16196928

Majumdar D (2021) How are the two most polluted metro-cities of India combating air pollution? Way forward after lifting of covid-19 lockdown. Aerosol Air Qual Res 21(1):200463. https://doi.org/10.4209/aaqr.2020.07.0463

Margaryan S (2021) Low emission zones and population health. J Health Econ 76:102402. https://doi.org/10.1016/j.jhealeco.2020.102402

Martinėnas B, Špakauskas V, Jasaitis D (2011) Semi-empirical model of the simulation of traffic pollution dispersion near roadways. Balt J Road Bridg Eng 6(1):17–22. https://doi.org/10.3846/bjrbe.2011.03

Matkovic V, Mulić M, Azabagić S, Jevtić M (2020) Premature adult mortality and years of life lost attributed to long-term exposure to ambient particulate matter pollution and potential for mitigating adverse health effects in Tuzla and Lukavac, Bosnia and Herzegovina. Atmosphere 11(10):1107. https://doi.org/10.3390/atmos11101107

McCarron A, Semple S, Braban CF, Swanson V, Gillespie C, Price HD (2023) Public engagement with air quality data: using health behaviour change theory to support exposure-minimising behaviours. J Eposure Sci Environ Epidemiol 33(3):321–331. https://doi.org/10.1038/s41370-022-00449-2

Meng Q, Zhang JW, Wang Y, Chang HL, Su CW (2023) Green household technology and its impacts on environmental sustainability in China. Sustainability 15(17):12919. https://doi.org/10.3390/su151712919

Meo SA, Almutairi FJ, Abukhalaf AA, Usmani AM (2021) Effect of green space environment on air pollutants PM2.5, PM10, CO, O_3, and incidence and mortality of SARS-CoV-2 in highly green and less-green countries. Int J Environ Res Public Health 18(24):13151. https://doi.org/10.3390/ijerph182413151

Monks PS, Williams ML (2020) What does success look like for air quality policy? A perspective. Phil Trans R Soc A 378(2183):20190326. https://doi.org/10.1098/rsta.2019.0326

Noh J, Sohn J, Cho J, Cho SK, Choi YJ, Kim C, Shin DC (2016) Short-term effects of ambient air pollution on emergency department visits for asthma: an assessment of effect modification by prior allergic disease history. J Prev Med Public Health 49(5):329. https://doi.org/10.3961/jpmph.16.038

Nosek R, Holubčík M, Jandacka J, Vician P (2016) Formation of emission from combustion of biomass in small heat source. Key Eng Mater 688:65–73. https://doi.org/10.4028/www.scientific.net/kem.688.65

Ogwu MC, Izah SC (2024) Air pollution and the sustainable development goals. In: Ogwu MC, Izah SC (eds) Sustainable strategies for air pollution mitigation: development, economics and technologies. Part of the book series: the handbook in environmental chemistry. Springer

Ogwu MC, Izah SC, Imarhiagbe O, Lori T, Aliu OO (2024a) Effects of air pollutants on biodiversity. In: Izah SC, Ogwu MC, Shahsavani A (eds) Air pollutants in the context of one health: fundamentals, sources and impacts. The handbook in environmental chemistry. Springer. https://doi.org/10.1007/698_2024_1135

Ogwu MC, Lori T, Aliu OO, Febnteh EB, Izah SC, Abdelkhalek ST (2024b) Agricultural air pollution: impacts, sources, and mitigation strategies. In: Izah SC, Ogwu MC, Shahsavani A (eds) Air pollutants in the context of one health: fundamentals, sources and impacts. The handbook in environmental chemistry. Springer. https://doi.org/10.1007/698_2024_1137

Oyebanji MO, Kirikkaleli D (2023) Green technology, green electricity, and environmental sustainability in Western European countries. Environ Sci Pollut Res 30(13):38525–38534. https://doi.org/10.1007/s11356-022-24793-w

Palacio LC, Pachajoa DC, Echeverri-Londoño CA, Saiz J, Tobón C (2023) Air pollution and cardiac diseases: a review of experimental studies. Dose-Response 21(4):15593258231212793. https://doi.org/10.1177/15593258231212793

Pérez IA, García MÁ, Sánchez ML, Pardo N, Fernández-Duque B (2020) Key points in air pollution meteorology. Int J Environ Res Public Health 17(22):8349. https://doi.org/10.3390/ijerph17228349

Peters JF, Burguillo M, Arranz JM (2021) Low emission zones: Effects on alternative-fuel vehicle uptake and fleet CO_2 emissions. Transp Res Part d: Transp Environ 95:102882. https://doi.org/10.1016/j.trd.2021.102882

Quarmby S, Santos G, Mathias M (2019) Air quality strategies and technologies: a rapid review of the international evidence. Sustainability 11(10):2757. https://doi.org/10.3390/su11102757

Raaschou-Nielsen O, Andersen ZJ, Beelen R, Samoli E, Stafoggia M, Weinmayr G, Hoffmann B, Fischer P, Nieuwenhuijsen MJ, Brunekreef B, Xun WW, Hoek G (2013) Air pollution and lung cancer incidence in 17 European cohorts: prospective analyses from the European Study of Cohorts for Air Pollution Effects (ESCAPE). Lancet Oncol 14(9):813–822. https://doi.org/10.1016/s1470-2045(13)70279-1

Ramírez AS, Ramondt S, Van Bogart K, Perez-Zuniga R (2019) Public awareness of air pollution and health threats: challenges and opportunities for communication strategies to improve environmental health literacy. J Health Commun 24(1):75–83. https://doi.org/10.1080/10810730.2019.1574320

Ramondt S, Ramírez AS (2020) Media reporting on air pollution: health risk and precautionary measures in national and regional newspapers. Int J Environ Res Public Health 17(18):6516. https://doi.org/10.3390/ijerph17186516

Rappold AG, Cascio WE, Kilaru VJ, Stone SL, Neas LM, Devlin RB, Diaz-Sanchez D (2012) Cardio-respiratory outcomes associated with exposure to wildfire smoke are modified by measures of community health. Environ Health 11:1–9. https://doi.org/10.1186/1476-069x-11-71

Relvas H, Ferreira J, Lopes D, Rafael S (2018) Improving air quality and human health: an approach based on artificial neural networks. WIT Trans Ecol Environ 230:205–212. https://doi.org/10.2495/air180191

Rizki M, Irawan MZ, Dirgahayani P, Belgiawan PF, Wihanesta R (2022) Low emission zone (LEZ) Expansion in Jakarta: acceptability and restriction preference. Sustainability 14(19):12334. https://doi.org/10.3390/su141912334

Roy R, D'Angiulli A (2024) Air pollution and neurological diseases, current state highlights. Front Neurosci 18:1351721. https://doi.org/10.3389/fnins.2024.1351721

Rumambi FJ (2023) Environmental effects on health: the role of sustainable environmental management. Jurnal Aisyah: Jurnal Ilmu Kesehatan 8(2). https://doi.org/10.30604/jika.v8i2.2028

Sakina B, Setiawan D (2023) Housing satisfaction and preferences towards indoor quality related to the environmental factors in the context of the Covid-19 pandemic. In: IOP conference series: earth and environmental science, vol 1169, No 1. IOP Publishing, p 012076. https://doi.org/10.1088/1755-1315/1169/1/012076

Shi T, Si S, Chan J, Zhou L (2021) The carbon emission reduction effect of technological innovation on the transportation industry and its spatial heterogeneity: evidence from China. Atmosphere 12(9):1169. https://doi.org/10.3390/atmos12091169

Singh A (2023) Design and development of air pollution monitoring using smartphone. Int J Eng Technol Manag Sci 7(3):489–493. https://doi.org/10.46647/ijetms.2023.v07i03.68

Skirienė AF, Stasiškienė Ž (2021) COVID-19 and air pollution: measuring pandemic impact on air quality in five European countries. Atmosphere 12(3):290. https://doi.org/10.3390/atmos12030290

Soesanti F, Uiterwaal CS, Meliefste K, Chen J, Brunekreef B, Idris NS, Grobbee DE, Klipstein-Grobusch K, Hoek G (2023) The effect of exposure to traffic related air pollutants in pregnancy on birth anthropometry: a cohort study in a heavily polluted low-middle income country. Environ Health 22(1):22. https://doi.org/10.1186/s12940-023-00973-0

Sompornrattanaphan M, Thongngarm T, Ratanawatkul P, Wongsa C, Swigris JJ (2020) The contribution of particulate matter to respiratory allergy. Asian Pac J Allergy Immunol 38(1):19–28. https://doi.org/10.12932/ap-100619-0579

Syuhada G, Akbar A, Hardiawan D, Pun V, Darmawan A, Heryati SHA, Siregar AYM, Kusuma RR, Driejana R, Ingole V, Kass D, Mehta S (2023) Impacts of air pollution on health and cost of illness in Jakarta, Indonesia. Int J Environ Res Public Health 20(4):2916. https://doi.org/10.3390/ijerph20042916

Tonne C, Wilkinson P (2013) Long-term exposure to air pollution is associated with survival following acute coronary syndrome. Eur Heart J 34(17):1306–1311. https://doi.org/10.1093/eurheartj/ehs480

Tsocheva I, Scales J, Dove R, Chavda J, Kalsi H, Wood HE, Colligan G, Cross L, Newby C, Hall A, Keating M, Griffiths CJ (2023) Investigating the impact of London's ultra-low emission zone on children's health: children's health in London and Luton (CHILL) protocol for a prospective parallel cohort study. BMC Pediatr 23(1):556. https://doi.org/10.1186/s12887-023-04384-5

Tunno BJ, Shields KN, Lioy P, Chu N, Kadane JB, Parmanto B, Pramana G, Zora J, Davidson C, Holguin F, Clougherty JE (2012) Understanding intra-neighborhood patterns in PM 2.5 and PM 10 using mobile monitoring in Braddock, PA. Environ Health 11:1–15. https://doi.org/10.1186/1476-069x-11-76

Uzoekwe SA, Izah SC, Aigberua AO (2021) Environmental and human health risk of heavy metals in atmospheric particulate matter (PM 10) around gas flaring vicinity in Bayelsa State, Nigeria. Toxicol Environ Health Sci 1–13. https://doi.org/10.1007/s13530-021-00085-7

Venter ZS, Hassani A, Stange E, Schneider P, Castell N (2024) Reassessing the role of urban green space in air pollution control. Proc Natl Acad Sci 121(6):e2306200121. https://doi.org/10.1073/pnas.2306200121

Waidyatillake NT, Campbell PT, Vicendese D, Dharmage SC, Curto A, Stevenson M (2021) Particulate matter and premature mortality: a Bayesian meta-analysis. Int J Environ Res Public Health 18(14):7655. https://doi.org/10.3390/ijerph18147655

Wang Q, Zhang F (2021) The effects of trade openness on decoupling carbon emissions from economic growth—evidence from 182 countries. J Clean Prod 279:123838. https://doi.org/10.1016/j.jclepro.2020.123838

Wang M, Li Y, Wang Z, Li J (2022) The heterogeneous relationship between pollution charges and enterprise green technology innovation, based on the data of Chinese industrial enterprises. Energies 15(5):1663. https://doi.org/10.3390/en15051663

Wang F, Harindintwali JD, Yuan Z, Wang M, Wang F, Li S, Yin Z, Huang L, Fu Y, Li L, Chang SX, Chen JM (2021) Technologies and perspectives for achieving carbon neutrality. Innovation 2(4). https://doi.org/10.1016/j.xinn.2021.100180

Westervelt DM, Ma CT, He MZ, Fiore AM, Kinney PL, Kioumourtzoglou MA, Wang S, Xing J, Ding D, Correa G (2019) Mid-21st century ozone air quality and health burden in China under emissions scenarios and climate change. Environ Res Lett 14(7):074030. https://doi.org/10.1088/1748-9326/ab260b

Wolch JR, Byrne J, Newell JP (2014) Urban green space, public health, and environmental justice: the challenge of making cities 'just green enough.' Landsc Urban Plan 125:234–244. https://doi.org/10.1016/j.landurbplan.2014.01.017

Woods ND, Potoski M (2010) Environmental federalism revisited: second-order devolution in air quality regulation. Rev Policy Res 27(6):721–739. https://doi.org/10.1111/j.1541-1338.2010.00468.x

World Health Organization (2021) WHO global air quality guidelines: particulate matter (PM2.5 and PM10), ozone, nitrogen dioxide, sulfur dioxide and carbon monoxide. World Health Organization. https://www.ncbi.nlm.nih.gov/books/NBK574587/

Wright WJ, Neitlich PN, Shiel AE, Hooten MB (2022) Mechanistic spatial models for heavy metal pollution. Environmetrics 33(8):e2760. https://doi.org/10.1002/env.2760

Wu M, Zhang G, Wang L, Liu X, Wu Z (2022) Influencing factors on airflow and pollutant dispersion around buildings under the combined effect of wind and buoyancy—a review. Int J Environ Res Public Health 19(19):12895. https://doi.org/10.3390/ijerph191912895

Wu Y, Shiledar A, Luo Y, Wong J, Chen C, Bai B, Zhang Y, Tamamitsu M, Ozcan A (2018) Spatial mapping and analysis of aerosols during a forest fire using computational mobile microscopy. In: Optics and biophotonics in low-resource settings IV, vol 10485. SPIE, pp 78–86. https://doi.org/10.1117/12.2288889

Wu Y, Zeng S, Zhong Z (2023) Study on the emission reduction effect of green technology innovation. Adv Manag Appl Econ 13(6):277–295. https://doi.org/10.47260/amae/13614

Xiao Q, Liu Y, Mulholland JA, Russell AG, Darrow LA, Tolbert PE, Strickland MJ (2016) Pediatric emergency department visits and ambient air pollution in the US State of Georgia: a case-crossover study. Environ Health 15:1–8. https://doi.org/10.1186/s12940-016-0196-y

Xu BL, Lin H, Cui WN, Hu Y, Zhu J, Tang S (2011) CUGrid and VGE based air pollution dispersion simulation. Adv Mater Res 143:1305–1310. https://doi.org/10.4028/www.scientific.net/amr.143-144.1305

Yadav R, Nagori A, Mukherjee A, Singh V, Lodha R, Kabra SK, Yadav G, Saini JK, Singhal KK, Jat KR, Madan K, George MP, Mani K, Mrigpuri P, Kumar R, Guleria R, Pandey RM, Sarin R, Dhaliwal RS (2022) Geographic information system-based mapping of air pollution and emergency room visits of patients for acute respiratory symptoms in Delhi, India (March 2018–February 2019). Indian J Med Res 156(4&5):648–658. https://doi.org/10.4103/ijmr.IJMR_136_21

Yamineva Y, Romppanen S (2017) Is the law failing to address air pollution? Reflections on international and EU developments. Rev Eur, Comp Int Environ Law 26(3):189–200. https://doi.org/10.1111/reel.12223

Yan W (2023) Effect of PM2.5 air pollution on the incidence of respiratory diseases: a Python-based data analysis. Theor Nat Sci 8:70–75. https://doi.org/10.54254/2753-8818/8/20240361

Yazid AWM, Sidik NAC, Salim SM, Saqr KM (2014) A review on the flow structure and pollutant dispersion in urban street canyons for urban planning strategies. Simulation 90(8):892–916. https://doi.org/10.1177/0037549714528046

Yu Z, Wang Y, Deng J, Shen Z, Wang K, Zhu J, Gan M (2017) Dynamics of hierarchical urban green space patches and implications for management policy. Sensors 17(6):1304. https://doi.org/10.3390/s17061304

Yudego EA, Candás JC, Álvarez EÁ, López MS, García L, Fernández-Pacheco VM (2018) Computational tools for analysing air pollutants dispersion: a comparative review. In: Proceedings, vol 2, No 23. MDPI, p 1408. https://doi.org/10.3390/proceedings2231408

Zhang Y, Van den Berg AE, Van Dijk T, Weitkamp G (2017) Quality over quantity: contribution of urban green space to neighborhood satisfaction. Int J Environ Res Public Health 14(5):535. https://doi.org/10.3390/ijerph14050535

Zhu T, Shang J, Zhao D (2011) The roles of heterogeneous chemical processes in the formation of an air pollution complex and gray haze. Sci China Chem 54:145–153. https://doi.org/10.1007/s11426-010-4181-y

Chapter 3
Water Quality Management: Processes Influencing Waterborne Diseases and Sustainable Solutions

Wisdom Ebiye Sawyer, Kurotimipa Frank Ovuru, Nsikak Godwin Etim, and Mohamed Azab El-Liethy

Abstract This chapter highlights the processes influencing waterborne diseases and the sustainable solutions for effective water quality management. Key processes include the presence of chemical contaminants, microbial pathogens, and physical factors, all of which contribute to the prevalence of waterborne diseases. Understanding these factors is critical for developing robust prevention and control strategies, particularly concerning environmental changes and emerging pathogens. The chapter also focuses on the diverse sources of water pollution, such as industrial discharge, agricultural runoff, and urban wastewater, emphasizing the need for targeted interventions to address both natural and human-induced pollution. The profound impact of waterborne diseases on public health, particularly in vulnerable populations, stresses the urgency of improved water management strategies as climate change intensifies these challenges. Sustainable water quality solutions, such as advanced purification technologies, robust policy frameworks, and community-based management practices, are discussed as critical strategies for improving water safety. However, the chapter acknowledges the economic, technical, and social barriers that hinder the implementation of these solutions. Through case studies of successful and unsuccessful interventions, the chapter illustrates how leadership, collaboration, and innovation can overcome these barriers. Hence, there is a need for an integrated,

W. E. Sawyer
Department of Community Medicine, Faculty of Clinical Sciences, Bayelsa Medical University, Yenagoa, Bayelsa State, Nigeria

K. F. Ovuru
Department of Disease Control and Immunization, Bayelsa State Primary Health Care Board, Yenagoa, Bayelsa State, Nigeria

N. G. Etim
Department of Medical Laboratory Science, Faculty of Basic Medical Sciences, Niger Delta University, Wilberforce Island, Bayelsa State, Nigeria

M. A. El-Liethy (✉)
Environmental Microbiology Lab., Water Pollution Research Department, National Research Centre, Dokki, Giza 12622, Egypt
e-mail: ma.el-liethy@nrc.sci.eg; mohamedazabr@yahoo.com

S. Chibueze Izah and M. Chidozie Ogwu (eds.), *Innovative Approaches in Environmental Health Management*, Environmental Science and Engineering,
https://doi.org/10.1007/978-3-031-81966-7_3

multidisciplinary approach to water quality management to protect public health, ecosystems, and water resources for future generations.

Keywords Waterborne diseases · Water quality management · Chemical contaminants · Emerging pathogens · Public health impact · Sustainable solutions · Water purification technologies

3.1 Introduction

Water quality management is a critical aspect of environmental science that focuses on maintaining and improving the quality of water resources to ensure their sustainability and safety for human consumption and ecological balance. The significance of effective water quality management cannot be overstated, as it directly impacts public health, environmental integrity, and economic development. The management practices involve monitoring, assessing, and regulating water quality parameters to meet established standards and protect water bodies from pollution and degradation. Sustainable water resource management is essential for preventing water scarcity and ensuring that future generations have access to clean water, a fundamental human right (Sudia et al. 2021; Pramaningsih et al. 2022).

The link between water quality and waterborne diseases is well-documented in the literature (Ogamba et al. 2021; Izah et al. 2022, 2023a, b, 2024a, b, c). Poor water quality significantly contributes to the prevalence of waterborne diseases, which pose a substantial public health risk, particularly in developing regions. Contaminated water can harbor pathogens such as bacteria, viruses, and protozoa, leading to diseases like cholera, dysentery, and typhoid fever. The World Health Organization (WHO) estimates that contaminated drinking water causes approximately one million diarrheal deaths yearly (WHO 2023). Therefore, adequate water quality management practices are crucial in mitigating these health risks by ensuring that water sources are adequately treated and monitored for contaminants (Linker et al. 2013). Water quality assessment is a vital component of water quality management, as it provides the necessary data to evaluate the safety and suitability of water for various uses, including drinking, agriculture, and recreation. Multiple indices and methodologies have been developed to assess water quality, such as the Water Quality Index (WQI), which aggregates various water quality parameters into a single score that reflects overall water quality (Escher and Neale 2021; Ban et al. 2014). These assessments help identify pollution sources, track changes, and inform policy decisions to improve water quality (Singh et al. 2019). In many regions, surface water sources are more susceptible to contamination than groundwater sources, primarily due to runoff from agricultural and urban areas. Studies have shown that fecal coliform bacteria are critical indicators of potential public health risks associated with surface water contamination (Tango and Batiuk 2013). Regular monitoring and assessment of surface water quality are essential to promptly detect and address contamination issues, reducing the risk of waterborne diseases (Xiao et al. 2020).

Developing water quality standards and regulations is another critical aspect of water quality management. These standards protect public health and the environment by establishing acceptable limits for various contaminants in water bodies. For instance, the United States Environmental Protection Agency (EPA) has set standards for drinking water quality, including maximum contaminant levels for pathogens, chemicals, and other pollutants (Saraswati et al. 2019). Compliance with these standards is essential for safeguarding public health and ensuring the safety of drinking water supplies (Javanmardi et al. 2022). In addition to regulatory frameworks, community engagement, and public awareness are vital for effective water quality management. Educating communities about the importance of water quality and the potential health risks associated with contaminated water can foster greater public participation in monitoring and protecting local water resources (Delgado et al. 2020). Initiatives that involve regional stakeholders in water quality assessments can enhance the effectiveness of management strategies and promote sustainable practices (Hasan 2023). Technological advancements have also played a significant role in improving water quality management practices. Integrating remote sensing, geographic information systems (GIS), and artificial intelligence in water quality monitoring has enabled more efficient and accurate assessments across large geographic areas (Ghafil et al. 2022). These technologies facilitate real-time monitoring and data analysis, allowing timely interventions for water quality issues (Delgado et al. 2019).

Furthermore, the impact of climate change on water quality is an emerging concern that necessitates adaptive management strategies. Changes in precipitation patterns, increased temperatures, and extreme weather events can exacerbate water quality issues by altering hydrological cycles and increasing pollutant runoff (Amori et al. 2015). Addressing these challenges requires a holistic approach to water quality management that considers the interplay between climate, land use, and water resources (Roy 2019).

The chapter focuses on critical processes influencing waterborne diseases by examining the role of chemical contaminants, biological pathogens, and physical factors in water quality. It also explores the sources of water pollution, distinguishing between natural and anthropogenic causes, and highlights the public health impact through case studies of outbreaks and vulnerable populations. Sustainable solutions, such as advanced water purification technologies, regulatory frameworks, and community-based management, are also discussed, alongside the challenges of implementing these measures due to economic, technical, and social barriers.

3.2 Key Processes Influencing Waterborne Diseases

Waterborne diseases pose a significant threat to public health globally, driven by a complex interplay of chemical, biological, and physical factors (Fig. 3.1 and Table 3.1). The presence of chemical contaminants, such as heavy metals and pesticides, in water sources can lead to severe health consequences. Heavy metals like lead, mercury, and cadmium are notorious for their toxic effects on human health, including neurological damage and developmental issues, particularly in children (Gibson 2014). Often used in agricultural practices, pesticides can leach into water supplies, leading to acute and chronic health effects, including endocrine disruption and carcinogenicity (Craun et al. 2010). The contamination of drinking water with these chemicals not only affects individual health but also burdens healthcare systems, as the economic costs associated with waterborne diseases are estimated to be in the billions annually (Ramírez-Castillo et al. 2015).

Biological pathogens, including bacteria, viruses, and parasites, are the primary agents responsible for waterborne diseases. Viruses are particularly concerning due to their high transmission efficiency and low infectious dose. For instance, the risk of illness from viral pathogens can be significantly higher than that from bacterial pathogens at similar exposure levels, with estimates suggesting a risk increase of 10–10,000 times (Gibson 2014). Pathogens such as *Cryptosporidium* and *Giardia* are prevalent in contaminated water and can cause severe gastrointestinal illnesses (Labana 2018). Detecting these pathogens in water sources is critical for public health, as they can lead to widespread outbreaks, particularly in vulnerable populations such as children and immunocompromised individuals (Nemati et al. 2023). Waterborne diseases are also influenced by physical factors such as turbidity and temperature. Turbidity can hinder the effectiveness of water treatment processes, allowing pathogens to survive and proliferate in surface water and drinking water

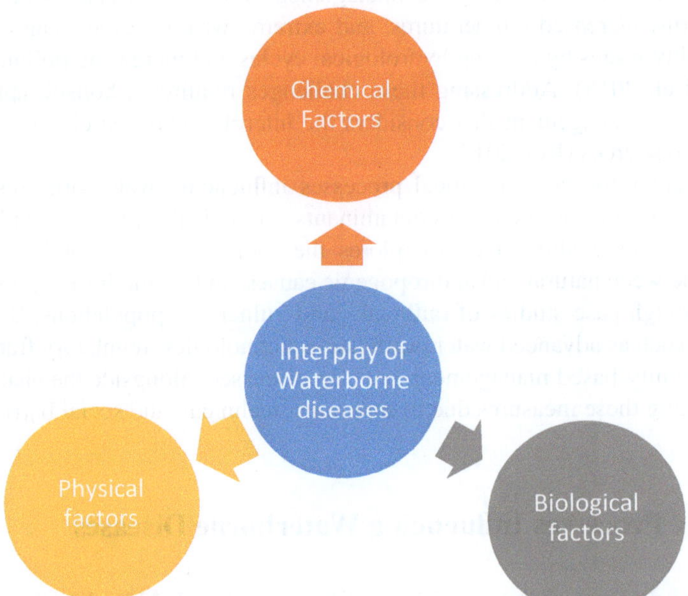

Fig. 3.1 Interplay of waterborne diseases

Table 3.1 Key processes influencing waterborne diseases

Process	Associated activities	Impacts on water quality	Examples of waterborne diseases	Mitigation strategies	Refs.
Chemical contaminants	Harmful substances like heavy metals, polycyclic aromatic hydrocarbons (PAHs), and pesticides enter water systems	It can lead to toxicity in aquatic organisms and humans	Lead poisoning and gastrointestinal issues from pesticide exposure	We are implementing stricter regulations on chemical discharges and improving waste management	Goutam and Mukherjee (2022)
Biological pathogens	Microorganisms that cause disease include bacteria, viruses, and parasites	Degrades water quality and poses health risks to humans and wildlife	Cholera (*Vibrio cholerae*), giardiasis (*Giardia lamblia*), and hepatitis A virus	We are enhancing water treatment processes and promoting sanitation and hygiene	Gerba (2009)
Physical factors	Environmental characteristics like turbidity and temperature influence pathogen survival and transport	Increased turbidity can shield pathogens from disinfection and impact ecosystem health	Increased risk of infections due to high turbidity levels facilitating pathogen survival	Monitoring and managing water temperature and turbidity levels in aquatic systems	Bradley and Lockaby (2023)

supplies (Hind et al. 2020; Ibrahim et al. 2019). High turbidity levels are often associated with heavy rainfall and flooding, which can lead to the runoff of contaminants into water sources (Buse and Ashbolt 2011). Temperature plays a crucial role in the survival and replication of pathogens; warmer temperatures can enhance the growth of certain bacteria and viruses, increasing the risk of waterborne disease outbreaks (Buse and Ashbolt 2011; El-Liethy et al. 2022). For example, *Legionella pneumophila*, a bacterium responsible for legionellosis, exhibits differential growth in various temperature ranges, which can affect its presence in water systems (Buse and Ashbolt 2011). In addition to the direct health impacts, the economic implications of waterborne diseases are substantial. The annual costs associated with waterborne illnesses in the United States alone are estimated to exceed $1 billion, with global losses reaching nearly $12 billion (Ramírez-Castillo et al. 2015). These costs arise from healthcare expenditures, lost productivity, and the economic burden on families affected by waterborne diseases. Furthermore, pathogens in water can increase treatment costs for municipalities as they strive to meet safety standards and protect public health (Ramírez-Castillo et al. 2015).

The interplay between chemical, biological, and physical factors complicates water quality management and the prevention of waterborne diseases. For instance, the presence of chemical contaminants can affect the survival of pathogens in water, potentially leading to increased risks of disease transmission (Bibby et al. 2019). Additionally, the use of microbial indicators, such as coliform bacteria, to assess water quality has limitations, as these indicators do not always correlate with the presence of pathogenic organisms (Payment and Locas 2011; Shaheen et al. 2024). This underscores the need for more comprehensive monitoring strategies incorporating advanced molecular techniques to detect a broader range of pathogens in water sources (Faillace et al. 2017). Emerging research highlights the importance of understanding the dynamics of waterborne pathogens in various environments. For example, studies have shown that viruses can persist in wastewater and surface waters, posing risks to human health through recreational water use and agricultural irrigation (Gruber et al. 2014). Detecting enteric viruses in urban wastewater emphasizes the need for effective surveillance and management strategies to mitigate the risks associated with contaminated water (Ozkay et al. 2022).

Furthermore, the co-infection of enteric viruses in populations with gastroenteritis illustrates the complex interactions between different pathogens and the challenges in diagnosing and treating waterborne diseases (Haramoto et al. 2013). The role of climate change in influencing waterborne diseases cannot be overlooked. Changes in precipitation patterns, increased flooding, and rising temperatures can exacerbate the contamination of water sources and enhance the survival of pathogens (Filipić et al. 2021). For instance, heavy rainfall events can lead to the overflow of sewage systems, introducing pathogens into drinking water supplies (Mohamed 2022). As climate change continues to impact water quality, public health strategies must adapt to address the evolving risks associated with waterborne diseases (He et al. 2018).

3.3 Sources of Water Pollution

Water pollution is a multifaceted issue that arises from various sources, significantly impacting ecosystems and human health. The primary contributors to water pollution are industrial discharge, agricultural runoff, and urban wastewater. Table 3.2 provides an overview of the sources of Water Pollution. Industrial activities often release many pollutants directly into water bodies, including heavy metals, organic compounds, and other hazardous materials. For instance, the textile industry has been noted for its detrimental effects on water quality due to the discharge of effluents rich in dyes and chemicals, which can lead to increased biological oxygen demand (BOD) and chemical oxygen demand (COD) in receiving waters (Makwana 2020; Yadav et al. 2016). Furthermore, studies indicate that industrial wastewater contributes significantly to the degradation of drinking water sources, with up to 62% of water resources in certain regions being affected by pollution from industrial activities (Xu et al. 2019).

Agricultural runoff is another critical source of water pollution, primarily due to using fertilizers and pesticides. These chemicals can leach into nearby water bodies during rainfall or irrigation, leading to nutrient pollution and eutrophication. The introduction of excess nutrients, particularly nitrogen and phosphorus, can result in harmful algal blooms, which deplete oxygen levels in water and produce toxins dangerous to aquatic life and humans (Ag et al. 2013; Ogedengbe and Akinbile 2011). Research has shown that agricultural practices significantly alter the chemical composition of rivers, as seen in the Qiantang River, where agrarian runoff has been identified as a significant contributor to water quality deterioration (Chen et al. 2018). The cumulative effects of these pollutants can lead to severe ecological consequences, including the loss of biodiversity and the disruption of aquatic food webs.

Urban wastewater also significantly contributes to water pollution, particularly in densely populated areas. Discharging untreated or inadequately treated sewage can introduce pathogens, nutrients, and various chemicals into water bodies, posing risks to public health and the environment (Jakubiak and Bojarski 2021; An et al. 2018). Urban wastewater treatment plants (WWTPs) play a crucial role in mitigating these impacts; however, many facilities struggle with efficiency and capacity, leading to the overflow of untreated sewage during peak periods (An et al. 2018). The increasing urbanization and population density exacerbate the challenges faced by wastewater management systems, necessitating improved infrastructure and regulatory frameworks to ensure effective treatment and discharge of urban wastewater (Wang and Islam 2022). The distinction between natural and anthropogenic sources of water pollution is essential for understanding the complexities of this environmental issue. Natural sources, such as sedimentation and the leaching of minerals from rocks, can contribute to water quality changes; however, anthropogenic activities have significantly intensified these effects (Fig. 3.2). For instance, introducing heavy metals from industrial processes poses a more significant threat than naturally occurring levels, leading to bioaccumulation in aquatic organisms and potential health risks for humans consuming contaminated fish (Li et al. 2022; Kamesh et al. 2023). The

Table 3.2 Sources of water pollution

Source of pollution	Description	Types of pollutants	Impacts on water quality	Examples	Mitigation strategies	Refs.
Industrial discharge	Release of waste and pollutants from manufacturing processes into water bodies	Heavy metals, organic compounds, and thermal pollution	Degrading water quality harms aquatic life and poses health risks	Discharge from factories and power plants	We are implementing wastewater treatment and stricter regulations	Zhang et al. (2023) El-Liethy et al. (2023)
Agricultural runoff	Surface runoff from agricultural land that carries fertilizers, pesticides, and sediments	Nutrients (nitrogen and phosphorus), pesticides, and sediments	Causes eutrophication, algal blooms, and sedimentation	Runoff from fields after rainstorms	We are adopting best management practices and buffer zones	Ogwu et al. (2024a, b)
Urban wastewater	Sewage and other waste from urban areas are discharged directly into water bodies, with, partially, or without any treatment	Pathogens, nutrients, and pharmaceuticals	Polluted water sources increase disease risks and deplete oxygen levels	Wastewater from sewage treatment plants and combined sewer overflows	Upgrading infrastructure and promoting water recycling	Xie et al. (2022), Iyiola et al. (2023)
Natural pollution sources	Pollutants originating from natural events or processes	Sediment, nutrients from erosion, and organic matter	It can contribute to nutrient loading but is less harmful than anthropogenic sources	Erosion from rivers and natural runoff	Monitoring natural sources and managing land use	Coffey et al. (2019)
Anthropogenic pollution sources	Pollution, resulting from human activities, impacts water bodies directly or indirectly	Chemicals, plastics, and waste products	This leads to habitat destruction, biodiversity loss, and human health issues	Plastic waste in rivers and lakes	Implementing pollution control measures and public awareness campaigns	Ogwu et al. (2014)

(continued)

Table 3.2 (continued)

Source of pollution	Description	Types of pollutants	Impacts on water quality	Examples	Mitigation strategies	Refs.
Atmospheric deposition	Pollutants that settle into water bodies from the atmosphere due to precipitation	Acids (sulfur and nitrogen compounds) and heavy metals	Alters pH levels and can lead to toxic conditions for aquatic life	Acid rain affects lakes and rivers	Reducing emissions from industrial and vehicular sources	Prakash et al. (2023)
Landfills and waste disposal	Leachate from landfills contaminates nearby water sources	Organic waste, heavy metals, and pathogens	Degrades groundwater and surface water quality	It contaminated groundwater near landfill sites	Proper landfill management and leachate treatment systems	Parvin and Tareq (2021)
Mining activities	Runoff and discharge from mining operations affecting nearby water bodies	Heavy metals, sediments, and acid mine drainage	Detrimental to aquatic ecosystems and water quality	Mining operations near rivers	Implementing responsible mining practices and rehabilitation	Wei et al. (2018)

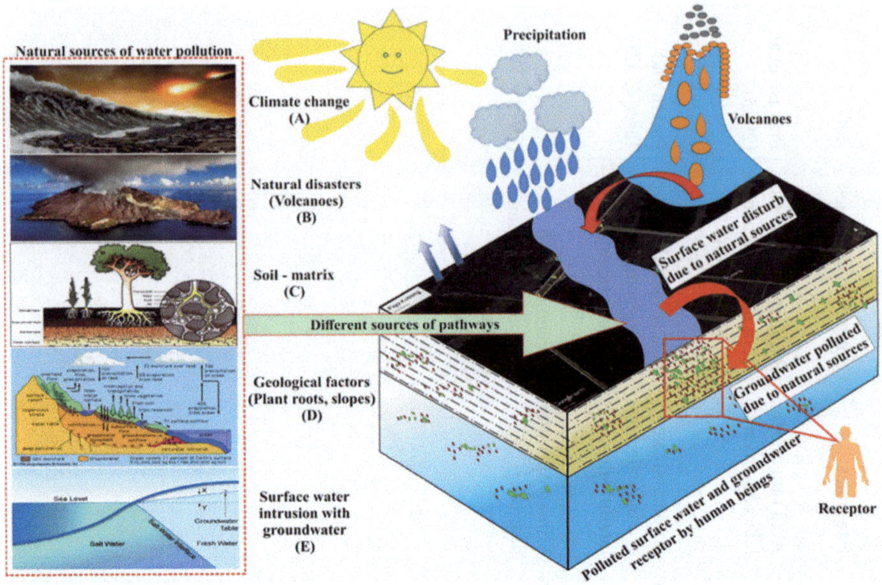

Fig. 3.2 Natural factors driving water pollution. *Source* Akhtar et al. (2021). https://creativecommons.org/licenses/by/4.0/

anthropogenic influence on water quality is evident in various studies, which highlight the correlation between industrialization and increased levels of pollutants in water bodies (Wang and Wang 2023).

Moreover, the interplay between different pollution sources complicates the assessment and management of water quality. For example, urban runoff can carry pollutants from industrial sites and agricultural fields into rivers and lakes, creating a complex mixture of contaminants that are challenging to remediate (Ilyas et al. 2019). The cumulative effects of these pollutants can lead to synergistic interactions, where the combined impact of multiple contaminants is greater than the sum of their individual effects, further complicating environmental management efforts (Yücesoy et al. 2021). Therefore, a comprehensive understanding of the sources and interactions of water pollutants is crucial for developing effective strategies to combat water pollution. Efforts to mitigate water pollution must consider the specific characteristics of each pollution source. Implementing stricter regulations and promoting cleaner production technologies for industrial discharges can significantly reduce the volume and toxicity of wastewater generated (Zowail et al. 2010).

Additionally, enhancing the efficiency of wastewater treatment processes and investing in advanced treatment technologies can help address the shortcomings of existing systems (An et al. 2018; Damian 2013). In the agricultural sector, promoting sustainable farming practices, such as integrated pest management and precision agriculture, can minimize the reliance on chemical inputs and reduce runoff into water bodies (Ag et al. 2013; Ogedengbe and Akinbile 2011). In urban areas, improving stormwater management practices is essential to mitigate the impacts of urban runoff. This can include implementing green infrastructure, such as rain gardens and permeable pavements, which can help absorb and filter pollutants before they enter water bodies (Jakubiak and Bojarski 2021; Meng et al. 2020). Furthermore, public education and community engagement are vital components of effective water pollution management, as they can foster a culture of environmental stewardship and encourage responsible practices among residents and businesses (Wang and Islam 2022; Yadav et al. 2016).

The consequences of water pollution extend beyond environmental degradation; they also pose significant risks to public health. Contaminated water sources can lead to waterborne diseases, which disproportionately affect vulnerable populations, particularly in developing regions (Fig. 3.2; Wang and Wang 2023). The economic implications of water pollution are also substantial, as polluted water bodies can reduce the availability of clean water for drinking, agriculture, and recreation, leading to increased healthcare costs and loss of productivity (Xu et al. 2019; Wang and Islam 2022). Therefore, addressing water pollution is an environmental imperative and a critical public health and economic concern.

3.4 Impact of Waterborne Diseases on Public Health

Waterborne diseases represent a significant public health challenge globally, with their impact felt most acutely in vulnerable populations. These diseases are primarily caused by pathogens transmitted through contaminated water sources, leading to various health issues, from mild gastrointestinal discomfort to severe, life-threatening conditions. The World Health Organization estimates that waterborne diseases account for approximately 485,000 deaths annually due to diarrheal diseases alone, highlighting the urgent need for effective water management and sanitation strategies (Oon et al. 2023). The burden of these diseases is particularly pronounced in low- and middle-income countries, where inadequate water, sanitation, and hygiene (WASH) infrastructure exacerbates the risk of outbreaks (Machado et al. 2022). Table 3.3 provides an overview of the Impact of Waterborne Diseases on Public Health.

Studies of waterborne disease outbreaks provide critical insights into transmission dynamics and public health interventions' effectiveness. For instance, a notable outbreak of norovirus gastroenteritis in Eslamabad-e Gharb, Iran, in 2016 underscored the role of contaminated water sources in facilitating viral transmission (Hatami et al. 2020). Similarly, the analysis of drinking water-related outbreaks in the United States from 1971 to 2006 revealed that a significant proportion of these incidents were linked to water treatment and distribution system deficiencies (Craun et al. 2010). Such case studies illustrate the direct health impacts of waterborne pathogens and the systemic vulnerabilities within water management frameworks that can lead to widespread outbreaks. Vulnerable populations, including children, the elderly, and individuals with compromised immune systems, are disproportionately affected by waterborne diseases. Research indicates that children, particularly

Table 3.3 Impact of waterborne diseases on public health

Waterborne disease	Pathogen type	Transmission route	Symptoms	Health impacts	At-risk populations	Mitigation strategies	Refs.
Cholera	Bacteria (*Vibrio cholerae*)	Contaminated water and food	Severe diarrhea, vomiting, dehydration	High mortality rate without treatment	Children, elderly, and immunocompromised individuals	Improved water supply and sanitation, vaccination campaigns	Ahmed et al. (2024)
Giardiasis	Parasite (*Giardia lamblia*)	Contaminated water	Diarrhea, abdominal cramps, bloating	This can lead to chronic gastrointestinal issues	Children and travelers to endemic areas	Filtration and treatment of drinking water	Ali et al. (2021)
Hepatitis A	Virus (Hepatitis A virus)	Contaminated water and food	Fatigue, jaundice, abdominal pain	Liver inflammation, potential chronic liver disease	Unvaccinated individuals, travelers, and children	Vaccination and sanitation improvements	Girish et al. (2024)
Typhoid fever	Bacteria (*Salmonella enterica*)	Contaminated water and food	Prolonged fever, abdominal pain, diarrhea	Severe illness requiring hospitalization	Low-income communities with poor sanitation	Safe drinking water access, hygiene education	Kabwama et al. (2017)
Cryptosporidiosis	Parasite (*Cryptosporidium* spp.)	Contaminated water	Watery diarrhea, stomach cramps, nausea	It can be severe in immunocompromised individuals	People with HIV/AIDS, children, and the elderly	Water treatment and public awareness campaigns	Bouzid et al. (2013)

(continued)

Table 3.3 (continued)

Waterborne disease	Pathogen type	Transmission route	Symptoms	Health impacts	At-risk populations	Mitigation strategies	Refs.
Dysentery	Bacteria (*Shigella* spp. and *Entamoeba histolytica*)	Contaminated water and food	Diarrhea with blood, abdominal pain	This can lead to severe dehydration and malnutrition	Children, malnourished individuals	Improved sanitation and access to clean water	Dejkam and Hatam-Nahavandi (2021)
Leptospirosis	Bacteria (*Leptospira* spp.)	Contaminated water, especially in floods	Fever, chills, muscle aches, jaundice	It can cause severe liver and kidney damage	Individuals in flood-prone areas	Rodent control, sanitation, and public awareness	Haake and Levett (2015)

in developing countries, are at a heightened risk due to their lower immunity and higher exposure to contaminated water (Hussain et al. 2023).

Furthermore, socio-economic factors, such as poverty and lack of access to clean water, exacerbate this vulnerability, as seen in regions where unsafe drinking water is prevalent (Mogasale et al. 2018). The intersection of environmental justice and public health is critical, as marginalized communities often bear the brunt of waterborne disease outbreaks due to systemic inequities in resource allocation and infrastructure development (Kamanmalek et al. 2023). The role of climate change in exacerbating waterborne diseases cannot be overlooked. Extreme weather events, such as heavy rainfall and flooding, have been shown to correlate with increased incidence of waterborne diseases due to the disruption of water treatment processes and the contamination of water sources (Cann et al. 2013; Herrador et al. 2021). For example, a systematic review highlighted the association between extreme precipitation events and outbreaks of gastrointestinal illnesses, emphasizing the need for adaptive water management strategies in the face of climate variability (Levy et al. 2016). As climate change continues to impact weather patterns, the public health implications of waterborne diseases are likely to intensify, necessitating proactive measures to safeguard vulnerable populations.

The economic burden of waterborne diseases is substantial, with estimates suggesting that the annual costs associated with treatment and lost productivity range between $2.2 billion and $3.7 billion in the United States alone (Oon et al. 2023). This economic impact is compounded by the long-term health consequences of waterborne diseases, which can lead to chronic health issues and diminished quality of life for affected individuals (Ramírez-Castillo et al. 2015). Investments in water purification technologies and improved sanitation infrastructure have yielded significant returns in terms of reduced disease burden and healthcare costs, reinforcing the argument for prioritizing WASH initiatives (Beach et al. 2016). Public health surveillance plays a crucial role in understanding and mitigating the impact of waterborne diseases. Establishing comprehensive surveillance systems, such as the one implemented in France from 2010 to 2019, has proven effective in tracking outbreaks and identifying risk factors associated with waterborne diseases (Pouey et al. 2021). Such systems enable health authorities to respond swiftly to emerging threats and implement targeted interventions to protect at-risk populations.

Moreover, integrating advanced detection technologies, such as molecular pathogen identification methods, can enhance outbreak response accuracy and timeliness (Oon et al. 2023; Lanrewaju et al. 2022). The relationship between water quality and public health is complex and multifaceted. Poor water quality, often due to environmental degradation and inadequate infrastructure, is a primary driver of waterborne diseases. Studies have shown that regions with compromised water quality experience higher rates of gastrointestinal illnesses, necessitating a focus on improving water treatment processes and ensuring compliance with safety standards (Harper et al. 2011). Additionally, the role of recreational water sources in disease transmission is increasingly recognized, with outbreaks linked to contaminated beaches and swimming pools highlighting the need for vigilant monitoring and regulation of these environments (Shahraki et al. 2019; McClung et al. 2018).

Addressing the impact of waterborne diseases on public health requires a multi-faceted approach that encompasses education, infrastructure development, and community engagement. Public health campaigns to raise awareness about safe water practices and hygiene can empower communities to take proactive measures to protect their health (Mogasale et al. 2018). Furthermore, collaboration between public health officials, environmental agencies, and community organizations is essential to develop sustainable solutions that address the root causes of waterborne diseases, particularly in vulnerable populations (Kamanmalek et al. 2023; Levy et al. 2016).

3.5 Sustainable Solutions for Water Quality Management

Sustainable solutions for water quality management are crucial in addressing the myriad challenges posed by pollution, climate change, and increasing water demand. Water purification technologies, policy frameworks, and community-based practices are integral components of a comprehensive strategy to ensure clean and safe water. Table 3.4 shows the overview of Sustainable Solutions for Water Quality Management.

Water purification technologies have evolved significantly, offering various methods to enhance water quality (Table 3.5). Filtration remains one of the most widely used techniques, effectively removing particulate matter and pathogens from water sources. Advanced filtration systems, such as membrane filtration and reverse osmosis (RO), have demonstrated high efficiency in removing contaminants, including heavy metals and microorganisms, thus ensuring safe drinking water (Wang et al. 2020; Bortolini et al. 2017). Furthermore, ultraviolet (UV) treatment has gained popularity due to its ability to inactivate a broad spectrum of pathogens without chemicals, making it an environmentally friendly option (Samijayani et al. 2023). Combining these technologies can improve water quality, particularly in areas facing severe contamination challenges (Gad et al. 2022).

In addition to physical and chemical purification methods, innovative technologies such as capacitive deionization (CDI) are emerging as sustainable alternatives for treating saline and contaminated water. CDI utilizes electric fields to remove ions from water, offering a low-energy solution compared to traditional desalination methods (Wang et al. 2020). This technology is particularly relevant in regions where freshwater resources are scarce and the demand for clean water is increasing due to population growth and industrial activities (Daud et al. 2017). Integrating such technologies into existing water management systems can significantly enhance the sustainability of water quality management efforts. Policy frameworks and regulations are pivotal in shaping water quality management practices. Effective regulations not only set standards for water quality but also incentivize industries to adopt water-efficient technologies and practices (Atakhanova 2023). For instance, in Canada, stringent effluent quality regulations have been shown to reduce both water quality

Table 3.4 Sustainable solutions for water quality management

Solution type	Techniques	Examples	Benefits	Challenges	Refs.
Water purification technologies	Techniques to remove contaminants and improve water quality	Filtration, UV treatment, reverse osmosis (RO)	Reduces pathogens and pollutants, ensuring safe drinking water	High initial costs and ongoing maintenance needs	Ogwu et al. (2024c)
Policy frameworks and regulations	Laws and guidelines that govern water resource management	Clean Water Act, EU Water Framework Directive	Promotes sustainable water use and ecosystem protection	Varying enforcement levels and compliance challenges	Giakoumis and Voulvoulis (2018)
Community-based water management	Local initiatives that empower communities in water management	Rainwater harvesting, watershed management	Increases community engagement and resource ownership	Requires continuous education and local involvement	Bernedo Del Carpio et al. (2021)
Ecosystem restoration	Rehabilitating ecosystems to improve water quality and biodiversity	Wetland restoration, riparian buffer zones	Enhances natural filtration, biodiversity, and flood resilience	Requires long-term commitment and funding	Meli et al. (2014)
Green infrastructure	Use of natural systems to manage stormwater and improve infiltration	Green roofs, permeable pavements, rain gardens	Reduces runoff and improves water infiltration	High setup costs and maintenance requirements	Golden and Hoghooghi (2018)
Public education and awareness	Programs to inform and engage the public on water conservation	Workshops, school programs, community campaigns	Increases public participation and reduces pollution	It needs ongoing resources and community involvement	Nelson et al. (2021)
Integrated water resource management (IWRM)	A holistic approach that considers social, economic, and environmental water needs	River basin management, stakeholder engagement	Promotes sustainable and equitable water use	Complexity in coordination among various stakeholders	Iyiola et al. (2023, 2024)
Sustainable agriculture practices	Methods to reduce agricultural impacts on water quality	Crop rotation, organic farming, cover cropping	Reduces runoff and improves soil health	Barriers in adoption and initial transition costs	Pretty (2008)

Table 3.5 Water purification technologies

Technology	Technique	Mechanism	Advantages	Limitations	Applications	Refs.
Filtration	The process of removing particles and contaminants from water uses a physical barrier	It uses various media such as sand, activated carbon, or membranes	Effective for turbidity removal and particle filtration	It may not remove dissolved contaminants	Drinking water treatment, wastewater treatment	Razali et al. (2023)
Ultraviolet (UV) treatment	Uses UV light to kill or inactivate pathogens in water	Disrupts the DNA of microorganisms	Chemical-free method; effective against bacteria and viruses	It requires clear water for effectiveness; there is no residual disinfection	Drinking water disinfection, wastewater treatment	Kim et al. (2023)
Reverse osmosis (RO)	A process that uses a semipermeable membrane to remove ions, molecules, and larger particles from water	Applies pressure to force water through the membrane	Highly effective at removing salts, contaminants, and microorganisms	Generates wastewater; requires energy	Desalination, household water treatment	Obotey Ezugbe and Rathilal (2020)
Activated carbon adsorption	Utilizes activated carbon to adsorb organic compounds and contaminants	Contaminants adhere to the surface of activated carbon	Effective for removing chlorine, odors, and taste	Limited lifespan; needs regular replacement	Drinking water treatment, aquarium filtration	Blachnio et al. (2020)
Ozone treatment	It uses ozone gas to disinfect and purify water	Ozone is a powerful oxidant that kills pathogens	Effective against a wide range of microorganisms; no residual chemicals	Requires careful handling; short contact time is needed	Drinking water treatment, industrial wastewater treatment	Botondi et al. (2023)

(continued)

Table 3.5 (continued)

Technology	Technique	Mechanism	Advantages	Limitations	Applications	Refs.
Chlorination	The addition of chlorine to water for disinfection	It kills bacteria and inactivates viruses	Effective and low-cost method for large-scale water treatment	It can form harmful byproducts and taste and odor issues	Municipal water supply, swimming pools	Lanrewaju et al. (2022) Abdo et al. (2022)
Membrane filtration	It uses various types of membranes to separate contaminants from water based on size	Includes microfiltration, ultrafiltration, nanofiltration, and reverse osmosis	It can remove bacteria, viruses, and other particles	Costly equipment and maintenance; fouling issues	Drinking water treatment, food and beverage industry	Cevallos-Mendoza et al. (2022)
Distillation	A process that involves boiling water and condensing the vapor to remove impurities	Removes contaminants through phase change	It is highly effective for removing salts and many contaminants	Energy-intensive; slow process	Desalination, laboratory water purification	Saxena and Chandwankar (2024)
Biological treatment	Utilizes microorganisms to degrade organic pollutants in water	Biological processes such as activated sludge or biofilters	Effective for treating wastewater; natural process	Limited to biodegradable materials; slower than chemical methods	Wastewater treatment, septic systems	Sravan et al. (2024)
Electrocoagulation	It uses electrical currents to remove suspended solids, oils, and pathogens from water	Charges particles to aggregate and settle	Reduces the need for chemicals; effective for various pollutants	Equipment costs require regular maintenance	Wastewater treatment, industrial applications	Ebba et al. (2022) Mahmoud and El-Liethy (2022)

and quantity impacts from mining activities, encouraging the adoption of sustainable practices (Atakhanova 2023). Similarly, the Sustainable Development Goals (SDGs) established by the United Nations emphasize the need for robust regulatory frameworks to ensure the availability and sustainable management of water and sanitation for all (Sala-Garrido et al. 2021). Policymakers must prioritize developing and enforcing regulations that protect water resources while promoting sustainable practices across various sectors.

Community-based water management practices are essential for fostering local engagement and ownership in water quality management. Research has shown that communities that actively monitor and manage their water resources have better water quality outcomes (Withanachchi et al. 2018). For example, local governance structures can empower communities to address water quality issues effectively, as seen in various case studies across different countries (Kayser et al. 2015). Community-led initiatives, such as rainwater harvesting and wastewater recycling, can also contribute to sustainable water management by reducing reliance on external water sources and enhancing resilience to climate variability (Beard et al. 2019). Moreover, education and awareness-raising efforts are critical in promoting community involvement in water quality management. Studies indicate that increasing knowledge about water hygiene practices can improve water quality and health outcomes (Kuberan et al. 2015; Manar et al. 2018). For instance, educating households about the importance of boiling water before consumption can significantly reduce the incidence of waterborne diseases (Manar et al. 2018). By fostering a culture of responsibility and awareness around water quality, communities can better manage their resources and advocate for necessary policy changes. Integrating technology in monitoring water quality is another vital aspect of sustainable water management. Advances in remote sensing and data analytics have enabled more effective monitoring of water quality indicators across various water bodies (Nguyen and Huynh 2022). For example, spectral indices and multivariate models can provide comprehensive water quality assessments, allowing timely interventions when pollution levels exceed acceptable thresholds (Elsayed et al. 2021). These technological innovations enhance the accuracy of water quality assessments and facilitate data-driven decision-making in water management.

Furthermore, addressing the impacts of climate change on water quality is imperative for sustainable management. Climate change can exacerbate water quality issues by altering hydrological cycles and increasing the frequency of extreme weather events, leading to higher pollutant loads in water bodies (Xu et al. 2023). Policymakers and water managers must consider climate resilience in their strategies, implementing adaptive management practices that account for changing environmental conditions (Pieper et al. 2022). This approach can help mitigate the adverse effects of climate change on water quality and ensure the long-term sustainability of water resources.

3.6 Challenges in Implementing Sustainable Water Management Solutions

Implementing sustainable solutions faces numerous challenges that can be categorized into economic, technical, and social barriers (Fig. 3.3). Economic barriers often manifest as insufficient funding or resources allocated to sustainability initiatives. For instance, many projects struggle to secure the necessary financial backing to sustain their operations over the long term, which is critical for achieving lasting impacts. A systematic review highlighted that the sustainability of public health programs is significantly influenced by the availability of resources, including financial support, human capital, and infrastructure (Scheirer and Dearing 2011). Furthermore, the lack of economic incentives for stakeholders to adopt sustainable practices can hinder progress. For example, transitioning to sustainable energy, such as electric vehicle adoption, requires substantial changes in consumer behavior and policy frameworks, often not aligned with immediate economic interests (Novak and Mohammadian 2023).

Technical barriers also play a crucial role in implementing sustainable solutions. These barriers include the lack of appropriate technology or expertise to execute sustainable practices effectively. For instance, the successful implementation of nature-based solutions (NBS) relies heavily on the availability of innovative technologies and methodologies that can adapt to local contexts (Dick et al. 2019).

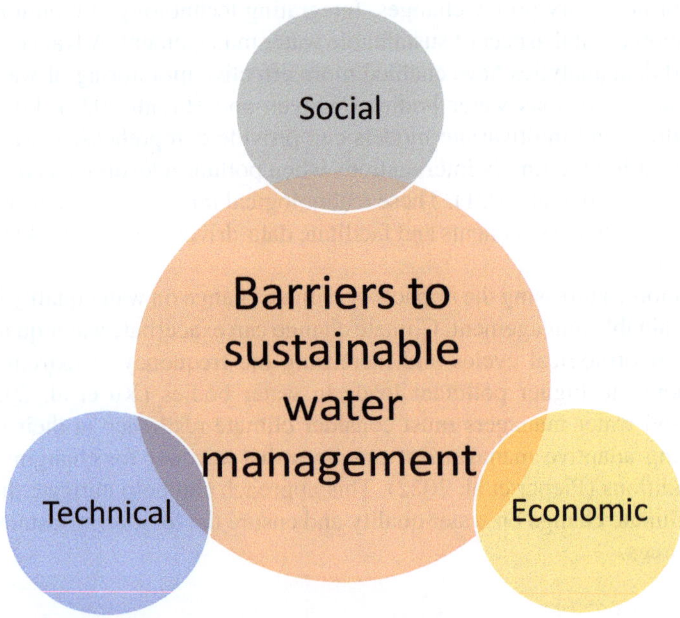

Fig. 3.3 Challenges of implementing sustainable solutions

Moreover, the complexity of integrating sustainable practices into existing systems can lead to resistance from stakeholders who may be unfamiliar with new technologies or processes. A study on the sustainability of a child mental health intervention emphasized the importance of aligning innovative practices with the existing organizational context to overcome these technical challenges (Pisciotta et al. 2023). Social barriers are equally significant in the discourse surrounding sustainable solutions. These barriers often stem from cultural attitudes, community engagement, and stakeholder buy-in. The success of sustainability initiatives frequently hinges on the level of community involvement and support. For example, a community-based obesity prevention project in Australia highlighted that effective community engagement is essential for the sustainability of health initiatives (Mathews et al. 2010).

Additionally, leadership commitment and solid partnerships among stakeholders are critical for fostering an environment conducive to sustainable practices (Ros et al. 2021). The absence of these social dynamics can lead to the failure of well-intentioned interventions, as seen in various case studies where a lack of community support resulted in unsustainable outcomes (Whelan et al. 2018). Studies of successful and unsuccessful interventions provide valuable insights into the complexities of implementing sustainable solutions. One notable success story is the Healthy Youth Healthy Communities initiative in Fiji, which demonstrated the importance of community participation and collaboration in achieving sustainable health outcomes (Waqa et al. 2013). The project faced numerous challenges, including logistical issues and resistance from some stakeholders, yet it ultimately succeeded due to strong community engagement and adaptive strategies. Conversely, the failure of specific public health programs can often be traced back to inadequate planning and insufficient consideration of the local context. For example, a systematic narrative synthesis review identified several factors that contributed to the sustainability of obesity prevention programs, including the need for adaptability and strong leadership (Whelan et al. 2018).

The role of leadership in promoting sustainable solutions cannot be overstated. Effective leadership is often the linchpin that holds together various components of sustainability initiatives. Leaders committed to sustainability can inspire and mobilize resources, foster collaboration among stakeholders, and create a culture of accountability (Ros et al. 2021). A study on the sustainment of patient flow interventions in healthcare settings found that leadership commitment significantly influenced the initiative's long-term success (Ros et al. 2021). This underscores the need for leaders to advocate for sustainability and actively engage with frontline staff and stakeholders to ensure that the benefits of interventions are recognized and valued. In addition to leadership, the adaptability of interventions plays a crucial role in their sustainability. The ability to modify programs in response to changing circumstances or feedback from stakeholders is essential for maintaining relevance and effectiveness. For instance, a systematic review of community-based obesity prevention interventions revealed that adaptability was one of the key factors contributing to sustainability (Whelan et al. 2018). This adaptability allows programs to evolve and respond to different contexts' unique challenges and opportunities, enhancing their chances of long-term success.

Furthermore, integrating sustainability into policy frameworks is vital for overcoming barriers to implementation. Policies that promote sustainable practices must address underlying causes rather than merely treating symptoms. A study on sustainable development diplomacy emphasized the importance of engaging diverse stakeholders in the policy-making process to ensure that solutions are tailored to the complexities of sustainability challenges (Moomaw et al. 2017). The likelihood of successful implementation increases significantly by fostering a sense of ownership among stakeholders and aligning policies with mutual benefits. The evaluation of sustainability measures is another critical aspect that influences the success of interventions. Understanding how to measure sustainability and its determinants can provide valuable insights into the effectiveness of various strategies. A systematic review highlighted the need for robust evaluation frameworks to assess sustainability initiatives' long-term impacts in public health and community settings (Hall et al. 2022). By establishing clear metrics and evaluation processes, stakeholders can better understand what works and what does not, facilitating continuous improvement and adaptation of interventions. Finally, the interplay between economic, technical, and social factors creates a complex landscape for implementing sustainable solutions. Addressing these challenges requires a multifaceted approach that considers the unique context of each initiative. For instance, the integration of nature-based solutions in urban planning must take into account not only the technical feasibility but also the economic implications and social acceptance (Dick et al. 2019). By fostering collaboration among diverse stakeholders and leveraging existing resources, it is possible to navigate these challenges and create sustainable solutions that are both effective and equitable.

3.7 Future Directions in Water Quality Management

Water quality management is an evolving field that increasingly integrates innovative technologies and methodologies to address the challenges posed by waterborne diseases. As urbanization and climate change exacerbate the risks associated with water quality, the need for advanced monitoring and treatment technologies becomes paramount. Innovations such as real-time water quality sensors and remote sensing technologies are being developed to enhance the detection of contaminants in water sources. For instance, integrating machine learning algorithms with sensor data allows continuous monitoring of water quality parameters, enabling timely interventions when contamination is detected (Maroju et al. 2023). Furthermore, the standardization of analytical procedures for waterborne contaminants, as highlighted by Avataneo et al. (2023), is critical to ensuring that data on water quality is reliable and comparable across different regions. This standardization is essential in emerging contaminants, such as asbestos in surface waters, which require rigorous analytical methods to assess their presence and impact on public health.

The role of artificial intelligence (AI) and data analytics in predicting and preventing waterborne diseases is becoming increasingly significant. AI can analyze vast datasets from various sources, including meteorological data, historical disease incidence, and water quality measurements, to identify patterns and predict outbreaks. For example, Pouey et al. (2021) demonstrated the utility of rainfall data in predicting waterborne disease outbreaks, as heavy rainfall often correlates with increased pathogen levels in drinking water. This predictive capability is essential for public health officials to implement preventive measures proactively. Additionally, wastewater-based epidemiology has emerged as a powerful tool for the early detection of waterborne pathogens, allowing for targeted interventions before outbreaks escalate (Zahedi et al. 2021). By leveraging AI and data analytics, water quality management can transition from reactive to proactive strategies, ultimately reducing the incidence of waterborne diseases.

Innovations in water treatment technologies are also crucial in the fight against waterborne diseases. Advanced treatment methods, such as photocatalytic disinfection, have shown promise in effectively inactivating waterborne pathogens, including viruses and bacteria (Zhao et al. 2023). These technologies improve the safety of drinking water and enhance the resilience of water supply systems against contamination events. Moreover, the application of nanotechnology in water treatment offers new avenues for developing efficient and cost-effective solutions to water quality challenges (Marembo and Xu 2020). Integrating these advanced treatment technologies into existing infrastructure can significantly reduce the burden of waterborne diseases, particularly in vulnerable populations. The impact of climate change on water quality and the prevalence of waterborne diseases cannot be overstated. Increased precipitation and extreme weather events, as noted by Herrador et al. (2021), are linked to higher incidences of gastroenteritis due to contaminated drinking water. As climate change continues to alter weather patterns, water quality management strategies must adapt accordingly. This includes enhancing stormwater management practices to mitigate the risks associated with flooding and runoff, which can introduce pathogens into drinking water supplies (Kamanmalek et al. 2023).

Furthermore, the design and implementation of resilient water infrastructure are essential to withstand the impacts of climate change, ensuring that communities have access to safe drinking water even in the face of environmental challenges. Community engagement and education are vital in water quality management and preventing waterborne diseases. Public awareness campaigns that inform communities about the importance of safe water practices and the risks associated with waterborne pathogens can empower individuals to take proactive measures. For example, promoting household water treatment methods, such as boiling or using filters, can significantly reduce the risk of waterborne diseases in areas with limited access to safe drinking water (Komba et al. 2022). Additionally, community-based management approaches that involve local stakeholders in decision-making processes can enhance the sustainability of water quality initiatives and ensure that interventions are culturally appropriate and effective. Integrating health data with water quality monitoring systems is another promising direction for future water quality management. Public health officials can better understand the relationships between waterborne diseases and environmental factors by linking health outcomes with water quality data. This approach allows for the identification of high-risk areas and the implementation of targeted interventions. For instance, epidemiological studies have shown a clear correlation between poor water quality and the prevalence of waterborne diseases, underscoring the need for comprehensive monitoring systems encompassing health and environmental data (Abdulkadir et al. 2019). Such integrated systems can facilitate timely responses to emerging threats and improve public health outcomes. In addition to technological innovations, regulatory frameworks, and policies must evolve to support effective water quality management.

Establishing clear guidelines for water quality standards and monitoring practices is essential to ensure all stakeholders are accountable for maintaining safe water supplies. As highlighted by Beach, investments in water infrastructure are directly linked to declines in waterborne diseases, emphasizing the need for sustained funding and commitment from governments and organizations (Beach 2022). Furthermore, addressing social determinants of health, such as access to clean water and sanitation, is crucial for reducing health disparities related to waterborne diseases, particularly in marginalized communities. The ongoing research and development of new technologies and methodologies will also shape the future of water quality management. Continuous innovation in biosensors, which can provide real-time monitoring of waterborne pathogens, is promising to enhance water safety (Marembo and Xu 2020). Machine learning and predictive analytics in water quality management can also lead to more effective and efficient responses to contamination events. As researchers continue to explore the potential of these technologies, it is essential to ensure that they are accessible and applicable to diverse contexts, particularly in low-resource settings where the burden of waterborne diseases is often highest.

3.8 Conclusion

In conclusion, water quality management is crucial in controlling waterborne diseases by addressing chemical, biological, and physical factors contributing to contamination. Key findings highlight that water pollution from industrial discharge, agricultural runoff, and urban wastewater significantly impacts public health, particularly in vulnerable populations. To mitigate these risks, it is imperative to implement robust monitoring systems, adopt advanced water purification technologies, and enforce stringent regulatory frameworks. Case studies of outbreaks emphasize the need for immediate intervention and adaptation to emerging pathogens and environmental changes. A multidisciplinary approach involving public health officials, environmental scientists, and policymakers can enhance waterborne disease prevention efforts while ensuring the sustainability of water resources.

Sustainable solutions must integrate technical innovations with policy reforms and community-based water management practices. Stakeholders must address the economic, social, and technical barriers that hinder effective implementation, particularly in developing regions where access to clean water remains challenging. Promoting filtration systems, UV treatments, and decentralized water purification technologies can provide immediate relief. At the same time, long-term strategies should focus on enhancing public awareness and encouraging community involvement. By learning from successful and failed interventions, governments and organizations can refine their approaches to sustainable water management. Collaboration at all levels—local, national, and international—will protect water resources, minimize disease burden, and ensure safe, clean water for future generations.

References

Abdo SM, El-Liethy MA, Doma HS, El Taweel GE, Ali GH (2022) Chlorine as an integrated approach for environmental health and hygiene: A case study on evaluation of waste stabilization ponds' performance at 11 governorates in Egypt. Emerg Contam 8:243–253

Abdulkadir N, Afolabi RO, Usman HM, Mustapha G, Abubakar UA (2019) Epidemiological studies of waterborne diseases in relation to bacteriological quality of water. Microbiol Res J Int 28(4):1–12. https://doi.org/10.9734/mrji/2019/v28i430141

Ag M, Am ES, Ms S (2013) Current situation of water pollution and its effect on aquatic life in Egypt. Egypt J Occup Med 37(1):95–115. https://doi.org/10.21608/ejom.2013.775

Ahmed AK, Sijercic VC, Akhtar MS, Elbayomy A, Marouf MA, Zeleke MS, Sayad R, Abdelshafi A, Laird NJ, El-Mokhtar MA, Ruthig GR, Hetta HF (2024) Cholera rages in Africa and the Middle East: a narrative review on challenges and solutions. Health Sci Rep 7(5):e2013. https://doi.org/10.1002/hsr2.2013

Akhtar N, Syakir Ishak MI, Bhawani SA, Umar K (2021) Various natural and anthropogenic factors responsible for water quality degradation: a review. Water 13(19):2660. https://doi.org/10.3390/w13192660

Ali Q, Ahmed S, Aamer S, Iqbal N, Mamoon N (2021) Giardiasis: an overlooked cause of recurrent abdominal pain. Cureus 13(9). https://doi.org/10.7759/cureus.17701

Amori AA, Odjegba EE, Adekitan AA (2015) Assessment of public perspectives on quality control measures in public water supply in Oyo state, Nigeria. J Appl Sci Environ Manag 19(1):13–18. https://doi.org/10.4314/jasem.v19i1.2

An M, He W, Degefu DM, Liao Z, Zhang Z, Yuan L (2018) Spatial patterns of urban wastewater discharge and treatment plants efficiency in China. Int J Environ Res Public Health 15(9):1892. https://doi.org/10.3390/ijerph15091892

Atakhanova Z (2023) Water impacts and effluent quality regulations of Canadian mining. Sustainability 15(22):16086. https://doi.org/10.3390/su152216086

Avataneo C, Capella S, Luiso M, Marangoni G, Lasagna M, De Luca DA, Bergamini M, Belluso E, Turci F (2023) Waterborne asbestos: good practices for surface waters analyses. Front Chem 11:1104569. https://doi.org/10.3389/fchem.2023.1104569

Ban X, Wu Q, Pan B, Du Y, Feng Q (2014) Application of composite water quality identification index on the water quality evaluation in spatial and temporal variations: a case study in Honghu Lake, China. Environ Monit Assess 186:4237–4247. https://doi.org/10.1007/s10661-014-3694-9

Beach B (2022) Water infrastructure and health in US cities. Reg Sci Urban Econ 94:103674. https://doi.org/10.1016/j.regsciurbeco.2021.103674

Beach B, Ferrie J, Saavedra M, Troesken W (2016) Typhoid fever, water quality, and human capital formation. J Econ Hist 76(1):41–75. https://doi.org/10.1017/s0022050716000413

Beard JE, Bierkens MF, Bartholomeus RP (2019) Following the water: characterising de facto wastewater reuse in agriculture in the Netherlands. Sustainability 11(21):5936. https://doi.org/10.3390/su11215936

Bernedo Del Carpio M, Alpizar F, Ferraro PJ (2021) Community-based monitoring to facilitate water management by local institutions in Costa Rica. Proc Natl Acad Sci 118(29):e2015177118. https://doi.org/10.1073/pnas.2015177118

Bibby K, Crank K, Greaves J, Li X, Wu Z, Hamza IA, Stachler E (2019) Metagenomics and the development of viral water quality tools. NPJ Clean Water 2(1):9. https://doi.org/10.1038/s41545-019-0032-3

Blachnio M, Derylo-Marczewska A, Charmas B, Zienkiewicz-Strzalka M, Bogatyrov V, Galaburda M (2020) Activated carbon from agricultural wastes for adsorption of organic pollutants. Molecules 25(21):5105. https://doi.org/10.3390/molecules25215105

Bortolini M, Gamberi M, Mora C, Pilati F, Regattieri A (2017) Design, prototyping, and assessment of a wastewater closed-loop recovery and purification system. Sustainability 9(11):1938. https://doi.org/10.3390/su9111938

Botondi R, Lembo M, Carboni C, Eramo V (2023) The use of ozone technology: an eco-friendly method for the sanitization of the dairy supply chain. Foods 12(5):987. https://doi.org/10.3390/foods12050987

Bouzid M, Hunter PR, Chalmers RM, Tyler KM (2013) Cryptosporidium pathogenicity and virulence. Clin Microbiol Rev 26(1):115–134. https://doi.org/10.1128/CMR.00076-12

Bradley EA, Lockaby G (2023) Leptospirosis and the environment: a review and future directions. Pathogens 12(9):1167. https://doi.org/10.3390/pathogens12091167

Buse HY, Ashbolt NJ (2011) Differential growth of Legionella pneumophila strains within a range of amoebae at various temperatures associated with in-premise plumbing. Lett Appl Microbiol 53(2):217–224. https://doi.org/10.1111/j.1472-765x.2011.03094.x

Cann KF, Thomas DR, Salmon RL, Wyn-Jones AP, Kay D (2013) Extreme water-related weather events and waterborne disease. Epidemiol Infect 141(4):671–686. https://doi.org/10.1017/s0950268812001653

Cevallos-Mendoza J, Amorim CG, Rodríguez-Díaz JM, Montenegro MDCB (2022) Removal of contaminants from water by membrane filtration: a review. Membranes 12(6):570. https://doi.org/10.3390/membranes12060570

Chen G, Luo J, Zhang C, Jiang L, Tian L, Chen G (2018) Characteristics and influencing factors of spatial differentiation of urban black and odorous waters in China. Sustainability 10(12):4747. https://doi.org/10.3390/su10124747

Coffey R, Paul MJ, Stamp J, Hamilton A, Johnson T (2019) A review of water quality responses to air temperature and precipitation changes 2: nutrients, algal blooms, sediment, pathogens. JAWRA J Am Water Resour Assoc 55(4):844–868. https://doi.org/10.1111/1752-1688.12711

Craun GF, Brunkard JM, Yoder JS, Roberts VA, Carpenter J, Wade T, Calderon RL, Roberts JM, Beach MJ, Roy SL (2010) Causes of outbreaks associated with drinking water in the United States from 1971 to 2006. Clin Microbiol Rev 23(3):507–528. https://doi.org/10.1128/cmr.00077-09

Damian C (2013) Environmental pollution in the petroleum refining industry. Ovidius Univ Ann Chem 24(2):109–114. https://doi.org/10.2478/auoc-2013-0018

Daud MK, Nafees M, Ali S, Rizwan M, Bajwa RA, Shakoor MB, Arshad MU, Chatha SAS, Deeba F, Murad W, Malook I, Zhu SJ (2017) Drinking water quality status and contamination in Pakistan. Biomed Res Int 2017:7908183. https://doi.org/10.1155/2017/7908183

Dejkam A, Hatam-Nahavandi K (2021) Dysentery in children. Iran J Public Health 50(9):1930–1931. https://doi.org/10.18502/ijph.v50i9.7082

Delgado A, Acuña M, Justano N, Llanos E, Puma I (2019) Water quality assessment in a watershed in Cusco, Peru using the grey clustering method. Int J Eng Adv Technol 9(2):5093–5098. https://doi.org/10.35940/ijeat.b3606.129219

Delgado A, Llanos E, Puma I, Acuña M, Justano N, Carbajal C (2020) Applying the grey clustering method to assess water quality in a watershed in Cusco, Peru. https://doi.org/10.18687/laccei2020.1.1.524

Dick J, Miller JD, Carruthers-Jones J, Dobel AJ, Carver S, Garbutt A, Hester A, Hails R, Magreehan V, Quinn M (2019) How are nature-based solutions contributing to priority societal challenges surrounding human well-being in the United Kingdom: a systematic map protocol. Environ Evid 8:1–11. https://doi.org/10.1186/s13750-019-0180-4

Ebba M, Asaithambi P, Alemayehu E (2022) Development of electrocoagulation process for wastewater treatment: optimization by response surface methodology. Heliyon 8(5). https://doi.org/10.1016/j.heliyon.2022.e09383

El-Liethy MA, Dakhil MA, El-Keblawy A, Abdelaal M, Halmy MWA, Elgarhy AH et al (2022) Temporal phytoremediation potential for heavy metals and bacterial abundance in drainage water. Sci Rep 12(1):8223

El-Liethy MA, Hemdan BA, El-Taweel GE (2023) New insights for tracking bacterial community structures in industrial wastewater from textile factories to surface water using phenotypic, 16S rRNA isolates identifications and high-throughput sequencing. Acta Trop 238:106806

Elsayed S, Gad M, Farouk M, Saleh AH, Hussein H, Elmetwalli AH, Elsherbiny O, Moghanm FS, Moustapha ME, Taher MA, Eid EM, Abou El-Safa MM (2021) Using optimized two and three-band spectral indices and multivariate models to assess some water quality indicators of Qaroun Lake in Egypt. Sustainability 13(18):10408. https://doi.org/10.3390/su131810408

Escher BI, Neale PA (2021) Effect-based trigger values for mixtures of chemicals in surface water detected with in vitro bioassays. Environ Toxicol Chem 40(2):487–499. https://doi.org/10.1002/etc.4944

Faillace CA, Lorusso NS, Duffy S (2017) Overlooking the smallest matter: viruses impact biological invasions. Ecol Lett 20(4):524–538. https://doi.org/10.1111/ele.12742

Filipić A, Dobnik D, Tušek Žnidarič M, Žegura B, Štern A, Primc G, Mozetič M, Ravnikar M, Žel J, Gutierrez Aguirre I (2021) Inactivation of pepper mild mottle virus in water by cold atmospheric plasma. Front Microbiol 12:618209. https://doi.org/10.3389/fmicb.2021.618209

Gad M, Saleh AH, Hussein H, Farouk M, Elsayed S (2022) Appraisal of surface water quality of Nile River using water quality indices, spectral signature and multivariate modeling. Water 14(7):1131. https://doi.org/10.3390/w14071131

Gerba CP (2009) Environmentally transmitted pathogens. In: Environmental microbiology. Academic Press, pp 445–484. https://doi.org/10.1016/B978-0-12-370519-8.00022-5

Ghafil JA, Zghair HK, Zgair AK (2022) Chemical and microbiological properties of drinking water in the city of Baghdad. Diyala J Med 23(1):44–51. https://doi.org/10.26505/djm.23016410301

Giakoumis T, Voulvoulis N (2018) The transition of EU water policy towards the water framework directive's integrated river basin management paradigm. Environ Manag 62(5):819–831. https://doi.org/10.1007/s00267-018-1080-z

Gibson KE (2014) Viral pathogens in water: occurrence, public health impact, and available control strategies. Curr Opin Virol 4:50–57. https://doi.org/10.1016/j.coviro.2013.12.005

Girish V, Grant LM, John S (2024, January) Hepatitis A. In: StatPearls. StatPearls Publishing, Treasure Island (FL). https://www.ncbi.nlm.nih.gov/books/NBK459290/

Golden HE, Hoghooghi N (2018) Green infrastructure and its catchment-scale effects: an emerging science. Wiley Interdiscip Rev Water 5(1):e1254. https://doi.org/10.1002/wat2.1254

Goutam Mukherjee A, Ramesh Wanjari U, Eladl MA, El-Sherbiny M, Elsherbini DMA, Sukumar A, Kannampuzha S, Ravichandran M, Renu K, Vellingiri B, Kandasamy S, Gopalakrishnan AV (2022) Mixed contaminants: occurrence, interactions, toxicity, detection, and remediation. Molecules 27(8):2577. https://doi.org/10.3390/molecules27082577

Gruber JS, Ercumen A, Colford JM Jr (2014) Coliform bacteria as indicators of diarrheal risk in household drinking water: systematic review and meta-analysis. PLoS ONE 9(9):e107429. https://doi.org/10.1371/journal.pone.0107429

Haake DA, Levett PN (2015) Leptospirosis in humans. In: Leptospira and leptospirosis, pp 65–97. https://doi.org/10.1007/978-3-662-45059-8_5

Hall A, Shoesmith A, Doherty E, McEvoy B, Mettert K, Lewis CC, Wolfenden L, Yoong S, Kingsland M, Shelton RC, Wiltsey Stirman S, Imad N, Sutherland R, Nathan N (2022) Evaluation of measures of sustainability and sustainability determinants for use in community, public health, and clinical settings: a systematic review. Implement Sci 17(1):81. https://doi.org/10.1186/s13012-022-01252-1

Haramoto E, Kitajima M, Kishida N, Konno Y, Katayama H, Asami M, Akiba M (2013) Occurrence of pepper mild mottle virus in drinking water sources in Japan. Appl Environ Microbiol 79(23):7413–7418. https://doi.org/10.1128/aem.02354-13

Harper SL, Edge VL, Schuster-Wallace CJ, Berke O, McEwen SA (2011) Weather, water quality and infectious gastrointestinal illness in two Inuit communities in Nunatsiavut, Canada: potential implications for climate change. EcoHealth 8:93–108. https://doi.org/10.1007/s10393-011-0690-1

Hasan F (2023) Evaluation of the performance of multiple water quality loading index systems for Sapgyo River. Doctoral dissertation, Toronto Metropolitan University. https://doi.org/10.32920/ryerson.14643678

Hatami H, Khorasani-Zavareh D, Shakiba E, Teimoori F (2020) Evaluation of waterborne norovirus gastroenteritis epidemic in Eslamabad-e Gharb in Kermanshah, Iran (2016). J Kermanshah Univ Med Sci 24(1). https://doi.org/10.5812/jkums.99892

He L, Liu D, Hu J, Sun W, Gao R, Shi L, He D, Li B, Wang X, Gu M, Hu S, Hu X, Hu Z, Peng D, Liu X (2018) A comprehensive comparison of the fifth-wave highly pathogenic and low-pathogenic H7N9 avian influenza viruses reveals potential threat posed by both types of viruses in mammals. Transbound Emerg Dis 65(6):1459–1473. https://doi.org/10.1111/tbed.12954

Herrador BG, Lund V, Fonahn W, Hisdal H, Hygen HO, Hyllestad S, Nordeng Z, Skaland RG, Sunde LS, Vold L, White R, Wong WK, Nygård K (2021) Heavy weather events, water quality and gastroenteritis in Norway. One Health 13:100297. https://doi.org/10.1016/j.onehlt.2021.100297

Hind H, Afkar H, Suhad J (2020) A review study of waterborne parasites in Iraq. GSC Biol Pharm Sci 10(3):022–026. https://doi.org/10.30574/gscbps.2020.10.3.0031

Hussain M, Cifci MA, Sehar T, Nabi S, Cheikhrouhou O, Maqsood H, Ibrahim M, Mohammad F (2023) Machine learning-based efficient prediction of positive cases of waterborne diseases. BMC Med Inform Decis Mak 23(1):11. https://doi.org/10.1186/s12911-022-02092-1

Ibrahim EME, El-Liethy MA, Abia ALK, Hemdan BA, Shaheen MN (2019) Survival of E. coli O157: H7, Salmonella Typhimurium, HAdV2 and MNV-1 in river water under dark conditions and varying storage temperatures. Sci Total Environ 648:1297–1304

Ilyas M, Ahmad W, Khan H, Yousaf S, Yasir M, Khan A (2019) Environmental and health impacts of industrial wastewater effluents in Pakistan: a review. Rev Environ Health 34(2):171–186. https://doi.org/10.1515/reveh-2018-0078

Iyiola AO, Valbo-Jørgensen J (2023) Intrinsic values of the African environment: a sustainable perspective. In: Sustainable utilization and conservation of Africa's biological resources and environment, pp 635–664

Iyiola AO, Kolawole AS, Ajayi FO, Ogidi OI, Ogwu MC (2024) Sustainable water use and management for agricultural transformation in Africa. In: Water resources management for rural development, pp 287–300

Izah SC, Aigberua AO, Srivastav AL (2022) Factors influencing the alteration of microbial and heavy metal characteristics of river systems in the Niger Delta region of Nigeria. In: Madhav S, Kanhaiya S, Srivastav AL, Singh VB, Singh P (eds) Ecological significance of river ecosystems: challenges and management. Elsevier, United Kingdom, pp 51–78. https://doi.org/10.1016/B978-0-323-85045-2.00005-4

Izah SC, Iyiola AO, Richard G (2023a) Impacts of pollution on the hydrogeochemical and microbial community of aquatic ecosystems in Bayelsa State, Southern Nigeria. In: Hydrogeochemistry of aquatic ecosystems. John Wiley & Sons Ltd, pp 283–305. https://doi.org/10.1002/9781119870562.ch13. Print ISBN: 9781119870531, Online ISBN: 9781119870562

Izah SC, Richard G, Stanley HO, Sawyer WE, Ogwu MC, Uwaeme OR (2023b) Integrating the one health approach and statistical analysis for sustainable aquatic ecosystem management and trace metal contamination mitigation. ES Food Agrofor 14(2):1012. https://doi.org/10.30919/esfaf1012

Izah SC, Richard G, Stanley HO, Ogwu MC, Sawyer WE, Uwaeme OR (2024a) Prospects and application of multivariate and reliability analyses to one health risk assessments of toxic elements. Toxicol Environ Health Sci 16(2):127–134. https://doi.org/10.1007/s13530-023-00199-0

Izah SC, Stanley HO, Richard G, Sawyer WE, Uwaeme OR (2024b) Environmental health risks of trace elements in sediment using multivariate approaches and contamination indices. Int J Environ Sci Technol 1–26. https://doi.org/10.1007/s13762-024-05974-1

Izah SC, Stanley HO, Richard G, Sawyer WE, Uwaeme OR (2024c) Surface water quality: a statistical perspective on the efficacy of environmental and human health assessment tools. Water Air Soil Pollut 235(3):205. https://doi.org/10.1007/s11270-024-06965-1

Jakubiak M, Bojarski B (2021) Impact of point source pollutants on the distribution of selected water parameters in the Vistula River in Puławy, Poland. J Water Land Dev. https://doi.org/10.24425/jwld.2021.139014

Javanmardi Z, Alighadri M, Parastar S, Sadeghi T (2022) Effects of the municipal solid waste landfill on the quality of water resources in Khalkhal: a case study. Avicenna J Environ Health Eng 9(1):25–34. https://doi.org/10.34172/ajehe.2022.04

Kabwama SN, Bulage L, Nsubuga F, Pande G, Oguttu DW, Mafigiri R, Kihembo C, Kwesiga B, Masiira B, Okullo AE, Kajumbula H, Matovu JKB, Makumbi I, Wetaka M, Kasozi S, Kyazze S, Dahlke M, Hughes P, Sendagala JN, Musenero M, Nabukenya I, Hill VR, Mintz E, Routh J, Gómez G, Bicknese A, Zhu BP (2017) A large and persistent outbreak of typhoid fever caused by consuming contaminated water and street-vended beverages: Kampala, Uganda, January–June 2015. BMC Public Health 17:1–9. https://doi.org/10.1186/s12889-016-4002-0

Kamanmalek S, Alamdari N, Blunt B, Hammond D (2023) The role of stormwater best management practices in controlling waterborne diseases across Florida with special attention to environmental justice. https://doi.org/10.21203/rs.3.rs-2596157/v1

Kamesh, Singh BP, Misra S, Verma KK, Singh CK, Kumar R (2023) An Emerging adsorption technology and its applicability on trees as an adsorbent for the remediation of water pollution: a review. Ecol, Environ Conserv 29(2):627–640. https://doi.org/10.53550/eec.2023.v29i02.014

Kayser GL, Amjad U, Dalcanale F, Bartram J, Bentley ME (2015) Drinking water quality governance: a comparative case study of Brazil, Ecuador, and Malawi. Environ Sci Policy 48:186–195. https://doi.org/10.1016/j.envsci.2014.12.019

Kim HJ, Yoon HW, Lee MA, Kim YH, Lee CJ (2023) Impact of UV-C irradiation on bacterial disinfection in a drinking water purification system. J Microbiol Biotechnol 33(1):106. https://doi.org/10.4014/jmb.2211.11027

Komba FE, Fabian C, Elimbinzi E, Shao GN (2022) Efficiency of common filters for water treatment in Tanzania. Bull Natl Res Cent 46(1):208. https://doi.org/10.1186/s42269-022-00888-9

Kuberan A, Singh AK, Kasav JB, Prasad S, Surapaneni KM, Upadhyay V, Joshi A (2015) Water and sanitation hygiene knowledge, attitude, and practices among household members living in rural setting of India. J Nat Sci, Biol, Med 6(Suppl 1):S69. https://doi.org/10.4103/0976-9668.166090

Labana RV (2018) Cryptosporidium in the Philippines. Int Ann Sci 6(1):18–27. https://doi.org/10.21467/ias.6.1.18-27

Lanrewaju AA, Enitan-Folami AM, Sabiu S, Swalaha FM (2022) A review on disinfection methods for inactivation of waterborne viruses. Front Microbiol 13:991856. https://doi.org/10.3389/fmicb.2022.991856

Levy K, Woster AP, Goldstein RS, Carlton EJ (2016) Untangling the impacts of climate change on waterborne diseases: a systematic review of relationships between diarrheal diseases and temperature, rainfall, flooding, and drought. Environ Sci Technol 50(10):4905–4922. https://doi.org/10.1021/acs.est.5b06186

Li Y, Wang Z, Wei Y (2022) Water lock-in within China's economic industry based on the input–output method and social network analysis. Environ Sci Pollut Res 1–13. https://doi.org/10.1007/s11356-021-17463-w

Linker LC, Batiuk RA, Shenk GW, Cerco CF (2013) Development of the Chesapeake Bay watershed total maximum daily load allocation. JAWRA J Am Water Resour Assoc 49(5):986–1006. https://doi.org/10.1111/jawr.12105

Machado A, Amorim E, Bordalo AA (2022) Spatial and seasonal drinking water quality assessment in a Sub-Saharan Country (Guinea-Bissau). Water 14(13):1987. https://doi.org/10.3390/w14131987

Mahmoud M, El-Liethy MA (2022) Three-dimensional, flow-through silver-magnetite nanocomposite-modified reactive electrochemical system with slow silver ions release for efficient bacterial disinfection of sewage. J Environ Chem Eng 10(1):106985

Makwana S (2020) Effect of textile industrial effluents on water quality of Bandi River (Pali) Rajasthan, India. Int J Res Appl Sci Eng Technol 8(4):21–28. https://doi.org/10.22214/ijraset.2020.4005

Manar MK, Singh SK, Bajpai PK, Singh S (2018) Drinking water hygiene practices in caregivers of under 6 year children: a cross sectional study in urban Lucknow, India. Int J Community Med Public Health 5(9):3829. https://doi.org/10.18203/2394-6040.ijcmph20183372

Marembo L, Xu W (2020) A graduated approach in the abolition of waterborne diseases in drinking water using an indicator based approach and nano based biosensors: a review. J Environ Earth Sci 10:64–71. https://core.ac.uk/download/pdf/304991776

Maroju RG, Choudhari SG, Shaikh MK, Borkar SK, Mendhe H (2023) Application of artificial intelligence in the management of drinking water: a narrative review. Cureus 15(11). https://doi.org/10.7759/cureus.49344

Mathews LB, Moodie MM, Simmons AM, Swinburn BA (2010) The process evaluation of It's Your Move!, an Australian adolescent community-based obesity prevention project. BMC Public Health 10:1–13. https://doi.org/10.1186/1471-2458-10-448

McClung RP, Roth DM, Vigar M, Roberts VA, Kahler AM, Cooley LA, Hilborn ED, Wade TJ, Fullerton KE, Yoder JS, Hill VR (2018) Waterborne disease outbreaks associated with environmental and undetermined exposures to water—United States, 2013–2014. Am J Transplant 18(1):262–267. https://doi.org/10.1111/ajt.14607

Meli P, Rey Benayas JM, Balvanera P, Martínez Ramos M (2014) Restoration enhances wetland biodiversity and ecosystem service supply, but results are context-dependent: a meta-analysis. PLoS ONE 9(4):e93507. https://doi.org/10.1371/journal.pone.0093507

Meng H, Xu T, Gao M, Bai J, Li C (2020) An oil-contamination-resistant PVP/PAN electrospinning membrane for high-efficient oil–water mixture and emulsion separation. J Appl Polym Sci 138(11):50043. https://doi.org/10.1002/app.50043

Mogasale VV, Ramani E, Mogasale V, Park JY, Wierzba TF (2018) Estimating typhoid fever risk associated with lack of access to safe water: a systematic literature review. J Environ Public Health 2018(1–14):9589208. https://doi.org/10.1155/2018/9589208

Mohamed ME (2022) Review article on cryptosporidiosis: epidemiology, clinical presentation, diagnosis, and treatment. Egypt Acad J Biol Sci, E Med Entomol Parasitol 14(2):149–157. https://doi.org/10.21608/eajbse.2022.271392

Moomaw WR, Bhandary RR, Kuhl L, Verkooijen P (2017) Sustainable development diplomacy: diagnostics for the negotiation and implementation of sustainable development. Global Pol 8(1):73–81. https://doi.org/10.1111/1758-5899.12350

Nelson S, Drabarek D, Jenkins A, Negin J, Abimbola S (2021) How community participation in water and sanitation interventions impacts human health, WASH infrastructure and service longevity in low-income and middle-income countries: a realist review. BMJ Open 11(12):e053320. https://doi.org/10.1136/bmjopen-2021-053320

Nemati S, Shalileh F, Mirjalali H, Omidfar K (2023) Toward waterborne protozoa detection using sensing technologies. Front Microbiol 14:1118164. https://doi.org/10.3389/fmicb.2023.1118164

Nguyen TG, Huynh THN (2022) Assessment of surface water quality and monitoring in southern Vietnam using multicriteria statistical approaches. Sustain Environ Res 32(1):20. https://doi.org/10.1186/s42834-022-00133-y

Novak D, Mohammadian HD (2023) Challenges of a sustainable energy and vehicle-related value chain for BEVs and FCEVs through the 5th wave theory. Prog Energy Fuels 12(1):3308–3308. https://doi.org/10.18282/pef.v12i1.3308

Obotey Ezugbe E, Rathilal S (2020) Membrane technologies in wastewater treatment: a review. Membranes 10(5):89. https://doi.org/10.3390/membranes10050089

Ogamba EN, Charles EE, Izah SC (2021) Distributions, pollution evaluation and health risk of selected heavy metal in surface water of Taylor creek, Bayelsa State, Nigeria. Toxicol Environ Heal Sci 13:109–121. https://doi.org/10.1007/s13530-020-00076-0

Ogedengbe K, Akinbile CO (2011) Comparative analysis of the impact of industrial and agricultural effluents on Ona stream in Ibadan, Nigeria. J Sci Technol (Ghana) 31(2). https://doi.org/10.4314/just.v31i2.69392

Ogwu IY, Tembe ET, Shomkegh SA (2014) Comparative analysis of calorific value of briquettes produced from sawdust particles of daniella oliveriand Afzelia africana combination at binary and tertiary levels with rice husk. J Res For Wildl Environ 6(2):13–21

Ogwu DA, Molua OC, Irunkwor TC, Ataman JO, Edobor M, Ukpene AO, Egheneji AO (2024a) Geo-electric assessment of groundwater potentials and vulnerability to contaminants for sustainable water management at Utue-Ogume, Delta State, Nigeria. Geol Geophy Russia South (in Russ) 14(1):91–102

Ogwu MC, Ahuekwe EF, Balogun D, Kwarpo Z, Shittu KA, Izah S C (2024b) Methods for assessing soil physicochemical and biological properties. In: Sustainable soil systems in global south. Springer Nature Singapore, Singapore, pp 49–82

Ogwu MC, Pamela BO, Dare AF, Ovuru KF, Iyiola AO (2024c) Traditional and conventional water treatment methods: a sustainable approach. In: Water crises and sustainable management in the global south. Springer Nature Singapore, Singapore, pp 461–486

Oon YL, Oon YS, Ayaz M, Deng M, Li L, Song K (2023) Waterborne pathogens detection technologies: advances, challenges, and future perspectives. Front Microbiol 14:1286923. https://doi.org/10.3389/fmicb.2023.1286923

Ozkay F, Tas I, Ozkandan H, Gunal AC (2022) Biological risks of wastewater for irrigation. Curr Trends Nat Sci 11(21):54–66. https://doi.org/10.47068/ctns.2022.v11i21.006

Parvin F, Tareq SM (2021) Impact of landfill leachate contamination on surface and groundwater of Bangladesh: a systematic review and possible public health risks assessment. Appl Water Sci 11(6):100. https://doi.org/10.1007/s13201-021-01431-3

Payment P, Locas A (2011) Pathogens in water: value and limits of correlation with microbial indicators. Groundwater 49(1):4–11. https://doi.org/10.1111/j.1745-6584.2010.00710.x

Pieper KJ, Pierce G, Dobbin K, Jones CN, Weiss S, Moloney K (2022) Impacts of regulated water service extension on water quality, perception, and affordability in Orleans, NY. Environ Sci Technol Lett 9(12):1068–1073. https://doi.org/10.1021/acs.estlett.2c00660

Pisciotta C, Hill-Kane R, Winters A, Hooley C, Lee KA, Green T, Gopalan G (2023) Sustainability of a child mental health intervention in child welfare services: case study. Med Res Arch 11(10). https://doi.org/10.18103/mra.v11i10.4524

Pouey J, Galey C, Chesneau J, Jones G, Franques N, Beaudeau P, groupe des referents regionaux EpiGEH, Mouly D (2021) Implementation of a national waterborne disease outbreak surveillance system: overview and preliminary results, France, 2010 to 2019. Eurosurveillance 26(34):2001466. https://doi.org/10.2807/1560-7917.ES.2021.26.34.2001466

Prakash J, Agrawal SB, Agrawal M (2023) Global trends of acidity in rainfall and its impact on plants and soil. J Soil Sci Plant Nutr 23(1):398–419. https://doi.org/10.1007/s42729-022-01051-z

Pramaningsih V, Purwanto AD, Yuliawati R, Kurniawan D (2022, December) Water quality status and water quality index of Selulu reservoir, East Kalimantan, Indonesia. In: IOP conference series: earth and environmental science, vol 1111, No 1, p 012039. https://doi.org/10.1088/1755-1315/1111/1/012039

Pretty J (2008) Agricultural sustainability: concepts, principles and evidence. Philos Trans R Soc B: Biol Sci 363(1491):447–465. https://doi.org/10.1098/rstb.2007.2163

Ramírez-Castillo FY, Loera-Muro A, Jacques M, Garneau P, Avelar-González FJ, Harel J, Guerrero-Barrera AL (2015) Waterborne pathogens: detection methods and challenges. Pathogens 4(2):307–334. https://doi.org/10.3390/pathogens4020307

Razali MC, Wahab NA, Sunar N, Shamsudin NH (2023) Existing filtration treatment on drinking water process and concerns issues. Membranes 13(3):285. https://doi.org/10.3390/membranes13030285

Ros E, Ros A, Austin EE, De Geer L, Lane P, Johnson A, Clay-Williams R (2021) Sustainment of a patient flow intervention in an intensive care unit in a regional hospital in Australia: a mixed-method, 5-year follow-up study. BMJ Open 11(6):e047394. https://doi.org/10.1136/bmjopen-2020-047394

Roy R (2019) An introduction to water quality analysis. Ess Int J Environ Rehabil Conserv 9(1):94–100. https://doi.org/10.31786/09756272.18.9.2.214

Sala-Garrido R, Mocholí-Arce M, Molinos-Senante M (2021) Assessing the quality of service of water companies: a 'benefit of the doubt' composite indicator. Soc Indic Res 155:371–387. https://doi.org/10.1007/s11205-020-02588-1

Samijayani ON, Saputra TP, Firdaus H, Mujadin A (2023) Solar powered wireless sensor network for water quality monitoring and classification. Green Intell Syst Appl 3(1):14–21. https://doi.org/10.53623/gisa.v3i1.244

Saraswati SP, Ardion MV, Widodo YH, Hadisusanto S (2019) Water quality index performance for river pollution control based on better ecological point of view (a case study in Code, Winongo, Gadjah Wong streams). J Civ Eng Forum 5(1):47–56. https://doi.org/10.22146/jcef.41165

Saxena GS, Chandwankar RR (2024) Distillation in water and used water purification. In: Handbook of water and used water purification. Springer International Publishing, Cham, pp 251–272. https://doi.org/10.1007/978-3-319-66382-1_100-1

Scheirer MA, Dearing JW (2011) An agenda for research on the sustainability of public health programs. Am J Public Health 101(11):2059–2067. https://doi.org/10.2105/ajph.2011.300193

Shaheen MN, Elmahdy EM, Rizk NM, Abdo SM, Hussein NA, Elshershaby A, Shahein YE, Fawzy ME, El-Liethy MA, Marouf MA, Abdel-Gawad FK, Hu A, Gad M (2024) Evaluation of physical, chemical, and microbiological characteristics of waste stabilization ponds, Giza, Egypt. Environ Sci Eur 36(1):170

Shahraki AH, Heath D, Chaganti SR (2019) Recreational water monitoring: nanofluidic qRT-PCR chip for assessing beach water safety. Environ DNA 1(4):305–315. https://doi.org/10.1002/edn3.30

Singh S, Ghosh NC, Krishan G, Kumar S, Gurjar S, Sharma MK (2019) Development of indices for surface and groundwater quality assessment and characterization for Indian conditions. Environ Monit Assess 191:1–20. https://doi.org/10.1007/s10661-019-7276-8

Sravan JS, Matsakas L, Sarkar O (2024) Advances in biological wastewater treatment processes: focus on low-carbon energy and resource recovery in biorefinery context. Bioengineering 11(3):281. https://doi.org/10.3390/bioengineering11030281

Sudia LB, Indriyani L, Yunus L, Mursidi B, Yasin A, Nurdin M (2021) Water quality in thirty freshwater springs and twenty-four brackish springs in the karst area to realize sustainable water resources management. Sustainability 13(5):2679. https://doi.org/10.3390/su13052679

Tango PJ, Batiuk RA (2013) Deriving the Chesapeake Bay water quality standards. JAWRA J Am Water Resour Assoc 49(5):1007–1024. https://doi.org/10.1111/jawr.12108

Wang G, Yan T, Zhang J, Shi L, Zhang D (2020) Trace-Fe-enhanced capacitive deionization of saline water by boosting electron transfer of electro-adsorption sites. Environ Sci Technol 54(13):8411–8419. https://doi.org/10.1021/acs.est.0c01518

Wang S, Islam MZ (2022) Legal institutional inefficiency and water pollution problem in Bangladesh. Resour Environ Econ 4(2):372–375. https://doi.org/10.25082/ree.2022.02.004

Wang X, Wang N (2023) Preparation of nanofiber affinity membrane by electrospinning and its application in metal ion recovery. Sci J Technol 5(3):91–94. https://doi.org/10.54691/sjt.v5i3.4492

Waqa G, Moodie M, Schultz J, Swinburn B (2013) Process evaluation of a community-based intervention program: Healthy Youth Healthy Communities, an adolescent obesity prevention project in Fiji. Glob Health Promot 20(4):23–34. https://doi.org/10.1177/1757975913501909

Wei W, Ma R, Sun Z, Zhou A, Bu J, Long X, Liu Y (2018) Effects of mining activities on the release of heavy metals (HMs) in a typical mountain headwater region, the Qinghai-Tibet Plateau in China. Int J Environ Res Public Health 15(9):1987. https://doi.org/10.3390/ijerph15091987

Whelan J, Love P, Millar L, Allender S, Bell C (2018) Sustaining obesity prevention in communities: a systematic narrative synthesis review. Obes Rev 19(6):839–851. https://doi.org/10.1111/obr.12675

WHO (World Health Organization) (2023) Drinking water, 13 September 2023. https://www.who.int/news-room/fact-sheets/detail/drinking-water

Withanachchi SS, Ghambashidze G, Kunchulia I, Urushadze T, Ploeger A (2018) A paradigm shift in water quality governance in a transitional context: a critical study about the empowerment of local governance in Georgia. Water 10(2):98. https://doi.org/10.3390/w10020098

Xiao L, Zhang Q, Niu C, Wang H (2020) Spatiotemporal patterns in river water quality and pollution source apportionment in the arid Beichuan River basin of northwestern China using positive matrix factorization receptor modeling techniques. Int J Environ Res Public Health 17(14):5015. https://doi.org/10.3390/ijerph17145015

Xie Y, Liu X, Wei H, Chen X, Gong N, Ahmad S, Lee T, Ismail S, Ni SQ (2022) Insight into impact of sewage discharge on microbial dynamics and pathogenicity in river ecosystem. Sci Rep 12(1):6894. https://doi.org/10.1038/s41598-022-09579-x

Xu R, Wu Y, Wang G, Zhang X, Wu W, Xu Z (2019) Evaluation of industrial water use efficiency considering pollutant discharge in China. PLoS ONE 14(8):e0221363. https://doi.org/10.1371/journal.pone.0221363

Xu C, Ren Z, Huang S, Li J, Zi Y, Hu X (2023) Simulation study on the impact of water flow regulation based on the MIKE 21 model in a river water environment. Sustainability 15(13):10313. https://doi.org/10.3390/su151310313

Yadav A, Rajhans KP, Ramteke S, Sahu BL, Patel KS, Blazhev B (2016) Contamination of industrial wastewater in central India. J Environ Prot 7(1):72–81. https://doi.org/10.4236/jep.2016.71007

Yücesoy H, Dönmez A, Atmaca-Aydın E, Yentür SP, Saruhan-Direskeneli G, Ankaralı H, Karagülle MZ (2021) Effects of balneological outpatient treatment on clinical parameters and serum cytokine levels in patients with chronic low back pain: a single-blind randomized controlled trial. Int J Biometeorol 65:1367–1376

Zahedi A, Monis P, Deere D, Ryan U (2021) Wastewater-based epidemiology—surveillance and early detection of waterborne pathogens with a focus on SARS-CoV-2, Cryptosporidium and Giardia. Parasitol Res 120(12):4167–4188. https://doi.org/10.1007/s00436-020-07023-5

Zhang P, Yang M, Lan J, Huang Y, Zhang J, Huang S, Yang Y, Ru J (2023) Water quality degradation due to heavy metal contamination: health impacts and eco-friendly approaches for heavy metal remediation. Toxics 11(10):828. https://doi.org/10.3390/toxics11100828

Zhao D, Lu H, Cheng Q, Huang Q, Ai J, Zhang Z, Liu H, He Z, Li Q (2023) Research progress on inactivation of bacteriophages by visible-light photocatalytic composite materials: a mini review. Materials 17(1):44. https://doi.org/10.3390/ma17010044

Zowail ME, Elezaby MM, Abdel-kareim AM, Yousef GA, Mahmoud AA (2010) Cytogenetic effect of industrial and agricultural wastes on Tilapia zilli fish. Egypt Acad J Biol Sci, B Zool 2(1):9–16. https://doi.org/10.21608/eajbsz.2010.15903

Chapter 4
Soil Health and Contamination: Processes, Remediation, and Impacts on Food Security

Ali Boularbah, Becky Nancy Aloo, Leila Benidire, Dora Neina, and Rachid Lahlali

Abstract Soil health is fundamental to sustainable agriculture and directly influences food security, environmental quality, and ecosystem services. However, soil contamination from anthropogenic activities such as industrialization, intensive farming, and improper waste management poses significant threats to these critical functions. The complex processes of soil contamination involve multiple sources, pathways, and impacts on soil health, affecting microbial activity, nutrient cycles, and ecosystem functioning. These disturbances can lead to reduced crop yields, compromised food quality, and long-term public health risks, exacerbating global food

A. Boularbah (✉) · L. Benidire
Laboratory of Bioresources and Food Safety, Department of Biology, Faculty of Science and Technics, Cadi Ayyad University, Marrakech, Morocco
e-mail: a.boularbah@uca.ac.ma

L. Benidire
e-mail: l.benidire@uca.ma

A. Boularbah
Center of Excellence for Soil and Fertilizer Research in Africa, College of Agriculture and Environmental Sciences, Mohammed VI Polytechnic University, Ben Guerir, Morocco

B. N. Aloo
Department of Biological Sciences, University of Eldoret, P. O. Box 1125-30100, Eldoret, Kenya
e-mail: baloo@uoeld.ac.ke

L. Benidire
Ecole Supérieure de Technologie d'El Kelâa Des Sraghna, Université Cadi-Ayyad, Route de Béni Mellal Km 8, BP 104, El Kelaa Des Sraghna, Morocco

D. Neina
Department of Soil Science, School of Agriculture, College of Basic and Applied Sciences, University of Ghana, P.O. Box LG 245, Legon, Accra, Ghana
e-mail: dneina@gmail.com

R. Lahlali
Phytopathology Unit, Department of Plant Protection, Ecole National d'Agriculture de Meknès, Km10, Rte Haj Kaddour, BP S/40, 50001 Meknès, Morocco
e-mail: rlahlali@enameknes.ac.ma

S. Chibueze Izah and M. Chidozie Ogwu (eds.), *Innovative Approaches in Environmental Health Management*, Environmental Science and Engineering,
https://doi.org/10.1007/978-3-031-81966-7_4

insecurity. This chapter focuses on the intricate dynamics between soil health and contamination, highlighting the mechanisms through which contaminants affect agricultural productivity and food security. It also explores various remediation strategies, including bioremediation and phytoremediation. It emphasizes the importance of integrating innovative technologies like artificial intelligence (AI) and predictive modeling to enhance the efficiency of these methods. Additionally, the chapter stresses the need for a comprehensive, multidisciplinary approach to soil management that involves farmers, researchers, policymakers, and communities working together to restore contaminated soils and secure the future of global food systems. By examining the socio-economic, environmental, and technological barriers to effective remediation, particularly in developing regions, this chapter offers insights into how sustainable soil management practices can be applied to ensure environmental justice and long-term agricultural viability.

Keywords Soil health · Soil contamination · Environmental impact · Food security · Sustainable agriculture · Remediation techniques · Ecosystem services

4.1 Introduction

Soil health is a critical component of sustainable agriculture, serving as the foundation for food production, supply of ecosystem services, and maintaining environmental quality. Healthy soils are characterized by their ability to support plant growth, conserve biodiversity, and regulate water and nutrient cycles. The importance of soil health cannot be overstated, as it directly influences agricultural productivity and resilience against climate change. Sustainable farming practices that enhance soil health are essential for ensuring food security and mitigating the adverse effects of farming practices on the environment (Ikram et al. 2022; Patil et al. 2019). Integrating soil health indicators, encompassing physical, chemical, and biological parameters, provides farmers with vital information to make informed decisions that promote sustainable practices and improve productivity (Ikram et al. 2022). However, soil contamination is a growing phenomenon that threatens soil health. It has far-reaching consequences, leading to reduced agricultural productivity, loss of biodiversity, and potential health risks for humans and wildlife (Bhise et al. 2019; Lee et al. 2020a). Soil contamination, defined as harmful substances in the soil that can adversely affect plant growth, human health, and the environment, poses a significant challenge to soil health and agricultural sustainability. Soil contaminants can originate from various sources, including industrial activities, farming practices, and urban development. Common pollutants include heavy metals (Boularbah et al. 2006a, b; Izah et al. 2017a, b, c, 2018, 2021), petroleum hydrocarbons (Aigberua et al. 2016a, b, 2017), and polycyclic aromatic hydrocarbons (PAHs), which can accumulate in the soil and disrupt its natural processes (Chuka et al. 2017; Lee et al. 2021; Wang et al. 2010).

The causes of soil contamination are multifaceted, often resulting from anthropogenic activities (Izah et al. 2021). Agricultural practices, such as the continuous overuse of fertilizers and pesticides, contribute to soil degradation and contamination. These practices can lead to nutrient imbalances, soil acidification, and the accumulation of toxic substances, which ultimately compromise soil health (Jena et al. 2022). Industrial activities, including mining and waste disposal, release heavy metals and organic pollutants into the soil, further exacerbating contamination (Lee et al. 2021; Sisay 2019). Urbanization also contributes to soil contamination through the introduction of pollutants from construction activities, vehicle emissions, and improper waste management (Lee et al. 2021; Wang et al. 2010), informal waste recycling such as e-wastes (Akese and Little 2018; Fosu-Mensah et al. 2017). Remediation of contaminated soils is essential to restore soil health and mitigate the impacts of contamination. Various remediation techniques have been developed, including physical, chemical, and biological methods. Physical methods, such as soil washing and thermal desorption, aim to remove contaminants from the soil matrix (Dike et al. 2013; Lee et al. 2021). Chemical methods involve the application of agents that stabilize or immobilize contaminants, while biological processes, such as bioremediation and phytoremediation, utilize microorganisms and plants to degrade or extract pollutants from the soil (Aboudrar et al. 2013; Teng et al. 2015; Zhu et al. 2010; Benidire et al. 2021). Each method has advantages and limitations, and the choice of remediation technique depends on the type and extent of contamination and the specific soil characteristics (Hu et al. 2023; Sisay, 2019).

The effectiveness of remediation strategies is often assessed through soil health indicators, which comprehensively understand the soil's physical, chemical, and biological properties. Monitoring these indicators is crucial for evaluating the success of remediation efforts and ensuring that soil health is restored (Lee et al. 2020a; Triantafyllidis et al. 2019). For instance, reducing pollutant concentrations alone does not guarantee improved soil health; ecotoxicity assessments are necessary to evaluate the overall health of the soil ecosystem (Lee et al. 2020a). Furthermore, the integration of innovative technologies, such as biochar application and the use of biosurfactants, has shown promise in enhancing the effectiveness of remediation efforts (Teng et al. 2015; Wang et al. 2018). In addition to remediation, sustainable agricultural practices play a vital role in preventing soil contamination and promoting soil health. Practices such as crop rotation, cover cropping, and reduced tillage can enhance soil structure, increase organic matter content, and improve nutrient cycling (Ikram et al. 2022; Patil et al. 2019). These practices contribute to soil health and enhance agricultural systems' resilience to climate change and other environmental stressors (Jena et al. 2022).

Furthermore, adopting precision agriculture technologies, including the Internet of Things (IoT), can help farmers monitor soil health in real time and make data-driven decisions to optimize inputs and minimize environmental impacts (Patil et al. 2019). The socio-economic implications of soil health and contamination cannot be overlooked. Contaminated soils can lead to decreased agricultural productivity, resulting in economic losses for farmers and increased consumer food prices (Lavee et al. 2012). Moreover, the costs associated with soil remediation can be substantial, often requiring significant investments in technology and labor (Lavee et al. 2012; Liu et al. 2011). Therefore, addressing soil contamination and promoting soil health is an environmental and economic necessity for sustainable agricultural development.

The chapter discusses the processes of soil contamination, highlighting industrial, agricultural, and urban sources and the pathways through which pollutants enter soils. It then explores the effects of contaminated soils on food security, focusing on crop quality, yield, and health risks through food chains. The chapter concludes by examining various soil remediation techniques, challenges to effective remediation, and future directions for sustainable soil management practices.

4.2 Processes of Soil Contamination

Soil contamination is a significant environmental issue that arises from various anthropogenic activities, including industrial, agricultural, and urban practices (Fig. 4.1). Industrial activities, such as manufacturing and waste disposal, often introduce hazardous substances into the soil, leading to contamination. For instance, the improper disposal of per- and polyfluorinated alkyl substances (PFAS) from consumer products has been shown to persist in soil environments for extended periods, indicating the long-term impacts of industrial waste on soil quality (Bugsel et al. 2022). Agricultural practices, mainly using fertilizers and pesticides, contribute to soil contamination by leaching chemicals into the soil matrix. Urban activities, including construction car traffic, urban and industrial wastewater, and solid waste, also exacerbate soil pollution by introducing heavy metals and organic pollutants (El Khalil et al. 2008a; Chacra et al. 2023) (Fig. 4.1). In urban areas, soils are also subjected to substantial human pressure, leading to the modification of their structure and functioning. These modifications affect mainly pH, aeration, water drainage, nutrient cycling, and temperature regime of soils (El Khalil et al. 2008b, 2013, 2016; Meuser 2010; Morel et al. 2005).

The cumulative effect of these sources highlights the urgent need for effective soil management and remediation strategies. Table 4.1 shows an overview of the processes of soil contamination.

The pathways through which pollutants enter soils can be categorized into chemical, biological, and physical mechanisms (Fig. 4.2). Chemical pathways involve directly deleting contaminants onto the soil surface or leaching from contaminated water sources. For example, heavy metals from industrial effluents can infiltrate the soil through precipitation and runoff, leading to significant bioaccumulation in soil organisms (Hamels et al. 2014). Biological pathways include the uptake of contaminants by plants and microorganisms, which can mitigate or exacerbate soil contamination depending on the species involved (Han et al. 2021). Physical pathways, such as erosion and sedimentation, can transport pollutants across landscapes, further complicating the contamination dynamics (Sarkar et al. 2021). Understanding these pathways is crucial for developing targeted remediation strategies that address the specific mechanisms of pollutant entry. The impact of soil contamination on microbial activity and nutrient cycles is profound and multifaceted. Contaminants

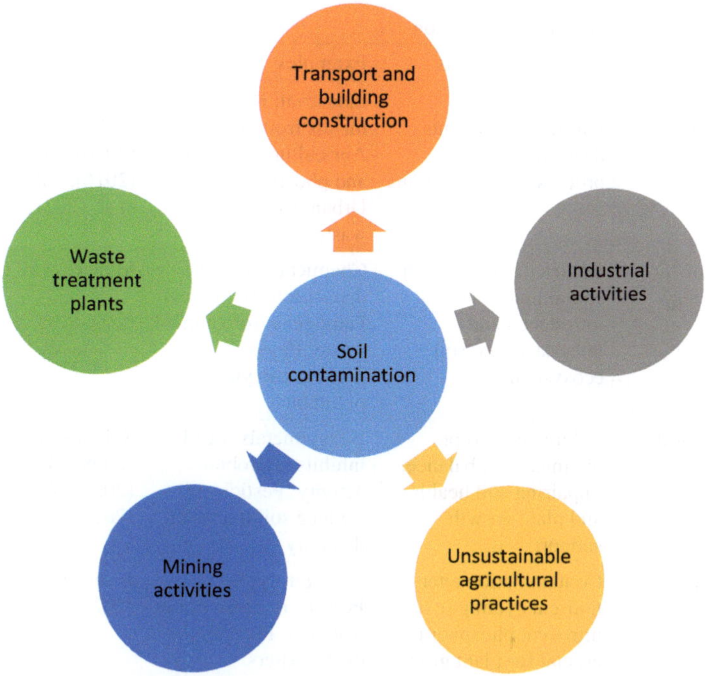

Fig. 4.1 Sources of soil contamination and environmental impact

can inhibit microbial growth, leading to a decline in microbial biomass and diversity, essential for maintaining soil health and fertility (Mészáros et al. 2022). For instance, heavy metals can disrupt microbial enzymatic activities, affecting nutrient-cycling processes such as nitrogen fixation and decomposition of organic matter (Naylo et al. 2019; Lee et al. 2020a).

Furthermore, the alteration of microbial communities due to contamination can lead to shifts in nutrient availability, impacting plant growth and ecosystem stability (Krumins et al. 2015). The interplay between soil contaminants and microbial dynamics underscores the importance of considering biological factors in soil remediation efforts. Contaminated soils can source heavy metals and other toxic substances that enter the food chain through agricultural practices (Jan et al. 2010; El Hamiani et al. 2015; Laaouidi et al. 2020). Consuming crops grown in contaminated soils poses significant health risks, particularly in regions where wastewater irrigation is common. Heavy metals such as lead and cadmium can accumulate in plant tissues, leading to adverse health effects in consumers (Jan et al. 2010). Therefore, understanding the contamination pathways and their implications for food safety is essential for developing effective public health policies.

Table 4.1 Processes of soil contamination

Process	Causes and effects	Examples	References
Sources of contamination	Human activities introduce pollutants, disrupting soil processes and functions	Industrial: Heavy metals from mining; Agricultural: Nitrate and phosphate runoff; Urban: Construction waste and emissions	Boularbah et al. (2006a, b), El Hamiani et al. (2010), El Khalil et al. (2016), Alengebawy et al. (2021)
Pathways of pollutant entry	Routes through which contaminants are introduced and integrated into soil ecosystems	Chemical: Airborne deposition; Biological: Pathogens from animal waste; Physical: Erosion carrying pollutants	Münzel et al. (2023)
Impacts on soil microbial activity	Pollutants disrupt the microbial balance, impairing soil health and plant growth support	Heavy metals: Lead inhibits microbial activity; Pesticides: Reduce soil microbial diversity	Sobolev and Begonia (2008), Benidire et al. (2020), Wang et al. (2022)
Impacts on nutrient cycles	Contamination alters nutrient cycling (e.g., nitrogen, phosphorus), crucial for plant growth	Nitrogen cycle: Fertilizers cause nitrate leaching; Phosphorus cycle: Alters availability, leading to eutrophication	Penuelas and Sardans (2023)

4.3 Effects of Soil Contamination on Food Security

Soil contamination significantly threatens food security, primarily through its adverse effects on crop yield and quality (Table 4.2). Contaminated soils can reduce agricultural productivity, as heavy metals and other pollutants disrupt the complex interactions within soil ecosystems. For instance, (Wu et al. 2021) highlighted that soil biodiversity is crucial for nutrient availability, directly influencing plant growth and crop yields. When soil is contaminated, the microbial communities that support these processes can be severely affected, leading to diminished soil health and reduced agricultural output (Zaib 2023). This decline in soil quality impacts not only the quantity of food produced but also the nutritional quality of crops, as contaminated soils can lead to the uptake of harmful substances by plants, further exacerbating food insecurity (Chen et al. 2018).

The influence of contaminated soils extends beyond crop yield to include the quality of the produce. Heavy metals such as cadmium, lead, and arsenic can accumulate in edible plant parts, posing severe health risks to consumers (Afzal et al. 2023). Moreover, the bioavailability of these contaminants is a critical factor; as highlighted by Hu (2010), the health risks associated with consuming contaminated vegetables are significant, particularly when considering long-term exposure to heavy

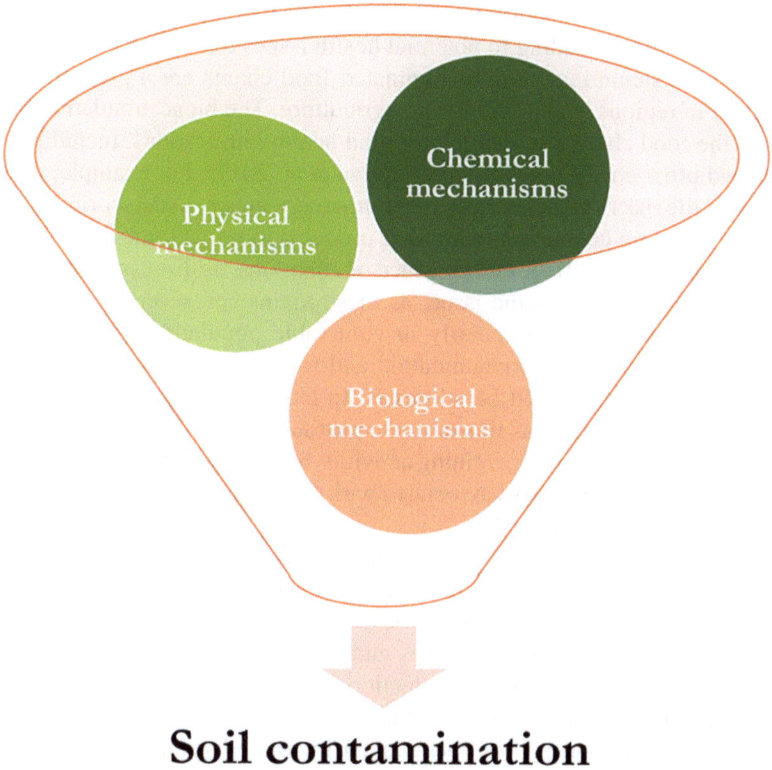

Fig. 4.2 Mechanistic processes resulting in soil contamination

Table 4.2 Effects of soil contamination on food security

Effect	Implications	Examples	References
Influence of contaminated soils on crop yield and quality	Contaminants reduce crop productivity and degrade nutritional quality	Heavy metals (Lead, cadmium, mercury reduce yields; Pesticides reduce nutrient cycling and crop quality	Ikhajiagbe et al. (2021), (2022)
Health risks from contaminated food chains	Soil contaminants accumulate in crops and livestock, posing health risks to consumers	Bioaccumulation of heavy metals causes cancer and neurological damage; Pesticides disrupt the endocrine system and weaken immunity	El Hamiani et al. (2015), Guo et al. (2019), Laaouidi et al. (2020)
Contamination impacts in agriculture-dependent regions	Soil contamination threatens food security and livelihoods in agricultural areas	China: 20% of arable land affected by industrial pollution; India: Heavy metals reduce yields	Zhang et al. (2015)

metals through the food chain. Thus, the quality of food produced in contaminated soils is compromised, leading to potential health issues for consumers.

Health risks stemming from contaminated food chains are a pressing concern, particularly in regions heavily reliant on agriculture. The bioaccumulation of heavy metals in the food chain can lead to profound health implications, including neurotoxicity and other chronic conditions (Supriya et al. 2017). For example, the EFSA has reported the risks associated with methylmercury in fish, underscoring the importance of monitoring contaminants in food products to protect public health (EFSA Scientific Committee, 2015). The presence of aflatoxins, as discussed by Afzal et al. (2023), further complicates the issue, as these toxins can severely impact human health and development, particularly in vulnerable populations such as children. The interplay between soil contamination and food safety thus creates a complex web of health risks that must be addressed to ensure food security. Studies from agriculture-dependent regions illustrate the profound impacts of soil contamination on food security. In areas where mining activities have led to soil degradation, such as Nanjing, China, the health risks associated with consuming contaminated vegetables have been documented (Hu 2010).

Similarly, research indicates that regions with high levels of industrial pollution face significant challenges in maintaining food safety due to the accumulation of heavy metals in crops (Teshale 2019). These case studies underscore the urgent need for effective soil management practices and remediation strategies to mitigate the impacts of contamination on food production and public health. The role of agricultural practices in exacerbating or alleviating soil contamination cannot be overlooked. Practices such as excessive use of chemical fertilizers and pesticides contribute to the accumulation of harmful substances in the soil (Suvorov and Stancu 2021). Sainju et al. (2019) advocate for improved management practices, including crop rotation and organic amendments, to enhance soil quality and reduce contamination risks. Integrating sustainable agricultural practices is essential for maintaining soil health and ensuring food security in the face of contamination challenges (Odubo et al. 2024).

4.4 Soil Remediation Techniques

Soil remediation is critical to restoring contaminated soils to a safe state for human health and the environment. Various techniques have been developed over the years, categorized broadly into physical, chemical, and biological methods (Table 4.3). Physical remediation techniques include soil replacement, thermal desorption, and washing, which focus on removing or altering contaminants. Chemical methods involve stabilization, solidification, and electrokinetic remediation, which chemically alter the pollutants to reduce their bioavailability or toxicity. Biological methods utilize living organisms to degrade or stabilize pollutants, including bioremediation and phytoremediation. Each of these methods has advantages and limitations, and their effectiveness can vary based on the type of contaminants present and the specific site conditions (Gusiatin and Rouhani 2023; Kaparwan et al. 2020; Khalid et al. 2017). Remediation techniques for contaminated soils vary widely, reflecting the complexity of contamination scenarios. Bioremediation, which utilizes microbial processes to degrade pollutants, has gained popularity due to its cost-effectiveness and environmental friendliness (Qin et al. 2023). For example, the application of biosurfactants derived from microorganisms has been shown to enhance the biodegradation of hydrocarbons in contaminated soils, facilitating the removal of pollutants (Moldes et al. 2011).

Additionally, phytoremediation, which employs plants to absorb and detoxify contaminants, has emerged as a promising approach for managing heavy metal pollution (Benidire et al. 2021, 2022; Chacra et al. 2023; Madline et al. 2024b). Integrating these biological methods with physical and chemical remediation can lead to more effective and sustainable soil recovery strategies. The specific characteristics of the contaminants and the soil environment influence the effectiveness of various remediation methods. For instance, thermal desorption has been identified as an efficient method for removing volatile organic compounds from contaminated soils, as it utilizes heat to volatilize and subsequently capture contaminants (Araujo et al. 2015). Similarly, electrokinetic remediation techniques have shown promise in treating soils contaminated with heavy metals and organic pollutants by facilitating the movement of contaminants toward electrodes for extraction (Ng et al. 2014). The choice of remediation technology must consider factors such as contaminant type, soil composition, and the desired speed of remediation to ensure optimal outcomes (Trifunović 2021).

Phytoremediation emerges as a promising strategy for addressing soil contamination and its effects on food security. This bioremediation technique utilizes plants to extract, stabilize, or degrade contaminants in the soil, thereby reducing their bioavailability and potential crop uptake (Muehe et al. 2015). Studies have shown that certain plant species can effectively remediate heavy metal-contaminated soils, improving crop yield and quality (Zhao et al. 2022). However, the success of phytoremediation is contingent upon understanding the interactions between soil microorganisms and plant roots, as these relationships are crucial for enhancing the efficacy of the remediation process (Saleem et al. 2019). The implications of soil contamination extend to food safety regulations and public health policies (Yarkwan et al. 2024). Petković et al. (2016) highlight that controlling contaminants in food products (Izah et al. 2017a, 2022, 2024), mainly dairy, is vital for protecting vulnerable populations. Stringent monitoring and regulation of food products from contaminated regions is paramount to prevent health risks associated with heavy metal exposure and other contaminants (Majed et al. 2016). Policymakers must prioritize research and intervention strategies that address the root causes of soil contamination to safeguard food security and public health (Ogwu et al. 2024). Zhao et al. (2022) emphasize that phytoremediation techniques can mitigate the risks associated with cadmium-contaminated soils. Still, the effectiveness of these methods is highly dependent on the soil's microbial community and its ability to support plant growth.

Soil washing is another remediation technique that has been effectively employed to treat contaminated soils, particularly those affected by heavy metals and petroleum

Table 4.3 Soil remediation techniques

Remediation technique	Description	Examples	References
Physical remediation	Techniques that physically remove or isolate soil contaminants	Soil washing: Uses water or other solvents to extract contaminants Thermal Desorption: Heats soil to volatilize organic contaminants for removal	Zoghi and Mafigholami (2023)
Chemical remediation	Chemical agents are applied to transform or neutralize contaminants in soils	Chemical oxidation: Injects oxidants (e.g. hydrogen peroxide, ozone) to break down pollutants Soil Stabilization: Adds binding agents (e.g. lime) to immobilize contaminants and prevent their spread	Rada et al. (2019)
Biological remediation	Living organisms (microorganisms or plants) are used to degrade or absorb contaminants	Bioremediation: Microbes degrade organic pollutants, such as hydrocarbons Phytoremediation: Plants absorb and accumulate heavy metals or other contaminants from soils	Bala et al. (2022)
Advances in bioremediation and phytoremediation	Enhanced methods for using biological systems to restore contaminated soils	Genetically Engineered Microorganisms (GEMs): Engineered microbes are more effective at breaking down pollutants Hyperaccumulator Plants: Plants genetically modified or selected for their ability to hyperaccumulate heavy metals like cadmium or lead	Aboudrar et al. (2013), Saha et al. (2021)

(continued)

Table 4.3 (continued)

Remediation technique	Description	Examples	References
Role of AI and predictive modeling	AI and predictive models optimize remediation strategies by analyzing data and predicting outcomes	Predictive models: Machine learning algorithms predict the effectiveness of remediation techniques based on soil properties and contaminant types AI-Driven decision support: AI systems assist in selecting the most efficient remediation method by analyzing environmental factors and previous case studies	Wu and Zhao (2023)

hydrocarbons. This method involves using water or chemical solutions to extract contaminants from the soil matrix, thereby reducing pollutant concentrations (Effendi et al. 2020). The efficiency of soil washing can be enhanced by optimizing the washing solutions and solid/liquid ratios, which can significantly improve the recovery of contaminants (Ramadan et al. 2018). However, the potential for secondary contamination must be carefully managed to prevent the spread of pollutants during the washing process. The role of soil enzymes as indicators of contamination and remediation success has garnered increasing attention in recent research. Soil enzymes are crucial for nutrient cycling and organic matter decomposition, and their activity can be significantly affected by the presence of contaminants (Naylo et al. 2019; Lee et al. 2020b). Monitoring changes in enzymatic activity can provide valuable insights into the health of the soil ecosystem and the effectiveness of remediation efforts. For instance, a decline in enzyme activity may indicate ongoing contamination, while a recovery in enzymatic function can signal successful remediation (Lee et al. 2020b; Benidire et al. 2021; Madline et al. 2024a, b). This approach underscores the importance of integrating biological indicators into soil management practices. The relationship between soil contamination and human health is a critical aspect of environmental science that warrants attention.

Recent advances in bioremediation and phytoremediation have shown promising results in addressing soil contamination, particularly with heavy metals and organic pollutants. Bioremediation employs microorganisms, such as bacteria and fungi, to metabolize and detoxify contaminants, while phytoremediation uses plants to absorb, degrade, or stabilize pollutants in the soil. Studies have demonstrated that certain plant species can enhance the uptake of heavy metals when combined with soil amendments like chelating agents (Benidire et al. 2021, 2022; Madline et al.

2024a, b; Alori et al. 2022; Chibuike and Obiora 2014; Ng et al. 2019). Furthermore, integrating bioremediation with other techniques, such as soil washing or chemical stabilization, can enhance the effectiveness of remediation efforts (Adnan et al. 2022; Yuan 2023). This synergy is essential in complex contaminated sites where multiple pollutants are present. The role of artificial intelligence (AI) and predictive modeling in optimizing remediation strategies is increasingly recognized in environmental science. AI algorithms can analyze large datasets to identify patterns and predict the behavior of contaminants in soil, which can inform the selection of appropriate remediation techniques. Predictive modeling can also help assess the potential effectiveness of various remediation strategies under different environmental conditions, thereby enabling more efficient planning and resource allocation (Ahmed et al. 2022; Ouvrard et al. 2011). For instance, Machine Learning models can be trained on historical data to predict the outcomes of bioremediation efforts based on soil characteristics and contaminant types, leading to more targeted and effective interventions.

In addition to traditional remediation methods, innovative approaches are being explored to enhance soil remediation processes. For example, using biochar as a soil amendment has gained attention for immobilizing heavy metals and organic pollutants, reducing their bioavailability (Gusiatin and Rouhani 2023; Zheng et al. 2021). Biochar can be produced from various biomass sources and tailored to enhance its adsorption properties for specific contaminants. Moreover, the application of nanotechnology in remediation processes, such as the use of nanoparticles for targeted delivery of remediation agents, is an area of active research that holds the potential for improving the efficiency of soil cleanup efforts (Sridevi 2022; Zhou et al. 2022). Various environmental factors, including soil texture, moisture content, and other contaminants, can influence the effectiveness of remediation techniques. For instance, the remediation of polycyclic aromatic hydrocarbons (PAHs) in contaminated soils can be particularly challenging due to their hydrophobic nature and tendency to adsorb strongly to soil particles (Kuppusamy et al. 2017; Zhao et al. 2021). Understanding the interactions between contaminants and soil properties is crucial for developing effective remediation strategies. Recent studies have emphasized the importance of site characterization and the need for tailored remediation approaches that consider the specific conditions of each contaminated site (Ampofo 2022; Fatdillah and Pauzi 2018).

The economic aspects of soil remediation cannot be overlooked. The costs associated with different remediation techniques can vary significantly, influenced by factors such as site location, contaminant type, and the scale of the remediation project. Cost-effective strategies are essential for ensuring the feasibility of remediation efforts, particularly in resource-limited settings (Ahmed et al. 2022; Azubuike et al. 2016). Recent reviews have highlighted various cost-reduction strategies, including bioremediation techniques, often more economical than traditional physical or chemical methods (Chao et al. 2021). Stakeholders can achieve better outcomes in soil cleanup initiatives by optimizing the choice of remediation technology based on effectiveness and cost. Integrating multiple remediation techniques is increasingly recognized as a best practice in soil remediation. A combination of physical, chemical, and biological methods can address the diverse challenges of contaminated soils more effectively than any single approach (Fatdillah and Pauzi 2018; Yuan 2023). For example, the sequential application of soil washing followed by bioremediation can enhance the removal of heavy metals and organic pollutants, improving overall remediation outcomes. This integrated approach allows for leveraging the strengths of each technique while mitigating their limitations (Villa et al. 2010).

4.5 Challenges and Barriers to Effective Soil Remediation

The challenges and barriers to effective remediation involve economic, technological, and policy-related dimensions (Table 4.4). In developing regions, these challenges are often exacerbated by limited resources, inadequate infrastructure, and a lack of regulatory frameworks. Economic constraints are particularly pronounced, as remediation efforts usually require substantial financial investment, which may not be feasible for many developing countries. For instance, the deployment of advanced remediation technologies can be prohibitively expensive, and the costs associated with long-term monitoring and maintenance can further strain limited budgets (Diarra et al. 2021; Harrison et al. 2017). Additionally, the economic impact of remediation projects must be carefully assessed to ensure that they do not disproportionately burden local communities, particularly in regions where livelihoods depend on natural resources (Apitz et al. 2017).

Technological challenges also play a critical role in hindering effective remediation. Many developing regions lack access to the latest remediation technologies, limiting their ability to address complex contamination issues effectively. For example, while advanced methods such as phytoremediation and bioremediation have shown promise in treating contaminated sites, their implementation

Table 4.4 Barriers to effective remediation and potential solutions

Challenges and barriers	Cause and effects	Potential solutions
Economic challenges	Limited funding, high costs, and resource competition hinder remediation efforts	Pursue public–private funding, cost-sharing, and economic incentives
Technological challenges	Lack of advanced technology, training, and local expertise in remediation	Invest in training, promote tech transfer, and create user-friendly solutions
Policy challenges	Weak regulations, poor enforcement, and conflicting policies limit progress	Advocate for reforms, strengthen enforcement, and improve collaboration
Barriers to developing regions	Insufficient infrastructure, political instability, and low public engagement	Engage communities, partner with local groups, and coordinate stakeholders

often requires specific knowledge and expertise that may not be readily available in these regions (Adnan et al. 2022; Khan et al. 2015). Furthermore, the variability in soil and water conditions can complicate the application of standardized remediation techniques, necessitating tailored approaches that consider local environmental factors (Maqbool et al. 2019). The lack of technological infrastructure can also impede the adoption of innovative solutions, such as nanotechnology and bioengineering, which can potentially enhance remediation efficiency (Maqbool et al. 2019). Policy challenges further complicate the remediation landscape, particularly in developing regions where regulatory frameworks may be weak or poorly enforced. The absence of comprehensive environmental policies can lead to inconsistent remediation practices and a lack of accountability for polluters (O'Connell and Hou 2015).

Moreover, the engagement of stakeholders in the decision-making process is often limited, which can result in remediation strategies that do not adequately reflect the needs and values of affected communities (Apitz et al. 2017). Effective policy frameworks are essential for establishing clear guidelines for remediation practices, ensuring they are scientifically sound and socially equitable (O'Connell and Hou 2015). In many cases, integrating sustainability principles into remediation policies can help balance environmental, economic, and social considerations, fostering more effective and inclusive remediation efforts (O'Connell and Hou 2015).

In developing regions, barriers to implementation are particularly pronounced due to a combination of economic, technological, and policy-related factors. The limited availability of financial resources can restrict the ability of governments and organizations to invest in necessary remediation infrastructure and technologies (Diarra et al. 2021). Additionally, the lack of trained personnel and expertise in remediation practices can hinder the effective execution of projects, as local stakeholders may not possess the requisite knowledge to implement advanced techniques (Khan et al. 2015). Furthermore, logistical challenges, such as inadequate transportation networks and limited access to contaminated sites, can impede remediation efforts' timely and efficient implementation (Diarra et al. 2021). Cultural factors can also pose significant barriers to effective remediation in developing regions. Community perceptions of environmental issues and remediation practices can influence the acceptance and success of remediation initiatives (Jones et al. 2018). For instance, if local populations do not perceive contamination as a significant threat or do not trust the motives of external organizations, they may resist remediation efforts (Jones et al. 2018). Building trust and fostering community engagement are essential for overcoming these barriers, as they can help ensure that remediation strategies are culturally appropriate and aligned with local values (Apitz et al. 2017). Additionally, education and awareness-raising initiatives can play a crucial role in changing perceptions and promoting the importance of environmental remediation (Charzyński et al. 2022; Jones et al. 2018).

The intersection of climate change and remediation presents another layer of complexity in addressing environmental contamination. As climate change exacerbates environmental challenges, such as increased flooding and drought, the effectiveness of traditional remediation methods may be compromised (O'Connell and

Hou 2015). For example, rising sea levels and extreme weather events can lead to the recontamination of previously remediated sites, necessitating the development of more resilient remediation strategies that can adapt to changing environmental conditions (O'Connell and Hou 2015; Madline et al. 2021, 2024b). This requires a shift in focus towards sustainable remediation practices that address current contamination and consider future environmental challenges (O'Connell and Hou 2015).

4.6 Future Directions and Sustainable Soil Management

The future of sustainable soil management is increasingly critical as global food security becomes a pressing concern amid climate change and population growth. Innovative soil conservation practices are essential for enhancing soil health and productivity, foundational for sustainable agriculture. Research indicates that sustainable soil management can significantly improve food production, with estimates suggesting potential increases of up to 58% in yield through effective soil management strategies (Handayani and Hale 2022). These practices include the integration of cover crops, reduced tillage, and crop rotation, enhancing soil structure and fertility, and promoting biodiversity and resilience against pests and diseases (Edralin et al. 2017; LaCanne and Lundgren 2018). Moreover, integrating sustainable agriculture with soil health management is vital for creating a holistic approach to food production. Sustainable agriculture emphasizes maintaining soil health as a critical component of agricultural systems. For instance, agroforestry and precision agriculture have been shown to enhance soil organic matter (SOM) and improve nutrient cycling, which is crucial for long-term agricultural sustainability (Al-Juthery 2023; Sanjay 2023). The role of SOM in improving soil structure, water retention, and microbial activity cannot be overstated, as it directly influences crop productivity and ecosystem health (Hussain 2023). Thus, a concerted effort to promote practices that enhance SOM is essential for achieving sustainable agricultural outcomes. Policy recommendations for ensuring long-term food security must focus on establishing frameworks that support sustainable soil management practices. Current policies often lack comprehensive regulations that address soil degradation and its impacts on food production (Stankovics et al. 2018). To rectify this, developing policies promoting sustainable agricultural practices and incentivizing farmers to adopt soil conservation measures is crucial. For example, implementing financial support mechanisms for farmers who engage in conservation tillage and cover cropping can facilitate a transition towards more sustainable farming systems (Helming et al. 2018). Additionally, policies should encourage research and development in soil health management, ensuring that farmers have access to the latest scientific knowledge and technologies.

Implementing effective soil management policies requires a multifaceted approach that includes education and knowledge exchange among stakeholders. Farmers' understanding of soil degradation and sustainable practices is often influenced by their experiences and local conditions (Jaworski et al. 2024). Therefore, creating knowledge-sharing platforms involving farmers, researchers, and policymakers is essential. This can be achieved through workshops, training programs, and collaborative research initiatives emphasizing soil health's importance in sustainable agriculture (Krzywoszynska 2019). Furthermore, integrating local knowledge with scientific research can enhance the relevance and applicability of soil management practices in diverse agricultural contexts. Innovative soil conservation practices are also emerging as critical components of sustainable soil management. Techniques such as no-till farming, cover cropping, and agroecological approaches have improved soil health while maintaining agricultural productivity (Shaheb et al. 2021). These practices reduce soil erosion and compaction and improve water retention and nutrient availability, which are essential for crop growth. For instance, the use of winter cover crops can significantly enhance soil structure and fertility, leading to improved yields in subsequent cropping seasons (LaCanne and Lundgren 2018). Moreover, adopting integrated pest management strategies can reduce reliance on chemical inputs, further contributing to soil health and ecosystem sustainability (Miner et al. 2020).

The interplay between climate change and soil management practices is another critical area that requires attention. Climate variability has been shown to exacerbate soil degradation, impacting agricultural productivity and food security (Mensah-Brako et al. 2021; Bismark et al. 2021). Therefore, developing adaptive soil management strategies that consider the effects of climate change is essential for ensuring long-term sustainability. This includes practices that enhance soil resilience to extreme weather events, such as droughts and heavy rainfall, which can severely impact crop yields (Randall et al. 2020; Madline et al. 2021, 2024b). Research into the effects of different soil management practices on soil health under varying climatic conditions is necessary to inform adaptive strategies that can mitigate the impacts of climate change on agriculture.

Furthermore, the role of soil monitoring and assessment in sustainable soil management cannot be overlooked. Establishing a robust soil monitoring network can provide valuable data on soil health indicators, enabling policymakers and farmers to make informed decisions regarding soil management practices (O'Sullivan et al. 2017). Regular soil health assessments can help identify degradation areas and inform targeted interventions to restore soil quality. Additionally, integrating soil health indicators into agricultural policy frameworks can promote accountability and encourage farmers' adoption of sustainable practices (Cardoso et al. 2013). Innovative soil conservation practices, such as no-till farming, crop rotation, and cover cropping, are essential for maintaining soil health and preventing erosion. By integrating sustainable agriculture techniques, like organic farming and precision agriculture, with soil health management strategies, farmers can enhance soil fertility, improve water retention, and reduce reliance on chemical inputs. These practices support resilient agricultural systems and promote soil biodiversity, boosting its capacity to sequester carbon and mitigate climate change. To ensure long-term food security, policymakers should prioritize incentives for sustainable farming, invest in research on soil health, and develop regulations that promote responsible land use. Policies that encourage the adoption of conservation practices and sustainable agriculture can help safeguard soil resources, supporting a productive agricultural sector that meets the needs of future generations.

4.7 Conclusion

Soil health plays a critical role in ensuring food security, as it directly affects agricultural productivity, environmental quality, and the resilience of ecosystems. Healthy soils provide essential nutrients to crops, support microbial activity, regulate water cycles, and help sequester carbon, which is crucial for sustainable agriculture. However, the integrity of soil health is often compromised by contamination from industrial activities, agricultural chemicals, and waste mismanagement, leading to diminished soil quality and reduced crop yields. Contaminated soils can disrupt the delicate balance of nutrient cycles, impact soil biodiversity, and threaten food safety by introducing toxins into the food chain. The growing global demand for food highlights the urgency of addressing soil contamination to maintain food security and environmental sustainability. Overcoming the challenges posed by soil contamination requires a multifaceted approach. Remediation techniques, such as bioremediation, phytoremediation, and advanced soil management practices, offer promising solutions for restoring contaminated soils. By integrating these technologies with innovative tools such as AI and predictive modeling, it is possible to develop more precise and efficient strategies for soil restoration.

Effective policies, robust regulatory frameworks, and strong community involvement are essential to ensuring the widespread implementation of these practices. The collaboration of farmers, researchers, policymakers, and communities is critical to addressing contamination issues, promoting soil health, and safeguarding food security. Sustainable agricultural practices that prioritize soil health will enhance food production and contribute to environmental conservation and climate resilience.

Acknowledgements The authors Ali BOULARBAH and Benidire Leila) gratefully acknowledge the Ministry of Higher Education, Scientific Research and Innovation (MESRSI)/DRSI), Morocco, (PRIMA/0002/2020) under the framework of the PRIMA (Program for Research and Innovation solutions in the Mediterranean region) project ReCROP—Bioinocula and CROPping systems: an integrated biotechnological approach for improving crop yield, biodiversity and REsilience of Mediterranean agro-ecosystems. They also gratefully acknowledge the financial support of OCP Nutricrops under the project « Environmental friendly tools to improve Soil health: Development of novel and quick biological soil health », Flagship Sustainable Agriculture program.

References

Aboudrar W, Schwartz C, Morel JL, Boularbah A (2013) Effect of nickel-resistant rhizosphere bacteria on the uptake of nickel by the hyperaccumulator Noccea caerulesens under controlled conditions. J Soils Sediments 13:501–507

Adnan M, Xiao B, Xiao P, Zhao P, Li R, Bibi S (2022) Research progress on heavy metals pollution in the soil of smelting sites in China. Toxics 10(5):231. https://doi.org/10.3390/toxics10050231

Afzal A, Ashraf S, Syed S, Mustafa R, Khan M, Iqbal J, Javed U, Shakil Q (2023) What makes research on aflatoxin crucial for human health and development? Plant Protect 7:109–120. https://doi.org/10.33804/pp.007.01.4497

Ahmed I, Nwaichi E, Ugwoha E, Ugbebor J, Arokoyu S (2022) Cost reduction strategies in the remediation of petroleum hydrocarbon contaminated soil. Open Res Africa 5:21. https://doi.org/10.12688/openresafrica.13383.1

Aigberua AO, Ekubo AT, Inengite AK, Izah SC (2016a) Evaluation of total hydrocarbon content and polycyclic aromatic hydrocarbon in an oil spill contaminated soil in Rumuolukwu Community in Niger Delta. J Environ Treatm Tech 4(4):130–142

Aigberua AO, Ekubo AT, Inengite AK, Izah SC (2016b) Seasonal variation of nutrient composition in an oil spill contaminated soil: a case of Rumuolukwu, Eneka, Port Harcourt, Nigeria. Biotechnol Res 2(4):179–186

Aigberua AO, Ekubo AT, Inengite AK, Izah SC (2017) Assessment of some selected heavy metals and their pollution indices in an oil spill contaminated soil in the Niger Delta: a case of Rumuolukwu community. Biotechnol Res 3:11–19

Akese GA, Little PC (2018) Electronic waste and the environmental justice challenge in Agbogbloshie. Environ Just 11(2):77–83. https://doi.org/10.1089/env.2017.0039

Alengebawy A, Abdelkhalek ST, Qureshi SR, Wang MQ (2021) Heavy metals and pesticides toxicity in agricultural soil and plants: ecological risks and human health implications. Toxics 9(3):42. https://doi.org/10.3390/toxics9030042

Al-Juthery H (2023) Innovative soil, water and plant management promoting sustainable agriculture and environments: a review. IOP Conf Series: Earth Environ Sci 1259:12014. https://doi.org/10.1088/1755-1315/1259/1/012014

Alori E, Gabasawa A, Elenwo C, Agbeyegbe O (2022) Bioremediation techniques as affected by limiting factors in soil environment. Front Soil Sci 2:937186. https://doi.org/10.3389/fsoil.2022.937186

Ampofo JA (2022) Remediation technologies for polluted sites: review. J Environ Earth Sci 12(12):5–18. https://doi.org/10.7176/jees/12-12-02

Apitz SE, Fitzpatrick AG, McNally A, Harrison D, Coughlin C, Edwards DA (2017) Stakeholder value-linked sustainability assessment: evaluating remedial alternatives for the Portland Harbor Superfund site, Portland, Oregon, USA. Integr Environ Assess Manag 14(1):43–62. Advance online publication. https://doi.org/10.1002/ieam.1998

Araujo M, Ignatius S, Oliveira A, Oliveira S, Fertonani F, Paste I (2015) Thermal desorption of each. J Therm Anal Calorim 123(2):1019–1029. https://doi.org/10.1007/s10973-015-4810-y

Azubuike CC, Chikere CB, Okpokwasili GC (2016) Bioremediation techniques–classification based on site of application: principles, advantages, limitations and prospects. World J Microbiol Biotechnol 32:1–18

Bala S, Garg D, Thirumalesh BV, Sharma M, Sridhar K, Inbaraj BS, Tripathi M (2022) Recent strategies for bioremediation of emerging pollutants: a review for a green and sustainable environment. Toxics 10(8):484. https://doi.org/10.3390/toxics10080484

Benidire LSIA, Pereira A, Naylo PML, Castro AB (2020) Do metal contamination and plant species affect microbial abundance and bacterial diversity in the rhizosphere of metallophytes growing in mining areas in a semiarid climate? J Soils Sediments. https://doi.org/10.1007/s11368-019-02475-4

Benidire L, Madline A, Pereira SIA, Castro PML, Boularbah A (2021) Synergistic effect of organo-mineral amendments and plant growth promoting rhizobacteria (PGPR) on the establishment of vegetation cover and amelioration of mine tailings. Chemosphere. https://doi.org/10.1016/j.chemosphere.2020.127803

Benidire L, Pereira S, Aboudrar W, Hafdi M, Castro P, Boularbah A (2022) Remediation of metal-contaminated mine tailings by the application of organic and mineral amendments. J Soils Sediments. https://doi.org/10.1007/s11368-021-03081-z

Bhise PR, Kulkarni SB, Borole VY (2019) Preprocessing and statistical analysis of soil parameters using conventional laboratory techniques and non-imaging spectral techniques for Vaijapur Taluka. Int J Recent Technol Eng 8(2):3092–3096

Bismark MB, Mexoese N, Pearl YN, Samuel AK, Paul OK (2021) Interplay of climate variability, land use change, crop yield and farm households food security in Offin River Basin, Ghana. Int J Sustain Dev Res 12(3):102–116

Boularbah A, Schwartz C, Bitton G, Morel JL (2006a) Heavy metal contamination from mining sites in South Morocco: 1. Use of a biotest to assess metal toxicity of tailings and soils. Chemosphere 63:802–810

Boularbah A, Schwartz C, Bitton G, Aboudrar W, Ouhammou A, Morel JL (2006b) Heavy metal contamination from mining sites in South Morocco: 2. Assessment of metal accumulation and toxicity in plants. Chemosphere 63:811–817

Bugsel B, Bauer R, Herrmann F, Maier ME, Zwiener C (2022) LC-HRMS screening of per-and polyfluorinated alkyl substances (PFAS) in impregnated paper samples and contaminated soils. Anal Bioanal Chem 414:1217–1225

Cardoso E, Vasconcellos RLF, Bini D, Miyauchi M, Santos C, Alves P, Paula A, Nakatani AS, Pereira J, Nogueira MA (2013) Soil health: Looking for suitable indicators. What should be considered to assess the effects of use and management on soil health? Scientia Agricola 70(4):274–289. https://doi.org/10.1590/s0103-90162013000400009

Chacra A, Mustapha S, Salam D, El-Kayal W (eds) (2023) Phytoremediation of cadmium and nickel contaminated clay soil in Lebanon using poplar trees. In: Proceedings of the 8th world congress on civil, structural, and environmental engineering (CSEE'23), Lisbon, Portugal, March 29–31, 2023, Paper No. ICEPTP, 91

Chao L, Mo X, Meng J, Li Y (eds) (2021) Study on enhanced bioremediation effect of oil-bearing dredging sediment. E3S Web Conf 252:02031. EDP Sciences

Charzyński P et al (2022) A global perspective on soil science education at third educational level; knowledge, practice, skills and challenges. Geodrma. 425:116053. https://doi.org/10.1016/j.geoderma.2022.116053

Chen P, Wang HY, Zheng RL, Zhang B, Sun GX (2018) Long-term effects of biochar on rice production and stabilization of cadmium and arsenic levels in contaminated paddy soils. Earth Environ Sci Trans R Soc Edinb 109(3–4):415–420. https://doi.org/10.1017/s175569101800049x

Chibuike GU, Obiora SC (2014) Heavy metal polluted soils: effect on plants and bioremediation methods. Appl Environ Soil Sci 1–12. https://doi.org/10.1155/2014/752708

Chuka DB, Clement O, Monago CC (2017) Effect of giant rat's tail grass (Sporobolus pyramidalis P. Beauv.) on total petroleum hydrocarbon (TPH) and heavy metals content of crude oil polluted soils. J Appl Sci Environ Manag 20(4):1045. https://doi.org/10.4314/jasem.v20i4.17

Diarra I, Kotra KK, Prasad S (2021) Application of phytoremediation for heavy metal contaminated sites in the South Pacific: strategies, current challenges, and future prospects. Appl Spectrosc Rev 57(6):490–512. https://doi.org/10.1080/05704928.2021.1904410

Dike BU, Okoro BC, Nwakwasi NN, Agbo KC (2013) Remediation of used motor engine oil contaminated soil: a soil washing treatment approach. J Civ Environ Eng 3(2). https://doi.org/10.4172/2165-784x.1000129

Edralin DA, Sigua GC, Reyes MR, Mulvaney MJ, Andrews SS (2017) Conservation agriculture improves yield and reduces weeding activity in sandy soils of Cambodia. Agron Sustain Dev 37(5):1–11. https://doi.org/10.1007/s13593-017-0461-7

Effendi AJ, Lestari V, Irsyad M (2020) Optimizing soil washing remediation of mercury-contaminated soil using various washing solutions and solid/liquid ratios. E3S Web Conf 148:5004. https://doi.org/10.1051/e3sconf/202014805004

European Food Safety Authority (EFSA) Scientific Committee (2015) Statement on the benefits of fish/seafood consumption compared to the risks of methylmercury in fish/seafood. EFSA J 13(1):1–36. https://doi.org/10.2903/j.efsa.2015.3982

El Hamiani O, El Khalil H, Lounate K, Sirguey C, Hafidi M, Bitton G, Schwartz C, Boularbah A (2010) Toxicity assessment of garden soils in the vicinity of mining areas in Southern Morocco. J Hazardous Mater 177:755–761

El Hamiani O, El Khalil H, Sirgueyb C, Ouhammou A, Bitton B, Schwartz C, Boularbah A (2015) Metal concentrations in plants from mining areas in South Morocco: health risks assessment of consumption of edible and aromatic plants. Clean- Soil, Air Water 43:399–407

El Khalil H, Schwartz C, Elhamiani O, Kubiniok J, Morel JL, Boularbah A (2008a) Contribution of technic materials to the mobile fraction of metals in urban soils in Marrakech (Morocco). J Soils Sediment 1:17–22

El Khalil H, El Hamiani O, Bitton G, Ouazzani N, Boularbah A (2008b) Heavy metal contamination from mining sites in South Morocco: monitoring metal content and toxicity of soil runoff and groundwater. Environ Monit Assess 136:147–216

El Khalil H, Schwartz C, El Hamiani O, Kubiniok J, Morel JL, Boularbah A (2013) Distribution of major elements and trace metals as indicators of technosolisation of urban and suburban soils. J Soils Sediments 13:519–530

El Khalil H, Schwartz C, El Hamiani O, Sirguey C, Kubiniok J, Boularbah A (2016) How physical alteration of technic materials affects mobility and phytoavailabilty of metals in urban soils? Chemosphere 152:407–414

Fatdillah ELN, Pauzi NIM (2018) Soil remediation technologies for heavy metals—A review. Int J Eng Manag Res 8(6):171–175. https://doi.org/10.31033/ijemr.8.6.16

Fosu-Mensah BY, Addae E, Yirenya-Tawiah D, Nyame F (2017) Heavy metals concentration and distribution in soils and vegetation at Korle Lagoon area in Accra, Ghana. Cogent Environ Sci 3:1405887. https://doi.org/10.1080/23311843.2017.1405887

Guo W, Pan B, Sakkiah S, Yavas G, Ge W, Zou W, Tong W, Hong H (2019) Persistent organic pollutants in food: contamination sources, health effects and detection methods. Int J Environ Res Public Health 16(22):4361. https://doi.org/10.3390/ijerph16224361

Gusiatin MZ, Rouhani A (2023) Application of selected methods to modify pyrolyzed biochar for the immobilization of metals in soil: a review. Materials 16(23):7342

Hamels F, Malevé J, Sonnet P, Kleja D, Smolders E (2014) Phytotoxicity of trace metals in spiked and field-contaminated soils: linking soil-extractable metals with toxicity. Environ Toxicol Chem 33(11):2479–2487. https://doi.org/10.1002/etc.2693

Han M, Yang H, Ding N, You S, Yu G (2021) The role of plant-associated bacteria in the phytoremediation of heavy metal contaminated soils. E3S Web Conf 261:4006. https://doi.org/10.1051/e3sconf/202126104006

Handayani I, Hale C (2022) Healthy soils for productivity and sustainable development in agriculture. IOP Conf Series. Earth Environ Sci 1018:12038. https://doi.org/10.1088/1755-1315/1018/1/012038

Harrison D, Coughlin C, Hogan D, Edwards D, Smith B (2017) Regional economic impact assessment: evaluating remedial alternatives for the Portland Harbor Superfund site, Portland, Oregon, USA. Integr Environ Assess Manag 14:32–42. https://doi.org/10.1002/ieam.1996

Helming K, Daedlow K, Hansjürgens B, Koellner T (2018) Assessment and governance of sustainable soil management. Sustainability 10(12):4432. https://doi.org/10.3390/su10124432

Hu G, He Y, Zhu KF, Zhang Z, Lou W, Zhang KN, Chen YG, Wang Q (2023) Experimental study on injection of ferrous sulphate for remediation of a clayey soil contaminated with hexavalent chromium. Environ Earth Sci 82(7):185. https://doi.org/10.1007/s12665-023-10853-y

Hu X (2010) Bioavailability of heavy metals and their health risk to inhabitants through the consumption of vegetables around a lead-zinc mining plant in Nanjing, China. In: Proceedings of the international conference on bioinformatics and biomedical engineering, pp 1–4. https://doi.org/10.1109/icbbe.2010.5517437

Hussain A (2023) The importance of soil organic matter (SOM) on soil productivity and plant growth. Biol Agric Sci Res J 11. https://doi.org/10.54112/basrj.v2023i1.11

Ikhajiagbe B, Ehiarinmwian RI, Omoregie GO (2022) Influence of different anionic charges on lead compounds in the remediation of lead by *Eleusine indica* (Gaertn). J Appl Sci Environ Manag 26(1):121–129. https://ajol.info/index.php/jasem/article/view/222674

Ikhajiagbe B, Ogwu MC, Lato NF (2021) Growth and yield responses of soybean (*Glycine max* [L.] *Merr.*) accessions after exposure to cadmium. Vegetos 34(1):107–118. https://doi.org/10.1007/s42535-021-00189-y

Ikram M, Kumar A, Shah SK, Singh M (2022) Importance of soil health indicators and their implications for sustainable agriculture. J Survey Fish Sci 8:377–380. https://doi.org/10.53555/sfs.v8i3.2395

Izah SC, Etebu EN, Aigberua AO, Odubo TC, Iniamagha I (2022) A meta-analysis of microbial contaminants in selected ready-to-eat foods in Bayelsa State, Nigeria: public health implications and risk-reduction strategies. Hygiene Environ Health Adv 4:100017. https://doi.org/10.1016/j.heha.2022.100017

Izah SC, Bassey SE, Ohimain EI (2017a) Assessment of heavy metal in cassava mill effluent contaminated soil in a rural community in the Niger Delta region of Nigeria. EC Pharmacol Toxicol 4(5):186–201

Izah SC, Bassey SE, Ohimain EI (2017b) Assessment of pollution load indices of heavy metals in cassava mill effluents contaminated soil: a case study of small-scale cassava processing mills in a rural community of the Niger Delta region of Nigeria. Biosci Methods 8:1–17

Izah SC, Bassey SE, Ohimain EI (2017c) Geo-accumulation index, enrichment factor and quantification of contamination of heavy metals in soil receiving cassava mill effluents in a rural community in the Niger Delta region of Nigeria. Mol Soil Biol 8(2):7–20

Izah SC, Bassey SE, Ohimain EI (2018) Ecological risk assessment of heavy metals in cassava mill effluents contaminated soil in a rural community in the Niger Delta Region of Nigeria. Mol Soil Biol 9:1–11

Izah SC, Richard G, Aigberua AO, Ekakitie O (2021) Variations in reference values utilized for the evaluation of complex pollution indices of potentially toxic elements: a critical review. Environ Chall 5:100322. https://doi.org/10.1016/j.envc.2021.100322

Izah SC, Richard G, Stanley HO, Ogwu MC, Sawyer WE, Uwaeme OR (2024) Prospects and application of multivariate and reliability analyses to one health risk assessments of toxic elements. Toxicol Environ Health Sci 16(2):127–134. https://doi.org/10.1007/s13530-023-00199-0

Jan F, Ishaq M, Khan S, Ihsanullah I, Ahmad I, Shakirullah M (2010) A comparative study of human health risks via consumption of food crops grown on wastewater irrigated soil (Peshawar) and relatively clean water irrigated soil (Lower Dir). J Hazard Mater 179(1–3):612–621. https://doi.org/10.1016/j.jhazmat.2010.03.047

Jaworski C, Krzywoszynska A, Leake J, Dicks L (2024) Sustainable soil management in the United Kingdom: a survey of current practices and how they relate to the principles of regenerative agriculture. Soil Use Manag 40. https://doi.org/10.1111/sum.12908

Jena P, Lippe R, Stellmacher T (2022) Editorial: sustainable certification standards: environmental and social impacts. Front Sustain Food Syst 6:922672. https://doi.org/10.3389/fsufs.2022.922672

Jones D, Rodriguez V, Butts S, Arheart K, Zulu R, Chitalu N (2018) Increasing acceptability and uptake of voluntary male medical circumcision in Zambia: implementing and disseminating an evidence-based intervention. Transl Behav Med 8(6):907–916. https://doi.org/10.1093/tbm/iby078

Kaparwan D, Rana N, Dhyani B (2020) Heavy metals toxicity in agricultural soils: Critical review of possible sources, influence on soil health and remedial measures to remove, reduce and stabilize contaminants in soil. Int J Curr Microbiol Appl Sci 9(6):1467–1482. https://doi.org/10.20546/ijcmas.2020.906.182

Khalid S, Shahid M, Niazi N, Murtaza B, Bibi I, Dumat C (2017) A comparison of technologies for remediation of heavy metal contaminated soils. J Geochem Explor 182:247–268. https://doi.org/10.1016/j.gexplo.2016.11.021

Khan MU, Sessitsch A, Harris M, Fatima K, Imran A, Arslan M, Shabir G, Khan QM, Afzal M (2015) Cr-resistant rhizo- and endophytic bacteria associated with Prosopis juliflora and their potential as phytoremediation enhancing agents in metal-degraded soils. Front Plant Sci 5:755. https://doi.org/10.3389/fpls.2014.00755

Krumins J, Goodey N, Gallagher F (2015) Plant–soil interactions in metal-contaminated soil. Soil Biol Biochem 80:224–231. https://doi.org/10.1016/j.soilbio.2014.10.011

Krzywoszynska A (2019) Making knowledge and meaning in communities of practice: what role may science play? The case of sustainable soil management in England. Soil Use Manag 35:160–168. https://doi.org/10.1111/sum.12487

Kuppusamy S, Thavamani P, Venkateswarlu K, Lee YB, Naidu R, Megharaj M (2017) Remediation approaches for polycyclic aromatic hydrocarbons (PAHs) contaminated soils: technological constraints, emerging trends and future directions. Chemosphere 168:944–968. https://doi.org/10.1016/j.chemosphere.2016.10.115

Laaouidi YA, Bahmed A, Naylo HE, Khalil S, Ouvrard CS, Boularbah A (2020) Trace elements in soils and vegetables from market gardens of urban areas in Marrakech city. Biol Trace Elem Res. https://doi.org/10.1007/s12011-019-01849-6

LaCanne CE, Lundgren JG (2018) Regenerative agriculture: merging farming and natural resource conservation profitably. PeerJ 6:e4428. https://doi.org/10.7717/peerj.4428

Lavee D, Ash T, Baniad G (2012) Cost-benefit analysis of soil remediation in Israeli industrial zones. Nat Res Forum 36(4):285–299. https://doi.org/10.1111/j.1477-8947.2012.01462.x

Lee SH, Kim MS, Kim JG, Kim SO (2020a) Use of soil enzymes as indicators for contaminated soil monitoring and sustainable management. Sustainability 12(19):8209. https://doi.org/10.3390/su12198209

Lee SH, Lee JH, Jung WC, Park M, Kim MS, Lee SJ, Park H (2020b) Changes in soil health with remediation of petroleum hydrocarbon contaminated soils using two different remediation technologies. Sustainability 12(23):10078

Lee SH, Kim SO, Lee SW, Kim MS, Park H (2021) Application of soil washing and thermal desorption for sustainable remediation and reuse of remediated soil. Sustainability 13(22):12523. https://doi.org/10.3390/su132212523

Liu AP, Sun XN, Yang F, Yao XX (2011) The study of remediation standards of heavy metal-Cu contaminated soil based on risk assessment. Adv Mater Res 414:93–98. https://doi.org/10.4028/www.scientific.net/amr.414.93

Madline A, Benidire L, Boularbah A (2021) Alleviation of salinity and metal stress using plant growth-promoting rhizobacteria isolated from semiarid Moroccan copper-mine soils. Environ Sci Pollut Res 28:67185–67202. https://doi.org/10.1007/s11356-021-15168-8

Madline A, Benidire L, Pereira SIA, El Khalil H, Michalski A, Castro PML, Charzyński P, Boularbah A (2024a) Optimizing biological and physicochemical properties of acidic mine tailings through combined organo-mineral amendments and topsoil application. J Soils Sediments https://doi.org/10.1007/s11368-024-03812-y

Madline A, Benidire L, Pereira SIA, Castro PML, Boularbah A (2024b) Enhancing native plant establishment in mine tailings under drought stress conditions through the application of organo-mineral amendments and microbial inoculants. Plants 13:863. https://doi.org/10.3390/plants13060863

Majed N, Real MIH, Akter M, Azam HM (2016) Food adulteration and bio-magnification of environmental contaminants: a comprehensive risk framework for Bangladesh. Front Environ Sci 4:34. https://doi.org/10.3389/fenvs.2016.00034

Maqbool A, Hui W, Sarwar MT (2019) Nanotechnology development for in-situ remediation of heavy metal(loid)s contaminated soil. Environ Ecosyst Sci 3(2):9–11. https://doi.org/10.26480/ees.02.2019.09.11

Mensah-Brako B, Nyatuame M, Yakanu PN, Ahorsu SK, Oppong PK (2021) Interplay of climate variability, land use change, crop yield and farm households food security in Offin River Basin, Ghana. Int J Sustain Dev Res 12(3):102–116
Mészáros E, Bodor A, Szierer Á, Kovács E, Perei K, Tölgyesi C, Bátori Z, Feigl G (2022) Indirect effects of COVID-19 on the environment: How plastic contamination from disposable surgical masks affects early development of plants. J Hazard Mater 436:129255. https://doi.org/10.1016/j.jhazmat.2022.129255
Meuser H (2010) Contaminated urban soils, Environmental Pollution Ser. Springer, 2010th edition. ISBN 978-9048193271
Miner GL, Delgado JA, Ippolito JA, Stewart CE (2020) Soil health management practices and crop productivity. Agric Environ Lett 5:e20023. https://doi.org/10.1002/ael2.20023
Moldes AB, Paradelo R, Rubinos D, Devesa-Rey R, Cruz JM, Barral MT (2011) Ex situ treatment of hydrocarbon-contaminated soil using biosurfactants from Lactobacillus pentosus. J Agric Food Chem 59(17):9443–9447. https://doi.org/10.1021/jf201807r
Morel JL, Schwartz C, Florentin L, de Kimpe C (2005): Urban soils. In: Hillel D (ed), Encyclopedia of soils in the environment. Elsevier Ltd., Academic Press, London, pp 202–208
Muehe EM, Weigold P, Adaktylou IJ, Planer-Friedrich B, Kraemer U, Kappler A, Behrens S (2015) Rhizosphere microbial community composition affects cadmium and zinc uptake by the metal-hyperaccumulating plant Arabidopsis halleri. Appl Environ Microbiol 81(6):2173–2181. https://doi.org/10.1128/aem.03359-14
Münzel T, Hahad O, Daiber A, Landrigan PJ (2023) Soil and water pollution and human health: what should cardiologists worry about? Cardiovasc Res 119(2):440–449. https://doi.org/10.1093/cvr/cvac082
Naylo A, Pereira S, Benidire L, El Khalil H, Castro P, Ouvrard S, Schwartz C, Boularbah A (2019) Trace and major element contents, microbial communities, enzymatic activities of urban soils of Marrakech city along an anthropization gradient. J Soils Sediments 19:2153–2165
Ng CC, Boyce AN, Abas MR, Mahmood NZ, Han F (2019) Phytoassessment of vetiver grass enhanced with EDTA soil amendment grown in single and mixed heavy metal–contaminated soil. Environ Monit Assess 191(7):1–16. https://doi.org/10.1007/s10661-019-7573-2
Ng YS, Sen Gupta B, Hashim MA (2014) Stability and performance enhancements of electrokinetic-Fenton soil remediation. Rev Environ Sci Bio/technol 13(3):251–263. https://doi.org/10.1007/s11157-014-9335-5
O'Sullivan L, Bampa F, Knights K, Creamer RE (2017) Soil protection for a sustainable future: options for a soil monitoring network for Ireland. Soil Use Manag 33(2):346–363. https://doi.org/10.1111/sum.12351
O'Connell S, Hou D (2015) Resilience: a new consideration for environmental remediation in an era of climate change. Remediat J 26:57–67. https://doi.org/10.1002/rem.21449
Odubo TC, Izah SC, Ogwu MC (2024) Mycological safety in Tiger Nut (Cyperus esculentus L.) drinks: cultivation, contaminants, and compliance strategies. In: Ogwu MC, Izah SC, Ntuli NR (eds) Food safety and quality in the global south Springer, Singapore, pp 463–487. https://doi.org/10.1007/978-981-97-2428-4_15
Ogwu MC, Izah SC, Ntuli NR, Odubo TC (2024) Food security complexities in the global south. In: Ogwu MC, Izah SC, Ntuli NR (eds) Food safety and quality in the global south. Springer, Singapore, pp 3–33. https://doi.org/10.1007/978-981-97-2428-4_1
Ouvrard S, Barnier C, Bauda P, Beguiristain T, Biache C, Bonnard M, Caupert C, Cébron A, Cortet J, Cotelle S et al (2011) In situ assessment of phytotechnologies for multicontaminated soil management. Int J Phytoremed 13:245–263. https://doi.org/10.1080/15226514.2011.568546
Patil S, Chavan MVG, Patil P (2019) Social innovation through precision farming: an IoT-based precision farming system for examining and improving soil fertility and soil health. Int J Innov Technol Explor Eng 8(11):2877–2881. https://doi.org/10.35940/ijitee.k2421.0981119
Penuelas J, Sardans J (2023) Human-driven global nutrient imbalances increase risks to health. Eco-Environ Health 2(4):246–251. https://doi.org/10.1016/j.eehl.2023.08.003

Petković V, Stanić S, Đermanović M, Bojanić L, Lazić V (2016) The public health significance of controlling contaminants in samples of milk and dairy. Scripta Medica 47(2):94–99. https://doi.org/10.18575/msrs.sm.e.16.16

Qin B, Meng F, Ren L, Jiang J, Gong H (2023) Coupling leaching-bioremediation for petroleum-contaminated soil. Structure 12:14–19. https://doi.org/10.11159/icepr23.114

Rada EC, Andreottola G, Istrate IA, Viotti P, Conti F, Magaril ER (2019) Remediation of soil polluted by organic compounds through chemical oxidation and phytoremediation combined with DCT. Int J Environ Res Public Health 16(17):3179. https://doi.org/10.3390/ijerph16173179

Ramadan BS, Effendi AJ, Helmy Q (2018) Integrating electrokinetic and bioremediation process for treating oil contaminated low permeability soil. E3S Web Conf 31:3005. https://doi.org/10.1051/e3sconf/20183103005

Randall KC, Brennan F, Clipson N, Creamer RE, Griffiths BS, Storey S, Doyle E (2020) An assessment of climate-induced increase in soil water availability for soil bacterial communities exposed to long-term differential phosphorus fertilization. Front Microbiol 11:682. https://doi.org/10.3389/fmicb.2020.00682

Saha L, Tiwari J, Bauddh K, Ma Y (2021) Recent developments in microbe-plant-based bioremediation for tackling heavy metal-polluted soils. Front Microbiol 12:731723. https://doi.org/10.3389/fmicb.2021.731723

Sainju UM, Ghimire R, Pradhan GP (2019) Nitrogen fertilization II: Management practices to sustain crop production and soil and environmental quality. In: Nitrogen fixation. IntechOpen, pp 1–23

Saleem M, Hu J, Jousset A (2019) More than the sum of its parts: microbiome biodiversity as a driver of plant growth and soil health. Annu Rev Ecol Evol Syst 50:145–168. https://doi.org/10.1146/annurev-ecolsys-110617-062605

Sanjay M (2023) Environmental sustainability through soil conservation: an imperative for future generations. Int J Environ Clim Change 13(10):1700–1707. https://doi.org/10.9734/ijecc/2023/v13i102826

Sarkar B, Mukhopadhyay R, Ramanayaka S, Bolan N, Ok YS (2021) The role of soils in the disposition, sequestration, and decontamination of environmental contaminants. Philos Trans Royal Soc B: Biol Sci 376(1834):20200177. https://doi.org/10.1098/rstb.2020.0177

Shaheb MR, Venkatesh R, Shearer SA (2021) A review of the effect of soil compaction and its management for sustainable crop production. J Biosyst Eng 46(4):417–439. https://doi.org/10.1007/s42853-021-00117-7

Sisay A (2019) Contamination of heavy metals, source, effects on living things, and different remediation techniques in soil: a review. Am J Nanosci 5(4):67. https://doi.org/10.11648/j.ajn.20190504.17

Sobolev D, Begonia MF (2008) Effects of heavy metal contamination upon soil microbes: lead-induced changes in general and denitrifying microbial communities as evidenced by molecular markers. Int J Environ Res Public Health 5(5):450–456. https://doi.org/10.3390/ijerph5050450

Sridevi S (2022) Microbial ecological effects and remediation of heavy metal-contaminated soil estimated by simultaneous equations based on panel data. Acad J Environ Biol 3:48–55. https://doi.org/10.38007/ajeb.2022.030105

Stankovics P, Tóth G, Tóth Z (2018) Identifying gaps between the legislative tools of soil protection in the EU member states for a common European soil protection legislation. Sustainability 10(8):2886. https://doi.org/10.3390/su10082886

Supriya RA, Swetha CS, Babu AJ, Reddy VB, Rao TM (2017) Estimation of chromium and copper in chicken liver samples by inductively coupled plasma–optical emission spectrometry method. Asian J Dairy Food Res 36(4):341–344. https://doi.org/10.18805/ajdfr.dr-1254

Suvorov N, Stancu A (2021) Environmental residues and contaminants. Western Balkan J Agric Econ Rural Dev 3:51–66. https://doi.org/10.5937/wbjae2101051s

Teng Y, Wang X, Li L, Li Z, Luo Y (2015) Rhizobia and their bio-partners as novel drivers for functional remediation in contaminated soils. Front Plant Sci 6:32. https://doi.org/10.3389/fpls.2015.00032

Teshale E (2019) Review on utilization of plants for the phytoremediation purpose on heavy metal contaminated soil. J Nat Sci Res 9(6):17–23. https://doi.org/10.7176/jnsr/9-6-03

Triantafyllidis V, Kosma C, Patakas A (2019) An assessment of the soil quality index in a Mediterranean agro-ecosystem. Emirates J Food Agric 30(12):1042–1050. https://doi.org/10.9755/ejfa.2018.v30.i12.1886

Trifunović V (2021) Vitrification as a method of soil remediation. Zastita Materijala 62(3):166–179. https://doi.org/10.5937/zasmat2103166t

Villa R, Trovó A, Nogueira R (2010) Soil remediation using a coupled process: soil washing with surfactant followed by photo-Fenton oxidation. J Hazard Mater 174(1–3):770–775. https://doi.org/10.1016/j.jhazmat.2009.09.118

Wang J, Xia K, Waigi MG, Gao Y, Odinga ES, Ling W, Liu J (2018) Application of biochar to soils may result in plant contamination and human cancer risk due to exposure of polycyclic aromatic hydrocarbons. Environ Int 121:169–177. https://doi.org/10.1016/j.envint.2018.09.010

Wang W, Simonich S, Xue M, Zhao J, Zhang N, Wang R, Cao J, Tao S (2010) Concentrations, sources, and spatial distribution of polycyclic aromatic hydrocarbons in soils from Beijing, Tianjin, and surrounding areas, North China. Environ Pollut 158(5):1245–1251. https://doi.org/10.1016/j.envpol.2010.01.021

Wang Y, Narayanan M, Shi X, Chen X, Li Z, Natarajan D, Ma Y (2022) Plant growth-promoting bacteria in metal-contaminated soil: current perspectives on remediation mechanisms. Front Microbiol 13:966226. https://doi.org/10.3389/fmicb.2022.966226

Wu J, Zhao F (2023) Machine learning: an effective technical method for future use in assessing the effectiveness of phosphorus-dissolving microbial agroremediation. Front Bioeng Biotechnol 11:1189166. https://doi.org/10.3389/fbioe.2023.1189166

Wu Z, Zeng Q, Tan W (2021) Cover cropping impacts soil microbial communities and functions in mango orchards. Agriculture 11(4):343. https://doi.org/10.3390/agriculture11040343

Yarkwan B, Isaac TO, Okopi A, Izah SC (2024) Evidence of the toxic potentials of agrochemicals on human health and biodiversity: Carcinogens and Mutagens. In: Ogwu MC, Izah SC, Ntuli NR (eds) Food safety and quality in the global south. Springer, Singapore, pp 331–359. https://doi.org/10.1007/978-981-97-2428-4_11

Yuan X (2023) Current problems and countermeasures of soil pollution management. E3S Web Conf 424:4017. https://doi.org/10.1051/e3sconf/202342404017

Zaib M (2023) Soil contamination and human health: exploring the heavy metal landscape: a comprehensive review. J Health Rehabil Res 3(2):351–356. https://doi.org/10.61919/jhrr.v3i2.123

Zhang X, Zhong T, Liu L, Ouyang X (2015) Impact of soil heavy metal pollution on food safety in China. PLoS ONE 10(8):e0135182. https://doi.org/10.1371/journal.pone.0135182

Zhao L, Lyu C, Li Y (2021) Analysis of factors influencing plant–microbe combined remediation of soil contaminated by polycyclic aromatic hydrocarbons. Sustainability 13(19):10695. https://doi.org/10.3390/su131910695

Zhao X, Lei M, Gu R (2022) Knowledge mapping of the phytoremediation of cadmium-contaminated soil: a bibliometric analysis from 1994 to 2021. Int J Environ Res Public Health 19(12):6987. https://doi.org/10.3390/ijerph19126987

Zheng C, Yang Z, Si M, Zhu F, Yang W, Zhao F, ... Shi Y (2021) Application of biochars in the remediation of chromium contamination: fabrication, mechanisms, and interfering species. J Hazard Mater 407:124376. https://doi.org/10.1016/j.jhazmat.2020.124376

Zhou C, Sun QM, Cao Q, He JH, Lu JM (2022) Synergistic effect of Fe single-atom catalyst for highly efficient microwave-stimulated remediation of chloramphenicol-contaminated soil. Nano Micro Small 19(2):2205341. https://doi.org/10.1002/smll.202205341

Zhu L, Lü L, Zhang D (2010) Mitigation and remediation technologies for organic contaminated soils. Front Environ Sci Eng China 4(4):373–386. https://doi.org/10.1007/s11783-010-0253-7

Zoghi P, Mafigholami R (2023) Optimisation of soil washing method for removal of petroleum hydrocarbons from contaminated soil around oil storage tanks using response surface methodology. Sci Rep 13:15457. https://doi.org/10.1038/s41598-023-42777-9

Chapter 5
Biodiversity and Human Health: The Interconnections of Species Loss and Ecosystem Services

Matthew Chidozie Ogwu, Aliu Olugbemiga Ojo, and Amarachi Chekosiba Alaka

Abstract This chapter elucidates the critical relationship between biodiversity and human health, emphasizing how species diversity of life on Earth supports ecosystem functions essential for human survival. Biodiversity provides a wide array of ecosystem services, such as food production, clean water, disease regulation, and climate stabilization, all of which underpin human health and well-being. Beyond ecological contributions, biodiversity also supports physical and mental health, as exposure to biodiverse environments has been linked to reduced stress, enhanced mental health, and increased physical activity. The chapter examines how the degradation of biodiversity due to human activities, including habitat destruction, pollution, overexploitation, and climate change, poses significant risks to public health by diminishing these vital ecosystem services. Furthermore, the chapter introduces the "One Health" approach, which recognizes the interconnectedness of humans, biodiversity, and environmental health. This framework emphasizes the importance of conserving biodiversity to mitigate risks such as zoonotic diseases arising from increased human-wildlife interactions. By detailing the role of biodiversity in supporting ecosystem functions, including disease regulation, nutritional security, and mental well-being, the chapter underscores the intrinsic value of biodiversity for

M. C. Ogwu
Goodnight Family Department of Sustainable Development, Appalachian State University, 212 Living Learning Center, 305 Bodenheimer Drive, Boone, NC 28608, USA
e-mail: ogwumc@appstate.edu

A. O. Ojo
Department of Plant Biology and Biotechnology, Faculty of Life Sciences, University of Benin, PMB 1154, Ugbowo, Benin City, Edo State, Nigeria

Department of Environmental Biology, Auchi Polytechnic Okene Road, PMB 13, Auchi, Edo State, Nigeria

A. C. Alaka (✉)
Department of Agricultural and Bio-Environmental Engineering, Imo State Polytechnic Omuma, Omuma, Imo State, Nigeria
e-mail: alakamarachi@gmail.com; alaka.amarachi@imopoly.edu.ng

S. Chibueze Izah and M. Chidozie Ogwu (eds.), *Innovative Approaches in Environmental Health Management*, Environmental Science and Engineering,
https://doi.org/10.1007/978-3-031-81966-7_5

human health. In conclusion, the chapter advocates for policies and strategies prioritizing biodiversity conservation as a critical component of public health, calling for integrated approaches that enhance ecosystem resilience and promote sustainable human societies.

Keywords One health approach · Species loss · Climate change · Zoonotic diseases · Conservation strategies · Public health · Ecological resilience

5.1 Introduction

Biodiversity, encompassing the rich tapestry of life on Earth, including species diversity, genetic diversity, and ecosystem divers, serves as a fundamental pillar for the health of our planet and its diverse inhabitants. This multifaceted diversity is vital for maintaining ecological equilibrium and is critical in enhancing human health and well-being through many interrelated mechanisms. Biodiversity is integral to the myriad of ecosystem services defined as the benefits provided by natural ecosystems to humanity that underpin human existence, such as food production, purification of water, regulation of diseases, and stabilization of climate (Korn et al. 2019; Zhang et al. 2022). The nexus between biodiversity and human health stretches beyond mere ecological functions as it encapsulates its direct and indirect contributions to physical and mental health outcomes. Environments rich in biodiversity have been associated with enhanced mental health, mitigated stress levels, and increased physical activity, primarily attributed to access to green spaces (Roslund et al. 2020; Clark et al. 2014). Furthermore, the diversity of endogenous and exogenous microbial entities in human physiology has profound implications for immune regulation and overall health, underscoring the intricate interdependencies between biodiversity and the human microbiome (Roslund et al. 2020).

Biodiversity is a fundamental pillar of the ecosystem services critical to human health and survival (Izah et al. 2017, 2018; Izah and Seiyaboh 2018a, b; Izah 2018). These ecosystem services, delineated into provisioning, regulating, cultural, and supporting categories, provide fundamental resources such as food, clean air, and safe drinking water essentials for everyday living. For instance, the biodiversity found within agricultural landscapes enhances crop pollination, pest management, and soil fertility, while wetlands and forests contribute significantly to water purification and flood mitigation (Korn et al. 2019). These ecosystem services bolster the resilience of ecological systems, enabling them to adapt to environmental fluctuations and disturbances, which is vital for sustaining human populations. Alarmingly, the ongoing decline in biodiversity due to anthropogenic pressures such as habitat degradation, pollution, overexploitation, and climate change poses formidable threats to public health. As biodiversity wanes, ecosystems diminish their capacity to provide essential services, elevating food insecurity risks, increased vulnerability to infectious diseases, and heightened susceptibility to climate-induced natural disasters (Romanelli et al. 2014). Consequently, biodiversity conservation is fundamentally intertwined with public health imperatives, serving as a bedrock for sustaining healthy ecosystems and securing the services crucial for human survival.

The "One Health" paradigm highlights interconnections between human, animal, and environmental health. This integrative approach posits that the health of each domain is interdependent, thereby underscoring biodiversity conservation as a strategic public health imperative. For example, curbing ecosystem degradation can diminish the propensity for zoonotic disease emergence, often escalating through increased human-wildlife interactions resulting from habitat loss and ecosystem fragmentation (Sun et al. 2023; Jones et al. 2013). Moreover, biodiverse ecosystems are crucial in regulating pathogens and diseases, reducing infection transmission that poses significant threats to human health (Franzolin et al. 2021). Additionally, the mental health benefits associated with biodiversity are gaining increasing recognition. Access to diverse natural environments has been linked to improved mental health outcomes, including reductions in anxiety, depression, and stress (Clark et al. 2014). This cultural ecosystem service underscores the intrinsic value of biodiversity for mental well-being and bolsters the call for preserving green spaces, particularly within urban environments. Additionally, empirical studies demonstrate that exposure to bio-diverse surroundings can promote physical activities, enhancing health benefits (Roslund et al. 2020).

This chapter focused on elucidating the complex interrelations between biodiversity and human health, exploring how biodiversity underpins essential ecosystem services that promote well-being. It also investigated the consequences of biodiversity loss on public health. It examined innovative models, such as the One Health approach, which advocates for integrating biodiversity conservation into public health policies. The chapter culminated in a call for prioritizing biodiversity within policies and actions as a core component of sustainable and healthy societies, emphasizing its vital role in enhancing resilience against health and environmental challenges.

5.2 The Concept of Ecosystem Services

Ecosystem services are the benefits that humans derive from natural ecosystems. They are essential for sustaining human life and well-being. These services can be categorized into four main types: provisioning, regulating, cultural, and supporting services (Fig. 5.1). Each category plays a crucial role in maintaining the balance of ecosystems and human health.

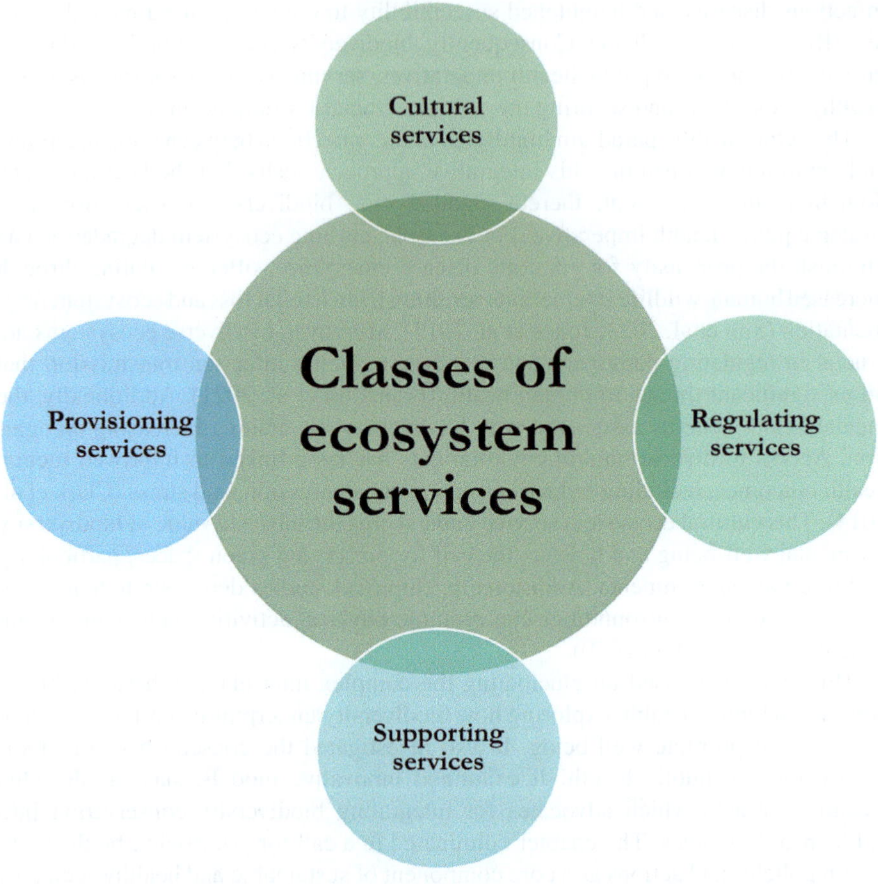

Fig. 5.1 Classes of ecosystem services

5.2.1 Classification of Ecosystem Services

Provisioning Services: These services provide tangible products humans utilize, including food, fresh water, timber, fiber, and genetic resources. For instance, agricultural biodiversity enhances food security by providing a variety of crops that can withstand pests and diseases, thereby ensuring a stable food supply (Haahtela 2019). Aquatic biodiversity is similarly vital, as diverse fish populations contribute to nutritional benefits through various seafood options rich in essential nutrients (Bernhardt and O'Connor 2021).

Regulating Services: These services involve regulating ecosystem processes, which are crucial for maintaining environmental quality. Examples include climate regulation, water purification, and disease control. Healthy ecosystems, such as wetlands, mangroves, and forests, significantly filter pollutants from water and sequester carbon, which helps mitigate climate change (Aerts et al. 2018). Air quality regulation is also notable, as vegetation absorbs carbon and releases oxygen, establishing a crucial balance in urban environments. Furthermore, biodiversity can influence disease transmission dynamics; for example, diverse ecosystems can dilute the prevalence of zoonotic diseases by reducing the contact between humans and wildlife that harbor pathogens (Salkeld et al. 2013).

Cultural Services: Cultural services encompass the non-material benefits contributing to ecosystems' cultural, aesthetic, and recreational values. Studies have shown that access to natural environments significantly correlates with improved mental health and overall well-being, as exposure to green spaces can alleviate stress while encouraging physical activities (Aerts et al. 2018). Additionally, cultural practices, spiritual beliefs, and traditional ecological knowledge are often rooted in biodiversity, demonstrating the intrinsic value of nature in human societies (Irvine et al. 2019). Preserving these cultural services is essential for maintaining the cultural heritage and identifying communities, particularly those dependent on natural landscapes for their livelihoods.

Supporting Services: Supporting services are necessary to produce all other ecosystem services. These processes include nutrient cycling, soil formation, and primary production. Biodiversity enhances these supporting services by maintaining the resilience and functionality of ecosystems. For example, diverse plant species contribute to soil health and fertility, essential for agricultural productivity (Haahtela 2019). Furthermore, healthy ecosystems contribute to pollination services, which are vital for food production systems; diverse habitats support a wide range of pollinator species, ensuring the productivity of crops. The stability of these supporting services is imperative in the face of ongoing environmental challenges, including land degradation and climate change, indicating an urgent need for integrated ecosystem management practices.

5.2.2 Role of Biodiversity in Supporting Ecosystem Functions

Biodiversity is fundamental in supporting ecosystem functions, the biological processes that sustain ecosystems, and the services they provide. The relationship between biodiversity and ecosystem functioning is well-documented; diverse ecosystems are generally more productive and resilient than those with low biodiversity (Zhang et al. 2022). This productivity is crucial for providing ecosystem services, ensuring ecosystems can adapt to changes and disturbances. Research has shown that ecosystems with higher species richness exhibit more excellent stability and resilience, enabling them to withstand environmental stressors such as climate change and habitat loss (Fenoglio et al. 2020). For instance, diverse crop rotations in agricultural systems can enhance soil health and reduce the need for chemical fertilizers and pesticides, leading to more sustainable farming practices (Bernhardt and O'Connor 2021).

Similarly, in marine ecosystems, various fish species can enhance nutrient cycling and contribute to the overall health of aquatic environments (Bernhardt and O'Connor 2021). Species interact with climate, resource availability, and disturbance patterns to influence ecological processes, shaping ecosystem complexity and enabling adaptive responses to environmental variability (Fig. 5.2) (Correia and Lopes 2023). Furthermore, incorporating multiple species into agricultural landscapes has optimized resource utilization, maximizing yields across different environmental conditions (Hector and Bagchi 2007).

Moreover, biodiversity contributes to the functioning of ecosystems through various mechanisms, including niche differentiation and facilitation. Different species often occupy unique ecological niches, allowing them to utilize resources more efficiently and reduce competition (Zhang et al. 2022). This functional diversity is essential for maintaining ecosystem processes and ensuring the delivery of services vital for human health and well-being. The connection between biodiversity, ecosystem services, and human health is evident in numerous examples. These services are crucial for sustaining life and significantly enhancing individuals' and communities' quality of life.

Studies have consistently shown that diverse pollinator communities can enhance crop yields and improve the nutritional quality of food (Haahtela 2019). Healthy ecosystems, particularly wetlands and riparian zones, act as natural filtration systems

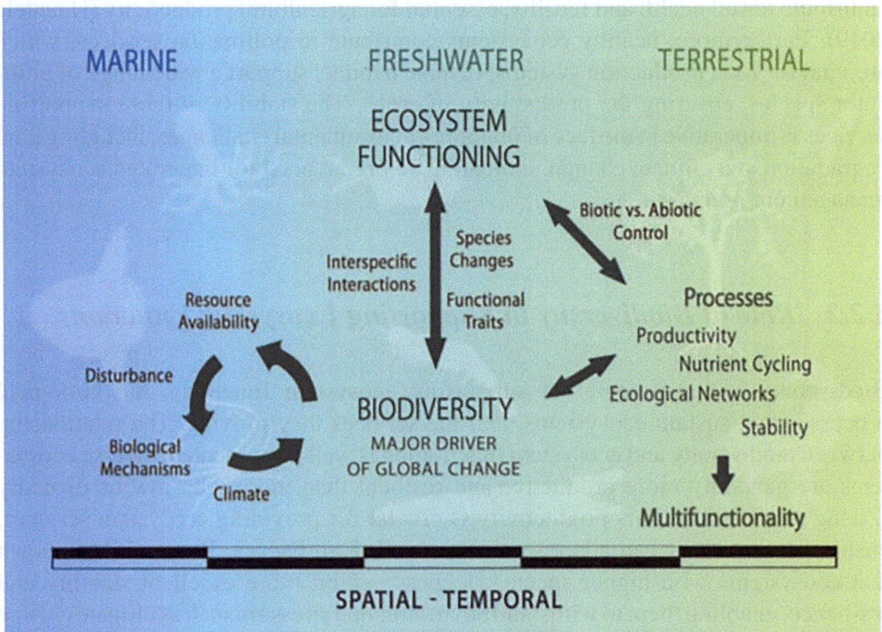

Fig. 5.2 Relationship between biodiversity and ecosystem functioning. *Source* Correia and Lopes (2023)

that purify water by removing pollutants and sediments. This service ensures access to clean drinking water and prevents waterborne diseases (Aerts et al. 2018). The degradation of these ecosystems can lead to increased water contamination and associated health risks. Biodiversity contributes to climate regulation through carbon sequestration and the maintenance of microclimates. Access to natural environments has been linked to improved mental health outcomes. Studies indicate that spending time in green spaces can reduce symptoms of anxiety and depression, enhance mood, and promote overall well-being (Jimenez et al. 2020). The therapeutic effects of nature are increasingly recognized in public health initiatives, emphasizing the importance of preserving biodiversity for mental health outcomes (Bratman et al. 2019). Notably, green spaces in urban areas support biodiversity and contribute to social cohesion and mental wellness, highlighting the multifaceted benefits of urban ecology (Kuo 2015). Biodiversity can also influence the dynamics of infectious diseases. Diverse ecosystems may reduce the risk of zoonotic diseases by dampening the presence of pathogens and their vectors (Salkeld et al. 2013). For example, maintaining healthy wildlife populations can help regulate disease transmission, such as Lyme disease and West Nile virus, protecting human health (Salkeld et al. 2013). Furthermore, landscape diversity can mitigate disease transmission by offering reduced habitat availability for vectors, aiding public health efforts (Keesing et al. 2006).

5.3 Causes/Impacts of Biodiversity Loss on the Ecosystem Services

5.3.1 Causes of Biodiversity Loss on the Ecosystem

Biodiversity loss is driven by various anthropogenic factors, which collectively erode the natural environment and threaten species across ecosystems. Understanding its causes is crucial for effective conservation strategies. The following (Fig. 5.3) are major proximate causes:

Habitat Destruction: Habitat destruction encompasses a variety of activities that fundamentally alter or degrade natural environments, leading to a precipitous decline in biodiversity. Deforestation, particularly in tropical regions, is a significant driver of habitat loss, with logging activities and agricultural conversion, such as establishing monoculture plantations (e.g., oil palm), notably detrimental. Urban expansion and infrastructure development further encroach upon wildlife habitats, exacerbating fragmentation and reducing viable living space for many species. Fragmentation can isolate populations, hindering gene flow and increasing vulnerability to extinction. A wide range of species, particularly those with specialized habitat requirements or limited geographic ranges, is particularly at risk. The loss of these habitats is particularly harmful to species with restricted ranges or specific habitat requirements (Foley

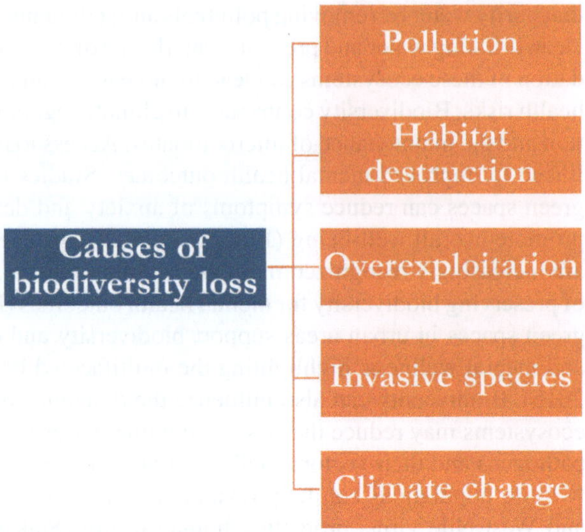

Fig. 5.3 Major proximate causes of biodiversity loss

et al. 2005), and this has been extensively documented in biodiversity hotspots like the Amazon rainforest, which is experiencing rapid deforestation (Barlow et al. 2016).

Pollution: Pollution is a critical stressor for biodiversity, impacting ecosystems through chemical contamination and degradation of habitat quality. Industrialization has introduced a variety of pollutants, including heavy metals and persistent organic pollutants (POPs), which infiltrate terrestrial and aquatic systems. Agricultural runoff leads to nutrient pollution, particularly from nitrogen and phosphorus fertilizers. It stimulates harmful algal blooms that deplete dissolved oxygen (hypoxia) in water bodies, rendering them inhospitable for aquatic organisms (Carpenter et al. 1998). Furthermore, plastic pollution poses significant risks to marine life, leading to ingestion, entanglement, and introduction of toxic substances into food webs (Doney et al. 2019). The cumulative effects of pollution diminish ecosystem resilience, complicating recovery efforts after other anthropogenic disturbances.

Climate Change: Climate change represents an existential threat to global biodiversity, driven by increasing concentrations of greenhouse gases. Altered temperature and precipitation patterns change species' distribution and communities' composition, challenging ecological interactions and mutualisms. Species with low adaptive capacity, such as those in specialized niches or restricted habitats, are at particular risk of extinction. For example, coral reefs, among the most biodiverse ecosystems, are undergoing widespread bleaching as sea surface temperatures rise, leading to the degradation of coral structures and the organisms that depend on them (Hoegh-Guldberg et al. 2019).

Overexploitation: Overexploitation occurs when renewable resources are harvested unsustainably, leading to the decline of populations of wildlife and plant species. This can manifest through unsustainable practices in hunting, fishing, logging, and extraction of natural resources. The repercussions extend beyond the targeted species, affecting entire ecosystems, as the balance of interactions within these systems is disturbed. An illustrative case is fisheries, where overfishing has led to drastic declines in fish populations. For instance, large predatory fish species have seen significant reductions, disrupting marine food webs and calling into question the viability of specific fisheries. Jackson et al. (2001) highlighted that the loss of fish populations compromises the stability and health of marine ecosystems like coral reefs and mangroves, which depend on fish for nutrient cycling and habitat structure. The decline of fish reduces genetic diversity within species and threatens the entire assemblage of organisms reliant on these fish, making ecosystems more susceptible to collapse (Worm et al. 2009).

5.3.2 Impacts of Species Loss on Ecosystem Services

The ongoing loss of biodiversity is an escalating global crisis that has far-reaching consequences for the sustainability of ecosystems and the services they provide. As species disappear, the ecosystems they support are disrupted, leading to a decline in the services that these ecosystems provide. Ecosystem services such as food, fresh water, rare species, climate moderation, disease regulation recreation, spiritual enrichment, nutrient cycling, and soil formation are essential for human survival, and their degradation threatens both ecological stability and public health (Chivian and Bernstein 2010). The extinction of species can severely compromise provisioning services, affecting food security and livelihoods. For instance, pollinators such as bees and butterflies are critical for the agricultural sector, contributing to the pollination of over 75% of global food crops (Klein et al. 2007). Their decline, driven by habitat loss, pesticides, and climate change, threatens crop yields and agricultural diversity, which in turn can exacerbate food scarcity and malnutrition (Garratt et al. 2016). Species loss also adversely impacts regulating services. For example, wetlands, which provide natural flood protection, water filtration, and carbon storage, become compromised when key species are removed (Zedler and Kercher 2005). Biodiversity is fundamental for maintaining the resilience and adaptability of ecosystems to changes, including climate change. The loss of keystone species can lead to altered nutrient cycling and increased vulnerability to invasive species, further destabilizing ecosystems (Díaz et al. 2020). Forests, for example, serve as significant carbon sinks, absorbing carbon dioxide from the atmosphere and mitigating climate change effects (Zhang et al. 2022). The loss of biodiversity can disrupt these processes, exacerbating climate-related health issues such as heat stress and respiratory problems due to poor air quality (Aerts et al. 2018).

The depletion of biodiversity diminishes cultural services essential for mental health and community well-being, thereby affecting community cohesion and individual mental health outcomes (Chan et al. 2006). Lastly, the loss of soil microbes and invertebrates can disrupt nutrient availability, impairing plant growth and ecosystem productivity (Bardgett and van der Putten 2014). This decline can result in reduced agricultural resilience and an increased need for chemical fertilizers, impacting environmental health and sustainability. Forests, grasslands, and marine ecosystems act as carbon sinks, storing carbon and helping regulate the Earth's climate. Deforestation and the loss of oceanic species that contribute to carbon sequestration, such as certain algae and seagrass, accelerate climate change by releasing stored carbon into the atmosphere (Pan et al. 2011). Deforestation releases this stored carbon into the atmosphere, contributing to climate change. For example, the Amazon rainforest, often called the "lungs of the Earth," has seen extensive deforestation due to agricultural expansion. The loss of tree species in the Amazon has not only reduced the forest's ability to sequester carbon. However, it has also impacted regional climate patterns, altered rainfall, and shifted temperatures that affect local agriculture and human settlements (Nobre et al. 2016).

5.4 Interconnections Between Biodiversity and Human Health

The interconnection between biodiversity and human health is a critical field of research that is underscored by various mechanisms revealing the dependencies of human populations on diverse biological systems. These mechanisms can be broadly categorized into three key areas: disease regulation, nutritional security, and mental health benefits. Understanding these connections is essential for developing effective strategies to mitigate the impacts of biodiversity loss on human health. Biodiversity is critical in regulating infectious diseases, particularly zoonotic diseases transmitted from animals to humans. The presence of diverse species can influence disease transmission dynamics, often acting as a buffer against outbreaks. This phenomenon is usually referred to as the "dilution effect," which posits that higher biodiversity can reduce the prevalence of pathogens by diluting the abundance of competent hosts Morand et al. (2014), Keesing et al. (2010). For example, non-competent hosts can reduce the likelihood of disease transmission in ecosystems with a rich species diversity. This is particularly relevant for vector-borne diseases, where the abundance of various host species can influence the population dynamics of disease vectors, such as mosquitoes and ticks (Rulli et al. 2017). The loss of biodiversity, therefore, can lead to an increase in the prevalence of infectious diseases, as fewer species are available to dilute the effects of pathogens. Lyme disease, caused by the bacterium Borrelia burgdorferi and transmitted by ticks, is a pertinent case study for understanding the dilution effect. Research has shown that areas with higher biodiversity, particularly those with a variety of small mammals, exhibit lower rates of Lyme disease transmission (Salkeld et al. 2013). In these biodiverse environments, non-competent hosts, such as certain species of birds, can reduce the likelihood of ticks acquiring the pathogen, lowering the risk of human infection.

Conversely, habitat fragmentation and biodiversity loss can increase Lyme disease cases. As natural habitats are altered or destroyed, the populations of certain host species, such as deer, may increase, leading to a higher density of ticks and a greater risk of transmission to humans (Salkeld et al. 2013). This relationship underscores the importance of maintaining biodiversity as a public health strategy to mitigate the risks associated with vector-borne diseases.

Biodiversity is essential for ensuring nutritional security, as it contributes to the availability of diverse food sources. A varied diet rich in fruits, vegetables, grains, and animal products is crucial for meeting human nutritional needs and preventing malnutrition (Haider et al. 2015). Biodiversity enhances food security by providing various crops that can adapt to changing environmental conditions, pests, and diseases (Haider et al. 2015; Andreazzi et al. 2023). The loss of agricultural biodiversity, particularly in monoculture systems, can lead to reduced resilience in food production and increased vulnerability to food shortages. For instance, the reliance on a limited number of crop species can result in significant losses during pest outbreaks or adverse weather conditions. In contrast, diverse agricultural systems incorporating multiple species can enhance soil health, improve pest resistance, and increase productivity (Haider et al. 2015; Andreazzi et al. 2023).

Moreover, biodiversity contributes to the nutritional quality of food. Plant and animal species provide essential vitamins, minerals, and other vital nutrients for human health. For example, consuming a diverse range of fruits and vegetables has been linked to lower risks of chronic diseases, such as heart disease and diabetes (Haider et al. 2015). Thus, preserving biodiversity is crucial for ecological balance and ensuring that populations can access nutritious food.

The mental health benefits of biodiversity are increasingly recognized in public health discourse. Access to natural environments and green spaces has been associated with numerous psychological benefits, including reduced stress, anxiety, and depression (Saha et al. 2021). Exposure to nature promotes physical activity, social interaction, and overall well-being, contributing to improved mental health outcomes. Studies have shown that individuals living in areas with abundant green spaces report higher levels of life satisfaction and lower levels of mental distress (Saha et al. 2021). Biodiversity in urban environments, such as parks and gardens, provides opportunities for recreation, relaxation, and connection with nature, which are essential for mental health.

Furthermore, the aesthetic and cultural values associated with biodiversity can enhance community cohesion and promote a sense of belonging, further contributing to mental well-being. The psychological benefits of biodiversity extend beyond individual health; they can also influence community health and resilience. Communities with access to diverse natural environments are better equipped to cope with stressors, such as economic challenges and social upheaval. Integrating green spaces into urban planning and community development can foster social ties and promote collective well-being (Saha et al. 2021).

5.5 Biodiversity and Emerging Infectious Diseases

As the interconnections between biodiversity and human health increasingly obvious, understanding how biodiversity impacts the emergence of infectious diseases is vital. Emerging infectious diseases (EIDs) are newly identified infections that have increased in incidence or geographical range over the past few decades. The relationship between biodiversity loss and the emergence of these diseases presents significant implications for global health, conservation, and public policy. Most emerging infectious diseases are zoonotic, originating in animals and spreading to humans. Biodiversity loss can increase the risk of zoonotic disease emerwell, with approximately 75% of emerging infectious diseases being zoonoses, and the rate of emergence appears to be rising, according to Smith and Wang (2013). In addition, the decline in biodiversity disrupts ecological balances and increases human-wildlife interactions, thereby enhancing the likelihood of zoonotic spillover events. As natural habitats are altered or destroyed, wildlife populations may become more vulnerable in smaller areas, increasing the chances of contact with humans and facilitating the transmission of pathogens (Olivero et al. 2017). The mechanisms through which biodiversity loss influences zoonotic diseases are multifaceted. Reduced biodiversity can lead to increased populations of certain species that are competent reservoirs for pathogens. For instance, when predators are removed from an ecosystem, populations of smaller mammals or birds that may harbor diseases can increase, leading to higher rates of disease transmission (Claude and Hawkes 2019). Furthermore, habitat fragmentation can disrupt animal movements and social structures and influence disease dynamics. In fragmented habitats, animals may be forced into closer contact with each other and humans, increasing the risk of pathogen spillover (Olivero et al. 2017).

Ebola virus disease (EVD) is a critical case study for understanding the relationship between biodiversity loss and the emergence of zoonotic diseases. The Ebola virus is believed to have a natural reservoir in fruit bats, and outbreaks have been linked to human interactions with wildlife, particularly in forested regions of Central and West Africa (Haahtela 2019). Research has shown that the recent loss of closed forests is associated with increased outbreaks of Ebola virus disease (Olivero et al. 2017). As forests are fragmented and degraded, the ecological dynamics that regulate wildlife populations are disrupted, leading to increased interactions between humans and potential disease reservoirs. In the Democratic Republic of the Congo (DRC), where multiple Ebola outbreaks have occurred, studies indicate that forest animals, such as bats and nonhuman primates, play a significant role in transmitting the virus to humans (Claude and Hawkes 2019). The proximity of human populations to forest edges increases the likelihood of contact with these animals, thereby heightening the risk of spillover events.

Moreover, local communities often rely on bushmeat hunting, which can further facilitate the transmission of the virus from infected animals to humans (Jalloh et al. 2017). The interplay between forest fragmentation, biodiversity loss, and human behavior underscores the importance of conserving natural habitats to mitigate the risk of zoonotic diseases like Ebola. The COVID-19 pandemic also underscores the intricate link between biodiversity and emerging infectious diseases. Research suggests that the virus that causes COVID-19 is likely of zoonotic origin, with bats and pangolins as potential intermediate hosts (Liu et al. 2020). The encroachment of human activity into wildlife habitats has promoted conditions that enable zoonotic spillovers.

The One Health and Planetary Health frameworks provide valuable perspectives for understanding the complex interconnections between biodiversity and human health. One Health emphasizes the interconnectedness of human, animal, and environmental health, advocating for a collaborative approach to address health challenges arising from these domains' interactions (Atlas et al. 2010). Given that over 60% of emerging infectious disease events are caused by zoonotic pathogens, a One Health approach is essential for reducing threats to global health (Atlas et al. 2010). This framework encourages interdisciplinary collaboration among public health officials, veterinarians, ecologists, and policymakers to develop comprehensive disease prevention and control strategies. On the other hand, Planetary Health focuses on the health of human civilization and the state of the natural systems on which it depends. This perspective recognizes that the degradation of ecosystems and biodiversity loss can directly and indirectly impact human health. For instance, the increasing frequency of zoonotic disease outbreaks is often linked to environmental changes driven by human activities, such as deforestation, urbanization, and climate change. By adopting a Planetary Health approach, we can better understand the systemic nature of health challenges and the need for sustainable practices that protect human health and the environment (Potter 2021). Both frameworks highlight the importance of biodiversity conservation as a strategy for promoting health. By maintaining healthy ecosystems, we can enhance the resilience of both natural systems and human populations. For example, preserving biodiversity can help regulate disease dynamics, ensuring that ecosystems provide essential services supporting human health. Additionally, these frameworks advocate for proactive management of emerging infectious diseases, emphasizing the need for surveillance, research, and community engagement to mitigate risks associated with zoonotic diseases (Pigott et al. 2017).

5.6 Nexus of Climate Change, Biodiversity, and Human Health

Climate change is increasingly recognized as a significant driver of biodiversity loss, affecting ecosystems and species across the globe. The rise in global temperatures, changes in precipitation patterns, and increased frequency of extreme weather events are altering habitats and threatening the survival of many species Leclère et al. (2020). As ecosystems become destabilized, the delicate balance that sustains biodiversity is disrupted, leading to shifts in species distributions, population declines, and, in some cases, extinction. Tropical regions, which are home to a significant portion of the world's biodiversity, are particularly vulnerable to the impacts of climate change. Studies indicate that tropical forests, despite experiencing lower conversion rates to human land use compared to other areas, are predicted to suffer substantial species losses due to future climate changes (Newbold 2018). The combination of habitat loss from land-use changes and the stressors imposed by climate change creates a synergistic effect that exacerbates biodiversity loss (Nunez and Alkemade 2021). For instance, species already under pressure from habitat destruction may find it increasingly difficult to adapt to changing climatic conditions, leading to further declines in their populations.

Moreover, climate change's impacts on biodiversity are not uniform; they vary spatially and temporally. In some regions, species may migrate to higher altitudes or latitudes in search of suitable habitats, while others may face local extinction if they cannot adapt quickly enough (Schipper et al. 2019). This uneven distribution of impacts can lead to changes in community composition and ecosystem functioning, further complicating the relationship between climate change and biodiversity.

The decline in biodiversity driven by climate change has direct and indirect implications for human health (Table 5.1). Ecosystems provide essential services that support human well-being, including clean air and water, food production, disease regulation, and climate stabilization. As biodiversity declines, the ability of ecosystems to deliver these services is compromised, leading to adverse health outcomes for human populations. For example, the loss of pollinator species due to climate change can threaten food security by reducing crop yields and the availability of diverse food sources (Jaureguiberry et al. 2022). Pollinators are crucial for reproducing many fruits, vegetables, and nuts, and their decline can lead to increased malnutrition and related health issues.

Similarly, the degradation of ecosystems that provide clean water can result in water scarcity and increased exposure to waterborne diseases, further exacerbating public health challenges (Roelvink et al. 2021). Climate change also influences the dynamics of infectious diseases. Changes in temperature and precipitation patterns can alter the habitats of disease vectors, such as mosquitoes and ticks, leading to shifts in the transmission patterns of vector-borne diseases (Teh et al. 2013). For instance, warmer temperatures may expand the range of mosquito well-being diseases like malaria and dengue fever, increasing the risk of outbreaks in previously unaffected areas (Aswani et al. 2015). This interplay between climate change, biodiversity loss, and human health underscores the urgent need for integrated approaches to address these interconnected challenges.

Coral reefs are among the most biodiverse ecosystems on the planet, providing critical habitat for a vast array of marine species. However, they are also highly susceptible to the impacts of climate change, mainly through coral bleaching events caused by elevated sea temperatures (Ainsworth and Mumby 2014). The loss of coral reefs threatens marine biodiversity and has profound implications for the human communities that depend on these ecosystems for their livelihoods and food security. In coastal regions, coral reefs serve as vital barriers against storm surges and coastal erosion, protecting communities from the impacts of extreme weather events (Dacks et al. 2018). The degradation of coral reefs due to climate change can increase

Table 5.1 Shared impacts of climate change on biodiversity and human health

Impact area	Climate change	Biodiversity	Human health
Temperature increase	Rising global temperatures lead to altered weather patterns, heatwaves, and increased frequency of extreme weather events	Species migration, habitat shifts, and increased extinction risks for temperature-sensitive species	Heat-related illnesses and deaths; exacerbation of chronic conditions, such as cardiovascular and respiratory diseases
Air quality degradation	Climate change can increase air pollutants like ozone and particulate matter	Loss of vegetation that filters pollutants from the air and fewer forests contribute to poorer air quality	Respiratory illnesses, such as asthma and lung disease, increase allergy incidences
Water scarcity	Reduced precipitation and altered rainfall patterns lead to droughts and water scarcity	Loss of freshwater habitats decreased aquatic biodiversity and reduced water quality	Waterborne diseases from contaminated water sources; food and nutritional insecurity
Ocean warming and acidification	Increased carbon dioxide levels cause ocean warming and acidification	Coral bleaching, loss of marine biodiversity, and disruption of aquatic food chains	Reduced seafood availability and loss of marine food sources; exposure to harmful algal blooms
Spread of infectious diseases	Warmer temperatures expand habitats suitable for disease vectors like mosquitoes	Disruption of ecosystems that help control disease vectors leads to increased risk of zoonotic diseases	Increased incidence of vector-borne diseases such as malaria, dengue, and Zika virus
Extreme weather events	I have increased the frequency and severity of hurricanes, floods, and wildfires	Habitat destruction, species displacement, and loss of biodiversity in affected areas	Direct injuries and fatalities; mental health impacts; infrastructure damage leading to health service disruption
Food security	Climate variability affects crop yields and growing seasons, leading to food scarcity	Reduced biodiversity in agricultural systems; loss of pollinators and soil organisms	Malnutrition and hunger; food insecurity-related health conditions, including stunted growth in children
Mental health and Well-being	Climate anxiety and stress from environmental changes and displacement due to extreme weather	Loss of natural spaces and green areas that promote mental health and well-being	Increased rates of anxiety, depression, and stress-related disorders

the vulnerability of coastal communities to flooding and other hazards, leading to displacement and loss of property.

Furthermore, the decline of reef ecosystems negatively impacts local fisheries, a primary protein source for millions of people worldwide (Rumagia 2023). As fish populations decline due to habitat loss wellbeing environmental conditions, food security is threatened, particularly in developing countries where reliance on fish is high (Cheal et al. 2017). The case of the Great Barrier Reef in Australia illustrates the interconnectedness of climate change, biodiversity loss, and human health. The reef has experienced significant coral bleaching events in recent years, leading to declines in fish populations and threatening the livelihoods of local fishers (Pratchett et al. 2011). The economic impacts of reef degradation extend beyond fisheries, affecting tourism and recreation, vital components of the local economy (Asner et al. 2010). As coral reefs continue to decline, the health and well-being of coastal communities are increasingly at risk.

5.7 Strategies for Conserving Biodiversity to Enhance Human Health

The conservation of biodiversity is essential not only for maintaining ecological balance but also for promoting human health and well-being. As the interconnections between biodiversity and health become increasingly evident, various strategies have emerged to conserve biodiversity while enhancing human health outcomes.

5.7.1 Sustainable Land Use and Conservation Strategies

Sustainable land use practices are critical for conserving biodiversity and ensuring that ecosystems provide essential services supporting human health. These practices involve managing land to balance ecological integrity with human needs, thereby reducing the negative impacts of land use on biodiversity.

Agroecology: Agroecological practices promote biodiversity by integrating ecological principles into agricultural systems. This approach emphasizes crop diversity, organic farming, and sustainable soil management, which can enhance ecosystem resilience and productivity (Zurba et al. 2019). By fostering diverse agroecosystems, agroecology can improve food security and nutritional outcomes while reducing reliance on chemical inputs that can harm biodiversity and human health. This diversity contributes to ecosystem stability and offers a more varied and nutritious food supply, directly enhancing human well-being (et al. 2019). By minimizing synthetic fertilizers and pesticides, organic farming practices reduce toxic inputs that can negatively impact biodiversity and human health (Reganold and Wachter 2016). Crop rotation, cover cropping, and reduced tillage promote soil health, fostering a diverse soil microbiome crucial for nutrient cycling and plant growth. By encouraging diverse agroecosystems, agroecology can improve food security and nutritional outcomes while reducing reliance on chemical inputs that can harm biodiversity and human health.

Sustainable Forestry: Implementing sustainable forestry practices can help conserve forest biodiversity while providing economic benefits to local communities. Practices such as selective logging, reforestation, and establishing protected areas within forest landscapes can maintain ecosystem functions and services (Bauch et al. 2015). Sustainable practices such as reducing-impact logging techniques can minimize damage to surrounding vegetation, promoting overall forest health. Reforestation can help restore degraded lands, reestablish native biodiversity, sequester carbon, and protect watershed functions. Establishing protected areas within forest landscapes is critical for preserving biodiversity hotspots. Such areas provide habitat for endangered species, allow for ecological processes to occur unimpeded, and serve as vital resources for education and ecotourism. Sustainable forest management preserves biodiversity and supports the livelihoods and cultural practices of communities that depend on forest resources.

Integrated Coastal Zone Management (ICZM): Coastal ecosystems, such as mangroves and coral reefs, are vital for biodiversity and human health. ICZM promotes the sustainable use of coastal resources while protecting habitats from degradation. This approach involves collaboration among stakeholders, including local communities, to balance ecological health with economic development (Salkeld et al. 2013). Healthy mangrove ecosystems also provide breeding grounds for a variety of marine species, thus underpinning local fisheries (Duke et al. 2007). By preserving coastal ecosystems, ICZM can enhance resilience to climate change and support fisheries crucial for food security.

5.7.2 *Community-Led and Indigenous Approaches to Biodiversity Conservation*

Community-led and Indigenous approaches to biodiversity conservation recognize the importance of local knowledge and practices in managing natural resources. These approaches often prioritize Indigenous peoples' and local communities' rights and involvement in conservation efforts.

Indigenous Protected and Conserved Areas (IPCAs): IPCAs are areas managed by Indigenous communities that prioritize the conservation of biodiversity while respecting cultural values and traditional knowledge (Zurba et al. 2019). These areas effectively maintain ecological integrity and enhance community well-being. By empowering Indigenous peoples to lead conservation efforts, IPCAs can foster a sense of stewardship and connection to the land, which is essential for biodiversity and cultural preservation.

Traditional Ecological Knowledge (TEK): TEK encompasses the knowledge and practices developed by Indigenous peoples over generations regarding the sustainable management of ecosystems. Incorporating TEK into conservation strategies can enhance biodiversity conservation by promoting practices attuning to local ecological conditions (Marentes et al. 2021). For example, traditional agroforestry systems can preserve biodiversity while providing food and resources for local communities.

Community-Based Conservation: Engaging local communities in conservation initiatives can lead to more effective and sustainable outcomes. Community-based conservation approaches involve collaboration between conservation organizations and stakeholders to develop management plans that reflect local needs and priorities (Rayne et al. 2020). This participatory approach can enhance community resilience, improve livelihoods, and foster a sense of ownership over well-being.

5.7.3 Policy Interventions: Global and Local Examples

Effective policy interventions are crucial for supporting biodiversity conservation and enhancing human health. These interventions can take various forms, from international agreements to local regulations.

International Agreements and Frameworks: Global initiatives, such as the Convention on Biological Diversity (CBD) and the Aichi Biodiversity Targets, aim to promote biodiversity conservation at the international level. These agreements encourage countries to develop national strategies for biodiversity conservation that align with global goals (Leiper et al. 2018). By committing to these frameworks, nations can enhance their efforts to protect biodiversity and promote sustainable development.

Protected Areas (PAs): Establishing protected areas is a widely recognized strategy for conserving biodiversity. PAs can safeguard critical habitats and ecosystems, providing refuge for endangered species and preserving ecosystem services (Vipat and Bharucha 2014). However, it is essential to ensure that the establishment and management of PAs consider the rights and needs of local communities to avoid conflicts and displacement (Goolmeer et al. 2022). Integrating community perspectives into PA management can enhance biodiversity conservation and human well-being.

Ecosystem-Based Approaches to Health: Policies that adopt ecosystem-based approaches to health recognize the interconnections between ecosystem health and human health. These approaches promote the conservation of biodiversity to enhance public health outcomes. For example, integrating health considerations into land use planning and environmental management can help mitigate the impacts of biodiversity loss on human health (Keune et al. 2013). By fostering collaboration between health and environmental sectors, these policies can create synergies that benefit both ecosystems and communities.

5.8 Technological and Scientific Innovations for Biodiversity and Health

The intersection of technological and scientific innovations with biodiversity conservation and human health is increasingly recognized as a vital area of research and action. Advances in conservation biology, biotechnology, and bioremediation are paving the way for new strategies that protect biodiversity and enhance human health outcomes.

5.8.1 Advances in Conservation Biology and Restoration Ecology

Conservation biology and restoration ecology have seen significant advances in recent years, driven by technological innovations and a deeper understanding of ecological processes. These advancements are crucial for addressing the ongoing biodiversity crisis and its implications for human health.

Genetic Tools and Techniques: The application of genetic tools, such as DNA barcoding and genomic sequencing, has revolutionized the field of conservation biology. These technologies enable researchers to identify species accurately, assess genetic diversity, and monitor population dynamics Howes et al. (2020). For instance, genetic analyses can help identify populations at risk of extinction and inform breeding programs to enhance genetic diversity, essential for species resilience.

Remote Sensing and Geographic Information Systems (GIS): Remote sensing technologies and GIS have become indispensable conservation planning and management tools. These technologies allow for monitoring habitat changes, assessing ecosystem health, and identifying critical habitats for conservation (Haahtela 2019). By providing spatial data, these tools facilitate informed decision-making and help prioritize conservation efforts where they are most needed.

Restoration Ecology: Advances in restoration ecology focus on rehabilitating degraded ecosystems to restore their ecological functions and services. Techniques such as assisted migration, where species are relocated to more suitable habitats in response to climate change, are gaining traction (Kang and Dai 2021). Additionally, using native plant species in restoration projects enhances biodiversity and improves ecosystem resilience, ultimately benefiting human health by maintaining ecosystem services such as clean air and water.

5.8.2 Role of Biotechnology in Preserving Biodiversity

Biotechnology plays a pivotal role in biodiversity conservation by providing innovative solutions for preserving genetic diversity and enhancing ecosystem resilience. Several biotechnological approaches are being explored to address the challenges of biodiversity loss.

Cryopreservation: Cryopreservation techniques allow for the long-term storage of genetic material from endangered species. By preserving seeds, sperm, and embryos at extremely low temperatures, researchers can safeguard genetic diversity for future restoration efforts (Young et al. 2014). This technology is precious for species at risk of extinction, as it provides a means to maintain genetic resources for breeding programs or reintroducing efforts.

Bio-banking: Biobanks are facilities that store biological samples, including DNA, tissues, and seeds, from various species. These repositories serve as a genetic insurance policy, allowing for the recovery of species that may face extinction due to habitat loss, climate change, or disease (Guarino and Pignatti 2010). Biobanks can also facilitate genetic diversity and adaptation research, providing crucial insights for effective conservation strategies.

Synthetic Biology: Synthetic biology is an emerging field that combines biology and engineering to design and construct new biological parts, devices, and systems. This technology has the potential to create genetically modified organisms that can enhance ecosystem functions or restore degraded habitats (Ward et al. 2022). For example, genetically engineered plants that can tolerate extreme environmental conditions may be used in reforestation, contributing to biodiversity conservation and ecosystem resilience.

Bioremediation: Bioremediation is a powerful tool for restoring contaminated environments and preserving biodiversity, with significant implications for human health. This process involves using microorganisms to degrade or remove environmental pollutants, restoring ecosystem health. Microbial bioremediation utilizes naturally occurring or engineered microorganisms to break down pollutants, such as petroleum hydrocarbons, heavy metals, and pesticides (Nduka et al. 2012). This environmentally friendly and cost-effective approach makes it an attractive option for cleaning up contaminated sites. For instance, bioremediation has been successfully employed in oil spill cleanups, where microbes metabolize hydrocarbons, reducing their toxicity and promoting ecosystem recovery (Koshlaf and Ball 2017). The successful application of bioremediation can lead to significant public health benefits. By removing harmful environmental pollutants, bioremediation reduces the risk of exposure to toxic substances that can cause health issues, such as respiratory problems, skin diseases, and cancers (Nduka et al. 2012).

Additionally, restoring contaminated ecosystems enhances biodiversity, which is essential for maintaining ecosystem services supporting human health, such as clean air and water. Engaging local communities in bioremediation efforts can enhance the effectiveness of these initiatives while promoting public health awareness. Community-led bioremediation projects empower residents to restore their environments actively, fostering a sense of ownership and responsibility for local ecosystems (Borca et al. 2023). By combining scientific expertise with local knowledge, these projects can achieve more sustainable outcomes and improve community health.

5.9 Policy and Governance for Biodiversity Protection and Human Health

The interconnected challenges of biodiversity loss and declining human health require coordinated policies and governance structures that can address these issues globally. Effective policy interventions are essential for conserving biodiversity, ensuring ecosystem services, and safeguarding human well-being.

5.9.1 International Agreements and Frameworks for Biodiversity Conservation

Several international agreements and frameworks provide guidance and set targets for biodiversity conservation. These frameworks reflect a growing understanding of the need to integrate biodiversity protection into broader policy goals related to human health and sustainable development. Key agreements include:

The Convention on Biological Diversity (CBD): Established in 1992, the CBD is a landmark international treaty focused on conserving biodiversity, sustainable use of its components, and the fair and equitable sharing of benefits from genetic resources. The CBD recognizes biodiversity as a global common good and promotes national action plans for conservation. It encourages countries to integrate biodiversity considerations into other sectors, such as agriculture, forestry, and health, fostering an approach that aligns biodiversity protection with human health and development goals (CBD Secretariat 2022). The CBD has been instrumental in establishing a comprehensive approach to biodiversity conservation and sustainable development. With 196 member countries, it provides a platform for countries to share knowledge, resources, and strategies for biodiversity protection. For example, under the CBD framework, countries like Costa Rica have implemented national biodiversity strategies that include forest conservation, protection of marine ecosystems, and promotion of sustainable tourism. These initiatives support biodiversity while generating economic benefits and improving public health through ecosystem services like water purification and climate regulation (CBD Secretariat 2022).

The Aichi Biodiversity Targets: Developed under the CBD in 2010, the Aichi Targets consist of 20 global goals to address the underlying causes of biodiversity loss and enhance conservation efforts by 2020. These targets include reducing habitat destruction, safeguarding ecosystems, and promoting sustainable practices across sectors. Although not all targets were fully met, they have influenced national biodiversity strategies and raised awareness about the importance of biodiversity conservation for health and well-being (Leadley et al. 2020). The Aichi Targets have influenced biodiversity policies worldwide by setting clear and measurable goals. Target 14, for instance, aimed to restore ecosystems that provide essential services, particularly those related to health, livelihoods, and well-being. In the European Union, efforts to achieve Target 14 led to the restoration of wetlands, which play a crucial role in flood mitigation, water purification, and habitat provision for species that contribute to agricultural pollination. While progress varied across countries, the Aichi Targets spurred actions that underscored the connection between biodiversity, ecosystem services, and human health (Leadley et al. 2020).

The Global Biodiversity Framework: In response to the lessons learned from the Aichi Targets, the Global Biodiversity Framework sets forth a comprehensive plan for biodiversity conservation by 2030. It emphasizes ambitious targets, such as protecting 30% of the planet's terrestrial and marine ecosystems, restoring degraded ecosystems, and promoting sustainable management practices. This framework also highlights the role of biodiversity in human health and the importance of integrating biodiversity conservation into global health agendas (Díaz et al. 2020). The Global Biodiversity Framework builds on the lessons from the Aichi Targets, setting ambitious new targets for 2030. One key aspect of this framework is the "30 by 30" wellbeing, which aims to protect 30% of Earth's land and ocean areas by 2030. Countries like Canada and the United Kingdom have already committed to this target by designating more protected areas and funding conservation programs. This framework also promotes the integration of biodiversity into health sectors, recognizing that healthy ecosystems contribute to public health by regulating air and water quality, supporting nutrition, and reducing disease transmission risks (Díaz et al. 2020).

5.9.2 *Role of Governments, NGOs, and the Private Sector in Promoting Biodiversity-Health Initiatives*

The successful implementation of biodiversity conservation policies requires the participation and collaboration of diverse stakeholders. Governments, non-governmental organizations (NGOs), and the private sector each play critical roles in promoting biodiversity-health initiatives:

Governments: National governments are responsible for creating policies, setting regulations, and allocating resources for biodiversity conservation. Governments can influence land use, conservation practices, and environmental protection through legislation, protected area designations, and financial incentives. Public health agencies also collaborate with environmental agencies to address the links between biodiversity and health, such as reducing pollution exposure and promoting access to green spaces (McMichael et al. 2008).

Non-Governmental Organizations (NGOs): NGOs, such as the World Wildlife Fund (WWF) and Conservation International, are pivotal in raising awareness, advocating for conservation policies, and conducting on-the-ground projects to protect biodiversity. NGOs work alongside governments to implement conservation programs, educate communities about sustainable practices, and support research that links biodiversity conservation with health outcomes. NGOs monitor policy effectiveness and hold governments accountable for international biodiversity commitments (Butchart et al. 2015).

Private Sector: Businesses and industries play a significant role in shaping biodiversity-health initiatives. Many companies are beginning to recognize the value of biodiversity and are implementing corporate sustainability programs to reduce their environmental footprint. Industries such as agriculture, forestry, and pharmaceuticals are directly impacted by biodiversity and thus are interested in supporting conservation efforts. For example, companies may engage in habitat restoration, adopt sustainable sourcing practices, and invest in biodiversity research. Additionally, public–private partnerships enable companies to collaborate with governments and NGOs to develop scalable solutions that benefit both biodiversity and human health (Garrett et al. 2019).

5.10 Conclusion

The interconnections between biodiversity, ecosystem services, and human health are complex and multifaceted. The classification of ecosystem services into provisioning, regulating, cultural, and supporting categories highlights ecosystems' diverse benefits. Biodiversity is critical in supporting these services and enhancing ecosystem functioning and resilience. As human activities continue to drive biodiversity loss, it is imperative to recognize the implications of this decline on ecosystem services and, consequently, human health and well-being. Effective conservation strategies that prioritize biodiversity are essential for ensuring ecosystem services' sustainability and future generations' health. Technological and scientific innovations are crucial in conserving biodiversity and enhancing human health. Advances in conservation biology and restoration ecology, the application of biotechnology, and the effective use of bioremediation contribute to a more sustainable future. As the challenges posed by biodiversity loss and environmental degradation continue to grow, it is imperative to leverage these innovations to protect ecosystems and promote human well-being. The mechanisms linking biodiversity to zoonotic diseases, as exemplified by the case of Ebola, underscore the importance of conserving natural habitats to reduce the risk of disease spillover. One Health and Planetary Health frameworks provide valuable insights into the interconnectedness of human, animal, and environmental health, emphasizing the need for collaborative approaches to address health challenges arising from biodiversity loss.

Additionally, biodiversity is crucial for ensuring nutritional security and providing access to diverse food sources essential for human health. The psychological and recreational value of natural environment and well-being underscores the importance of preserving biodiversity for mental well-being. As biodiversity continues to decline globally, it is imperative to recognize and address the implications of this loss on human health. Effective conservation strategies prioritizing biodiversity are essential for safeguarding ecosystem services and promoting public health. The influence of biodiversity loss on emerging infectious diseases is a critical concern for public health. The mechanisms linking biodiversity to zoonotic diseases, as exemplified by the case of Ebola, underscore the importance of conserving natural habitats to reduce the risk of disease spillover. One Health and Planetary Health frameworks provide valuable insights into the interconnectedness of human, animal, and environmental health, emphasizing the need for collaborative approaches to address health challenges arising from biodiversity loss. By recognizing the intrinsic links between biodiversity and health, we can develop effective strategies to mitigate the impacts of emerging infectious diseases and promote a healthier future for both people and the planet.

References

Aerts R, Honnay O, Nieuwenhuyse A (2018) Biodiversity and human health: mechanisms and evidence of the positive health effects of diversity in nature and green spaces. Br Med Bull 127(1):5–22. https://doi.org/10.1093/bmb/ldy021

Ainsworth C, Mumby P (2014) Coral–algal phase shifts alter fish communities and reduce fisheries production. Glob Change Biol 21(1):165–172. https://doi.org/10.1111/gcb.12667

Andreazzi C, Martinez-Vaquero L, Winck G, Cardoso T, Teixeira B, Xavier S, … D'Andrea P (2023) Vegetation cover and biodiversity reduce parasite infection in wild hosts across ecological levels and scales. Ecography 2023(5). https://doi.org/10.1111/ecog.06579

Asner G, Loarie S, Heyder U (2010) Combined effects of climate and land-use change on the future of humid tropical forests. Conserv Lett 3(6):395–403. https://doi.org/10.1111/j.1755-263x.2010.00133.x

Aswani S, Mumby P, Baker A, Christie P, McCook L, Steneck R, … Richmond R (2015) Scientific frontiers in the management of coral reefs. Front Marine Sci 2. https://doi.org/10.3389/fmars.2015.00050

Atlas R, Rubin C, Maloy S, Daszak P, Colwell R, Hyde B (2010) One health—attaining optimal health for people, animals, and the environment. Microbe Mag 5(9):383–389. https://doi.org/10.1128/microbe.5.383.1

Bardgett RD, van der Putten WH (2014) Belowground biodiversity and ecosystem functioning. Nature 515(7528):505–511. https://doi.org/10.1038/nature13855

Barlow J, Lennox G, Ferreira J et al (2016) Anthropogenic disturbance in tropical forests can double biodiversity loss from deforestation. Nature 535:144–147. https://doi.org/10.1038/nature18326

Bauch S, Birkenbach A, Pattanayak S, Sills E (2015) Public health impacts of ecosystem change in the Brazilian amazon. Proc Natl Acad Sci 112(24):7414–7419. https://doi.org/10.1073/pnas.1406495111

Bernhardt J, O'Connor M (2021) Aquatic biodiversity enhances multiple nutritional benefits to humans. Proc Natl Acad Sci 118(15). https://doi.org/10.1073/pnas.1917487118

Borca M, Ciobîcă A, Iordache A (2023) Modern aspects on the complex interactions between biodiversity and the main neuropsychiatric disorders. https://doi.org/10.20944/preprints202307.1722.v1

Bratman GN, Anderson CB, Baker SM, DeVries, S. (2019). Nature and mental health: An ecosystem service perspective. Sci Adv 5(7):eaax0522. https://doi.org/10.1126/sciadv.aax0522

Butchart SHM, Scharlemann JPW, Evans MI, Quader S, Aricò S, Arinaitwe J., ... Woodley S (2015) Toward a comprehensive database of the world's protected areas. Science 352(6285):35–36. https://doi.org/10.1126/science.351.6285.35

Carpenter SR, Caraco NF, Correll DL, Howarth RW, Sharpley AN, Smith VH (1998) Nonpoint pollution of surface waters with phosphorus and nitrogen. Ecol Appl 8(3):559–568. https://doi.org/10.1890/1051-0761(1998)008[0559:NPOSWW]2.0.CO;2

CBD Secretariat (2022) Global biodiversity outlook 5. Montreal: Convention on Biological Diversity. https://www.cbd.int/gbo/gbo5/publication/gbo-5-en.pdf

Chan KMA, Shaw MR, Cameron, David RC, Underwood, Emma CU, Gretchan CD (2006) Ecosystem services are not just for the rich. Nature 481(7380):317–328. https://doi.org/10.1371/journal.pbio.0040379

Cheal A, MacNeil M, Emslie M, Sweatman H (2017) The threat to coral reefs from more intense cyclones under climate change. Glob Change Biol 23(4):1511–1524. https://doi.org/10.1111/gcb.13593

Chivian E, Bernstein A (2010) Sustaining life: how human health depends on biodiversity. Oxford University Press

Clark N, Lovell R, Wheeler B, Higgins S, Depledge M, Norris K (2014) Biodiversity, cultural pathways, and human health: a framework. Trends Ecol Evol 29(4):198–204. https://doi.org/10.1016/j.tree.2014.01.009

Claude K, Hawkes M (2019) The cat that kills people:' community beliefs about ebola origins and implications for disease control in eastern democratic republic of the congo. Pathogens and Global Health 113(4):149–157. https://doi.org/10.1080/20477724.2019.1650227

Correia AM, Lopes LF (2023) Revisiting biodiversity and ecosystem functioning through the lens of complex adaptive systems. Diversity 15(8):895. https://doi.org/10.3390/d15080895

Dacks R, Ticktin T, Jupiter S, Friedlander A (2018) Drivers of fishing at the household scale in Fiji. Ecol Soc 23(1). https://doi.org/10.5751/es-09989-230137

Díaz S, Settele J, Brondízio ES, Ngo HT, Guèze M, Agard J, ... Zayas CN (2020) Pervasive human-driven decline of life on Earth points to the need for transformative change. Science 366(6471):1052–1054. https://doi.org/10.1126/science.aax3100

Doney SC, Fabry VJ, Feely RA, Kleypas JA (2009) Ocean acidification: the other CO2 problem. Annu Rev Mar Sci 169–192

Duke NC, Meynecke JO, Dittmann S, Ellison AM, Anger K, Berger U et al (2007) A world without mangroves? Science 317:41–42. https://doi.org/10.1126/science.317.5834.41b

Fenoglio M, Rossetti M, Videla M (2020) Negative effects of urbanization on terrestrial arthropod communities: a meta-analysis. Glob Ecol Biogeogr 29(8):1412–1429. https://doi.org/10.1111/geb.13107

Foley JA, Defries R, Asner GP, Barford C, Bonan G, Carpenter SR, ... Snyder PK (2005) Global consequences of land use. Science 309(5734):570–574. https://doi.org/10.1126/science.1111772

Franzolin F, Carvalho G, Santana C, Calegari A, Almeida, E., Soares, J., … Lemos, E. (2021). Students' interests in biodiversity: links with health and sustainability. Sustainability, 13(24), 13767. https://doi.org/10.3390/su132413767

Garrett RD, Carlson KM, Rueda X, Noojipady P (2019) Assessing the potential of supply-side sustainability initiatives to reduce deforestation in Amazonia. Glob Environ Chang 57:101931. https://doi.org/10.1016/j.gloenvcha.2019.101931

Garratt MPD et al (2016) Apple pollination: demand depends on variety and supply depends on pollinator identity. PLOS One 11:e0153889

Goolmeer T, Skroblin A, Grant C, Leeuwen S, Archer R, Gore-Birch C, … Wintle B (2022) Recognizing culturally significant species and indigenous-led management is key to meeting international biodiversity obligations. Conserv Lett 15(6). https://doi.org/10.1111/conl.12899

Guarino R, Pignatti S (2010) Diversitas and biodiversity: the roots of a twenty-first century myth. Rendiconti Lincei 21(4):351–357. https://doi.org/10.1007/s12210-010-0104-2

Haahtela T (2019) A biodiversity hypothesis. Allergy 74(8):1445–1456. https://doi.org/10.1111/all.13763

Haider M, Patel M, Bhattacharjee P, Bassa M (2015) The impacts of biotechnology on biodiversity in global health: a case study on avian influenza in Bangladesh. Int J Public Health Sci (Ijphs) 4(2):102. https://doi.org/10.11591/.v4i2.4719

Hector A, Bagchi R (2007) Biodiversity and ecosystem multifunctionality. Nature 448(7150):188–190. https://doi.org/10.1038/nature05947

Hoegh-Guldberg O et al (2019) The human imperative of stabilizing global climate change at 1.5°C. Science 365(6974). https://doi.org/10.1126/science.aaw6974

Howes M, Quave C, Collemare J, Tatsis E, Twilley D, Lulekal E, … Lughadha E (2020) Molecules from nature: reconciling biodiversity conservation and global healthcare imperatives for sustainable use of medicinal plants and fungi. Plants People Planet 2(5):463–481. https://doi.org/10.1002/ppp3.10138

Irvine K, Hoesly D, Bell-Williams R, Warber S (2019) Biodiversity and spiritual well-being, pp 213–247. https://doi.org/10.1007/978-3-030-02318-8_10

Izah SC, Angaye CN, Aigberua AO, Nduka JO (2017) Uncontrolled bush burning in the Niger Delta region of Nigeria: potential causes and impacts on biodiversity. Int J Mol Ecol Conserv 7(1):1–15

Izah SC (2018) Ecosystem of the Niger Delta region of Nigeria: potentials and threats. Biodivers Int J 2(4):338–345

Izah SC, Seiyaboh EI (2018a) Challenges of wildlife with therapeutic properties in Nigeria; a conservation perspective. Int J Avian Wildlife Biol 3(4):259–264

Izah SC, Seiyaboh EI (2018b) Changes in the protected areas of Bayelsa state, Nigeria. Int J Mol Evol Biodivers 8(1):1–11

Izah SC, Aigberua AO, Nduka JO (2018c) Factors affecting the population trend of biodiversity in the Niger Delta region of Nigeria. Int J Avian Wildlife Biol 3(3):206–214

Jackson JBC, Kirby MX, Berger WH, Bjorndal KA, Botsford LW, Bourque BJ, ... Warner RR (2001) Historical overfishing and the recent collapse of coastal ecosystems. Science 293(5530):629–637. https://doi.org/10.1126/science.1059199

Jalloh M, Sengeh P, Monasch R, Jalloh M, DeLuca N, Dyson M, … Bunnell R (2017) National survey of ebola-related knowledge, attitudes and practices before the outbreak peak in Sierra Leone: august 2014. BMJ Global Health 2(4):e000285. https://doi.org/10.1136/bmjgh-2017-000285

Jaureguiberry P, Titeux N, Wiemers M, Bowler D, Coscieme L, Golden A, … Purvis A (2022) The direct drivers of recent global anthropogenic biodiversity loss. Sci Adv 8(45). https://doi.org/10.1126/sciadv.abm9982

Jimenez MP, DeVille NV, Elliott EG, Schiff JE, Wilt GE, Hart JE, James P (2021) Associations between nature exposure and health: a review of the evidence. Int J Environ Res Public Health 18(9):4790. https://doi.org/10.3390/ijerph18094790

Kang Y, Dai R (2021) Toward a more expansive discourse in a changing world: an analysis of political leaders' speeches on biodiversity. Sustainability 13(5):2899. https://doi.org/10.3390/su13052899

Keesing F, Belden L, Daszak P, Dobson A, Harvell C, Holt R, … Ostfeld R (2010) Impacts of biodiversity on the emergence and transmission of infectious diseases. Nature 468(7324):647–652. https://doi.org/10.1038/nature09575

Keesing F, Holt RD, Ostfeld RS (2006) Effects of species diversity on disease risk. Ecol Lett 9(4):485–98. https://doi.org/10.1111/j.1461-0248.2006.00885.x

Keune H, Kretsch C, Blust G, Gilbert M, Flandroy L, Berge K, … Bauler T (2013) Science–policy challenges for biodiversity, public health and urbanization: examples from belgium. Environ Res Lett 8(2):025015. https://doi.org/10.1088/1748-9326/8/2/025015

Klein AM, Vaissière BE, Cane JH, Steffan-Dewenter I, Cunningham SA, Kremen C, Tscharntke T (2007) Importance of pollinators in changing landscapes for world crops. Proc Royal Soc b: Biol Sci 274(1608):303–313. https://doi.org/10.1098/rspb.2006.3721

Korn H, Stadler J, Bonn A (2019) Global developments: policy support for linking biodiversity, health and climate change, pp 315–328. https://doi.org/10.1007/978-3-030-02318-8_14

Koshlaf E, Ball A (2017) Soil bioremediation approaches for petroleum hydrocarbon polluted environments. Aims Microbiol 3(1):25–49. https://doi.org/10.3934/microbiol.2017.1.25

Kuo M (2015) How might contact with nature promote human health? Promising mechanisms and a possible central pathway. Front Psychol 6:1093. https://doi.org/10.3389/fpsyg.2015.01093

Leadley P, Krug C, Alkemade R, Pereira HM, Sumaila UR, Walpole M, ... Larigauderie A (2020) Progress towards the Aichi biodiversity targets: an assessment of biodiversity trends, policy scenarios, and key actions. Global Environ Change 40:41–55. https://doi.org/10.1016/j.gloenvcha.2015.10.012

Leclère D, Obersteiner M, Barrett M, Butchart S, Chaudhary A, Palma A, … Young L (2020) Bending the curve of terrestrial biodiversity needs an integrated strategy. Nature 585(7826):551–556. https://doi.org/10.1038/s41586-020-2705-y

Leiper I, Zander K, Robinson C, Carwadine J, Moggridge B, Garnett S (2018) Quantifying current and potential contributions of australian indigenous peoples to threatened species management. Conserv Biol 32(5):1038–1047. https://doi.org/10.1111/cobi.13178

Liu P, Jiang JZ, Wan XF, Hua Y, Li L, Zhou J, Wang X, Hou F, Chen J, Zou J, Chen J (2020) Are pangolins the intermediate host of the 2019 novel coronavirus (SARS-CoV-2)? PLoS Pathog. https://doi.org/10.1371/journal.ppat.1008421

Marentes M, Venturi M, Scaramuzzi S, Focacci M, Santoro A (2021) Traditional forest-related knowledge and agrobiodiversity preservation: the case of the chagras in the indigenous reserve of monochoa (colombia). Biodivers Conserv 31(10):2243–2258. https://doi.org/10.1007/s10531-021-02263-y

McMichael AJ, Woodruff RE, Hales S (2008) Climate change and human health: present and future risks. The Lancet 367(9513):859–869. https://doi.org/10.1016/S0140-6736(08)60454-6

Morand S, Jittapalapong S, Suputtamongkol Y, Abdullah M, Huan T (2014) Infectious diseases and their outbreaks in Asia-Pacific: biodiversity and its regulation loss matter. PLoS ONE 9(2):e90032. https://doi.org/10.1371/journal.pone.0090032

Nduka J, Umeh L, Okerulu I, Umedum L, Okoye H (2012) Utilization of different microbes in bioremediation of hydrocarbon-contaminated soils stimulated with inorganic and organic fertilizers. J Petrol Environ Biotechnol 3(2). https://doi.org/10.4172/2157-7463.1000116

Newbold T (2018) Future effects of climate and land-use change on terrestrial vertebrate community diversity under different scenarios. Proc Royal Soc B Biol Sci 285(1881):20180792. https://doi.org/10.1098/rspb.2018.0792

Nobre CA, Sampaio G, Borma LS, Castilla-Rubio JC, Silva JS, Cardoso M (2016) Land-use and climate change risks in the Amazon and the need of a novel sustainable development paradigm. Proc Natl Acad Sci 113(39):10759–10768. https://doi.org/10.1073/pnas.1605516113

Nunez S, Alkemade R (2021) Exploring interaction effects from mechanisms between climate and land-use changes and the projected consequences on biodiversity. Biodivers Conserv 30(12):3685–3696. https://doi.org/10.1007/s10531-021-02271-y

Olivero J, Fa J, Real R, Márquez A, Farfán M, Vargas J, … Nasi R (2017) Recent loss of closed forests is associated with ebola virus disease outbreaks. Sci Reports 7(1). https://doi.org/10.1038/s41598-017-14727-9

Pan Y, Birdsey RA, Fang J, Houghton R, Kauppi PE, Kurz WA, ... Hayes D (2011) A large and persistent carbon sink in the world's forests. Science 333(6045):988–993. https://doi.org/10.1126/science.1201609

Pigott D, Deshpande A, Letourneau I, Morozoff C, Reiner R, Kraemer M, … Hay S (2017) Local, national, and regional viral haemorrhagic fever pandemic potential in Africa: a multistage analysis. The Lancet 390(10113):2662–2672. https://doi.org/10.1016/s0140-6736(17)32092-5

Potter T (2021) Planetary health: an essential framework for nursing education and practice. Creat Nurs 27(4):226–230. https://doi.org/10.1891/cn-2021-0017

Pratchett M, Hoey A, Wilson S, Messmer V, Graham N (2011) Changes in biodiversity and functioning of reef fish assemblages following coral bleaching and coral loss. Diversity 3(3):424–452. https://doi.org/10.3390/d3030424

Rayne A, Byrnes G, Collier-Robinson L, Hollows J, McIntosh A, Ramsden M, … Steeves (2020) Centring indigenous knowledge systems to re-imagine conservation translocations. People Nat 2(3):512–526. https://doi.org/10.1002/pan3.10126

Reganold JP, Wachter JM (2016) Organic farming in the twenty-first century. Nat Plants 2(2):15221. https://doi.org/10.1002/pan3.10126

Roelvink F, Storlazzi C, Dongeren A, Pearson S (2021) Coral reef restorations can be optimized to reduce coastal flooding hazards. Front Marine Sci 8. https://doi.org/10.3389/fmars.2021.653945

Romanelli C, Corvalán C, Cooper H, Manga L, Maiero M, Campbell-Lendrum D (2014) From manaus to maputo: toward a public health and biodiversity framework. EcoHealth 11(3):292–299. https://doi.org/10.1007/s10393-014-0959-2

Roslund M, Puhakka R, Grönroos M, Nurminen N, Oikarinen S, Gazali A… Sinkkonen A (2020) Biodiversity intervention enhances immune regulation and health-associated commensal microbiota among daycare children. Sci Adv 6(42). https://doi.org/10.1126/sciadv.aba2578

Rulli M, Santini M, Hayman D, D'Odorico P (2017) The nexus between forest fragmentation in Africa and ebola virus disease outbreaks. Sci Reports 7(1). https://doi.org/10.1038/srep41613

Rumagia F (2023) Sustainability development index of reef fisheries in indonesia: a case study of reef fisheries at ternate island, north maluku province. Omni-Akuatika 19(1):15. https://doi.org/10.20884/1.oa.2023.19.1.731

Saha S, Hasan S, Haque M, Ahamed T (2021) Perception based assessment of ecosystem services of madhupur sal forest in bangladesh. Eur J Agric Food Sci 3(1):39–44. https://doi.org/10.24018/ejfood.2021.3.1.194

Salkeld D, Padgett K, Jones J (2013) A meta-analysis suggesting that the relationship between biodiversity and risk of zoonotic pathogen transmission is idiosyncratic. Ecol Lett 16(5):679–686. https://doi.org/10.1111/ele.12101

Schipper A, Hilbers J, Meijer J, Antão L, Benítez-López A, Jonge M, … Huijbregts M (2019) Projecting terrestrial biodiversity intactness with globio 4. Global Change Biol 26(2):760–771. https://doi.org/10.1111/gcb.14848

Smith I, Wang L (2013) Bats and their virome: an important source of emerging viruses capable of infecting humans. Curr Opin Virol 3(1):84–91. https://doi.org/10.1016/j.coviro.2012.11.006

Sun X, Liddicoat C, Tiunov A, Wang B, Zhang Y, Lu C, … Zhu Y (2023) Harnessing soil biodiversity to promote human health in cities. NPJ Urban Sustain 3(1). https://doi.org/10.1038/s42949-023-00086-0

Teh L, Teh L, Sumaila U (2013) A global estimate of the number of coral reef fishers. PLoS ONE 8(6):e65397. https://doi.org/10.1371/journal.pone.0065397

Vipat A, Bharucha E (2014) Sacred groves: the consequence of traditional management. J Anthropol 2014:1–8. https://doi.org/10.1155/2014/595314

Ward D, Melbourne-Thomas J, Pecl G, Evans K, Green M, McCormack P, … Layton C (2022) Safeguarding marine life: conservation of biodiversity and ecosystems. Rev Fish Biol Fish32(1):65–100. https://doi.org/10.1007/s11160-022-09700-3

Worm B, Hilborn R, Baum JK, Branch TA, Collie JS, Costello C, Fogarty MJ et al (2009) Rebuilding global fisheries. Science 325:578–585

Young J, Waylen K, Sarkki S, Albon S, Bainbridge I, Balian E, ... Watt A (2014) Improving the science-policy dialogue to meet the challenges of biodiversity conservation: having conversations rather than talking at one-another. Biodivers Conserv 23(2):387–404. https://doi.org/10.1007/s10531-013-0607-0

Zedler JB, Kercher S (2005) Wetland resources: status, trends, ecosystem services, and restorability. Annu Rev Environ Resour 30(1):39–74. https://doi.org/10.1146/annurev.energy.30.050504.144248

Zhang Y, Wang Z, Lü Y, Zuo L (2022) Editorial: biodiversity, ecosystem functions and services: interrelationship with environmental and human health. Front Ecol Evol 10. https://doi.org/10.3389/fevo.2022.1086408

Zurba M, Beazley K, English E, Buchmann-Duck J (2019) Indigenous protected and conserved areas (ipcas), aichi target 11 and canada's pathway to target 1: focusing conservation on reconciliation. Land 8(1):10. https://doi.org/10.3390/land8010010

Chapter 6
Waste Management and Health: Addressing the Processes Behind Hazardous Waste and Pollution

Sylvester Chibueze Izah, Matthew Chidozie Ogwu, and Milan Hait

Abstract Managing hazardous waste is a critical aspect of public health and environmental Sustainability. This chapter explores the processes behind dangerous waste generation, disposal, and its associated pollution, focusing on the impact on health and ecosystems. The classification of hazardous waste, including chemical, biological, and radioactive materials, highlights the complexities of addressing waste from industrial, agricultural, and urban sources. Improper waste disposal leads to air, water, and soil quality degradation, resulting in short-term and long-term health effects. Regulatory frameworks play a vital role in waste management, but challenges such as inadequate infrastructure and socioeconomic barriers hinder their effectiveness. The chapter also stresses the need for sustainable waste management practices, including adopting circular economy approaches, innovative technologies, and community engagement to mitigate the impacts of hazardous waste. Collaborative efforts from policymakers, researchers, and local communities are essential for overcoming waste management challenges. Prioritizing public health, environmental justice, and technological innovation, will help societies promote healthier environments and ensure that waste management systems are practical and equitable.

Keywords Hazardous waste · Pollution · Waste management · Public health · Sustainability · Disposal · Environmental health · Regulatory frameworks

S. Chibueze Izah (✉)
Department of Community Medicine, Faculty of Clinical Sciences, Bayelsa Medical University, Yenagoa, Bayelsa State, Nigeria
e-mail: sylvester.izah@bmu.edu.ng

Department of Microbiology, Faculty of Science, Bayelsa Medical University, Yenagoa, Bayelsa State, Nigeria

M. C. Ogwu
Goodnight Family Department of Sustainable Development, Appalachian State University, 212 Living Learning Center, 305 Bodenheimer Drive, Boone, NC 28608, USA
e-mail: ogwumc@appstate.edu

M. Hait
Department of Chemistry, Dr. C. V. Raman University, Bilaspur, Chhattisgarh, India

S. Chibueze Izah and M. Chidozie Ogwu (eds.), *Innovative Approaches in Environmental Health Management*, Environmental Science and Engineering,
https://doi.org/10.1007/978-3-031-81966-7_6

6.1 Introduction

The intricate relationship between waste management and public health has garnered significant attention in recent years, particularly as urbanization and industrialization continue to escalate globally. Effective waste management systems are essential not only for maintaining environmental integrity but also for safeguarding public health. Poor waste management practices can lead to the proliferation of hazardous materials, directly threatening human health and the environment. The link between waste management and public health is well-documented in the literature. Guerrero et al. (2012) highlighted that inadequate waste management systems in developing countries can exacerbate health issues, including respiratory infections and vector-borne diseases, due to the accumulation of waste in urban areas. Furthermore, Tang et al. (2020) provided empirical evidence indicating that stringent environmental regulations can significantly reduce public health costs associated with air and water pollution, thereby underscoring the importance of effective waste management practices. The systematic review by Fazzo et al. (2017) emphasized the causal relationships between hazardous waste exposure and various health outcomes, including cancers and congenital anomalies, reinforcing the need for robust waste management strategies.

The health impacts of poor waste management are particularly pronounced in low-income regions, where infrastructure for waste disposal is often lacking. Ngwuluka et al. (2011) discussed how pharmaceutical waste management practices in Nigeria have been integrated into broader health and safety protocols, demonstrating a proactive approach to mitigating health risks associated with hazardous waste. Similarly, the study by Zangina and Ali (2021) revealed that indiscriminate dumping of hazardous waste in Nigeria poses significant public health risks, necessitating urgent reforms in waste management policies. These findings highlight the critical need for comprehensive waste management systems to protect public health. Addressing hazardous waste is paramount for public health, as the improper disposal of such materials can lead to severe health consequences.

Moreover, the research conducted by Li et al. (2021) illustrated that industrial activities, particularly in developing nations, often result in significant environmental pollution, which in turn adversely affects public health. This relationship is further corroborated by the findings of Wu and Pu (2020), who argued that air pollution, exacerbated by inadequate waste management, leads to increased public health expenditures and worsens health outcomes (Wu and Pu 2020). The implications of hazardous waste on health are not limited to immediate physical effects; they also involve long-term health burdens. For instance, the systematic review by Fazzo et al. (2017) indicated that exposure to hazardous waste is linked to various chronic health conditions, necessitating the implementation of effective public health policies to manage these risks. Furthermore, the study by Yang et al. (2023) highlighted that environmental regulations can significantly improve public health outcomes by reducing pollution levels, reinforcing regulatory frameworks' importance in waste management.

Environmental regulations are crucial in shaping waste management practices and public health outcomes. The empirical analysis by Tang et al. (2020) demonstrated that effective ecological regulations can significantly improve air quality and reduce pollution-related public health costs. This is particularly relevant in developing countries, where regulatory frameworks may be weak or poorly enforced. The findings of Greenstone and Hanna (2014) further emphasized that local pollutant concentrations in developing nations can have dire health implications, including increased infant mortality rates. Moreover, the research by Xiao-Chun et al. (2022) indicated that the interplay between economic growth and environmental pollution necessitates a balanced approach to waste management, wherein economic development does not come at the expense of public health. This perspective is echoed in the work of Di and Xiao (2023), who argued that fiscal expenditures aimed at energy conservation and environmental protection are essential for enhancing public health outcomes. Thus, establishing and enforcing robust environmental regulations are critical for effective waste management and public health safety. Industrialization has significantly influenced waste management practices and public health outcomes. Li et al. (2021) revealed that industrial agglomeration can indirectly affect public health through environmental pollution, highlighting the need for integrated waste management strategies considering industrial activities.

Additionally, the research by Jamin and Mahmood (2015) pointed out that the management of scheduled waste in Malaysia is heavily influenced by the level of industrial development and the effectiveness of existing waste management systems. As industries expand, hazardous waste generation will likely increase, necessitating more stringent management practices. The findings of Wang et al. (2023) suggested that implementing life-cycle management concepts in hazardous waste management can enhance the Sustainability of industrial practices and mitigate health risks. This approach aligns with the broader goal of promoting environmental Sustainability while safeguarding public health.

The chapter focuses on the classification and sources of hazardous waste, including chemical, biological, and radioactive materials. It explores waste generation and disposal methods across industrial, agricultural, and urban settings. It also examines the health impacts of hazardous waste and pollution on humans and the environment alongside the regulatory frameworks governing waste management. Additionally, the chapter discusses sustainable waste management practices, addresses challenges, and offers recommendations for future innovations and policy development.

6.2 Types of Hazardous Waste

Classifying hazardous waste is crucial for effectively managing and mitigating its impacts on human health and the environment. Hazardous waste can be broadly categorized into several types, including chemical, biological, and radioactive waste (Table 6.1). Chemical waste involves a variety of substances that can be toxic, ignitable, corrosive, or reactive, posing significant risks if not managed properly. Biological waste, often generated from healthcare facilities, includes materials containing pathogens or infectious agents, necessitating stringent disposal protocols to prevent public health risks (Bedada et al. 2019; Derso et al. 2018). Radioactive waste from nuclear power generation and medical applications requires specialized handling and disposal due to its long-lived hazardous nature (Zangina and Ali 2022).

Each type of hazardous waste has distinct properties and requires tailored management strategies to minimize environmental and health impacts. Various industrial processes generate hazardous waste, including manufacturing, chemical production, and waste treatment. Industries such as pharmaceuticals, petrochemicals, and metal processing significantly contribute to hazardous waste generation, often producing large volumes of toxic byproducts that require careful management (Zakianis et al. 2023; Akpan and Olukanni 2020). For instance, the manufacturing sector in Spain has been identified as a significant source of hazardous waste, particularly in base-metal industries (Zakianis et al. 2023). Additionally, the improper disposal of hazardous waste from industrial activities can lead to severe environmental degradation and health risks for nearby communities, highlighting the need for effective regulatory frameworks and waste management practices (Basu and Chakraborty 2016; Gutberlet and Uddin 2017). Agricultural activities also generate hazardous waste, primarily through pesticides, fertilizers, and veterinary pharmaceuticals. These substances can contaminate soil and water resources, adversely affecting ecosystems and human health (Chen and Ogunseitan 2021). The accumulation of hazardous waste in agricultural settings often stems from inadequate disposal practices and a lack of awareness regarding the potential risks associated with these materials (Budiman 2023; Akpan and Olukanni 2020). Furthermore, the rise of urban agriculture has introduced new challenges, as urban environments may lack the infrastructure to manage hazardous agricultural waste effectively (Bedada et al. 2019).

Urban settings are increasingly recognized as significant sources of hazardous waste, mainly through the generation of electronic waste (e-waste) and municipal solid waste (MSW). E-waste, which includes discarded electronic devices, is one of the fastest-growing categories of hazardous waste globally, posing severe environmental and health risks due to the presence of toxic materials such as lead, mercury, and cadmium (Chen and Ogunseitan 2021; Ahmad et al. 2020). In urban areas, improper e-waste disposal can contaminate soil and water, affecting human populations and wildlife (Chen and Ogunseitan 2021). Moreover, urban MSW often contains hazardous components, such as batteries and chemicals, which can exacerbate environmental pollution if not managed appropriately (Nkuh et al. 2023). The management of hazardous waste is further complicated by socioeconomic factors, particularly in low- and middle-income countries where waste management systems may be underdeveloped. In many urban areas, inadequate infrastructure and limited resources hinder effective hazardous waste management, increasing health risks for vulnerable populations, including waste pickers and nearby communities (Gutberlet and Uddin 2017; Derso et al. 2018). For instance, in Ethiopia, studies have shown that a significant portion of healthcare waste is improperly disposed of, posing risks to public health and the environment (Bedada et al. 2019; Derso et al. 2018). This

Table 6.1 Types of hazardous waste

Type of hazardous waste	Classification	Constituents	Sources	Health risks	Management strategies	References
Chemical waste	Chemical	Waste containing harmful chemicals can pose risks to human health and the environment	Industrial manufacturing, laboratories, and pharmaceuticals	Respiratory issues, skin irritation, carcinogenic effects	Proper labeling, secure storage, and regulated disposal	Padmanabhan and Barik (2019)
Biological waste	Biological	Waste that contains or is derived from living organisms	Healthcare facilities, research laboratories, and veterinary clinics	Infectious diseases, allergic reactions	Autoclaving, incineration, and secure containment	Jacob et al. (2021)
Radioactive waste	Radioactive	Waste containing radioactive materials, which can emit radiation	Nuclear power plants, medical facilities, and research institutions	Cancer, genetic mutations, acute radiation syndrome	Containment, shielding, and specialized disposal sites	Folkers and Gunter (2022)
Electronic waste (E-Waste)	Electronic	Discarded electronic devices containing hazardous materials	Consumer electronics, I.T. equipment, and other appliances	Heavy metal exposure (lead, mercury), chemical hazards	Recycling, refurbishment, and proper disposal methods	Parvez et al. (2021)
Heavy metal waste	Chemical	Waste containing heavy metals like lead, cadmium, and arsenic	Battery manufacturing, metal plating , and mining activities	Neurological damage, kidney damage, and developmental issues	Specialized recycling and detoxification processes	Tchounnwou et al. (2012), Ikhajiagbe and Ogwu (2020)
Industrial waste	Chemical/physical	Waste generated from industrial processes can be hazardous	Factories, chemical plants, and oil refineries	Various health risks depend on the specific waste components	Waste minimization, treatment, and safe disposal practices	Kumar et al. (2023)

(continued)

Table 6.1 (continued)

Type of hazardous waste	Classification	Constituents	Sources	Health risks	Management strategies	References
Agricultural waste	Biological/chemical	Waste is generated from agricultural activities, including pesticides and fertilizers	Farms and food processing facilities	Chemical exposure, groundwater contamination, pest resistance	Integrated Pest Management (IPM), composting, and safe disposal	Pathak et al. (2022)
Construction and demolition waste	Physical	Waste generated during construction, renovation, or demolition projects	Construction sites and renovation projects	Exposure to hazardous materials (asbestos, lead)	Recycling materials, proper waste segregation, and safe disposal	Purchase et al. (2021)
Asbestos waste	Physical	Waste containing asbestos fibers, which are hazardous when disturbed	Construction sites and renovation projects	Lung diseases, including asbestosis and mesothelioma	Specialized removal and disposal procedures	Emmett (2021)

highlights the urgent need for improved waste management practices and community education to mitigate the impacts of hazardous waste in urban settings. Environmental justice is critical in hazardous waste management, as marginalized communities often bear disproportionate exposure to hazardous materials. Studies have indicated that industrial facilities generating hazardous waste are frequently located in urbanized districts with higher populations of disadvantaged groups, exacerbating existing inequalities (Basu and Chakraborty 2016; Laborde et al. 2015). This spatial distribution of hazardous waste sources underscores the importance of equitable waste management policies prioritizing the health and safety of all communities, particularly those historically affected by environmental injustices (Basu and Chakraborty 2016; Laborde et al. 2015).

The health impacts of hazardous waste exposure are well-documented, with numerous studies linking exposure to various health outcomes, including cancers, respiratory diseases, and reproductive health issues (Fazzo et al. 2017; Laborde et al. 2015). For example, exposure to endocrine-disrupting chemicals found in hazardous waste sites has been associated with increased incidences of breast and testicular cancers (Fazzo et al. 2017). Furthermore, hazardous waste in urban environments can lead to chronic health conditions among residents, particularly children, who are more susceptible to environmental toxins (Laborde et al. 2015). This necessitates a comprehensive approach to hazardous waste management, including public health considerations and community engagement. Innovative strategies for hazardous waste management are emerging, focusing on reducing waste generation and enhancing recycling efforts. For instance, industrial symbiosis promotes the exchange of byproducts between industries, reducing the overall volume of hazardous waste generated (Ažman Momirski et al. 2021). Additionally, advancements in waste treatment technologies, such as incineration and bioremediation, offer promising solutions for managing hazardous waste in an environmentally sound manner (Šomplák et al. 2022; Jamin and Mahmood 2015). However, successfully implementing these strategies requires collaboration among stakeholders, including government agencies, industries, and local communities, to ensure sustainable waste management practices.

6.3 Waste Generation and Disposal Processes

Waste generation and disposal processes are critical for environmental management and public health. The methods of waste generation vary widely, influenced by factors such as urbanization, population growth, and economic activities. In urban settings, domestic waste is predominantly generated from households, which includes organic waste, plastics, and other materials. According to research, improper disposal practices, such as open dumping and burning, are prevalent in many developing countries, leading to significant environmental and health risks (Yoada et al. 2014; Lema et al. 2019; Frempong 2013). Table 6.2 shows the overview of waste generation and disposal processes. The philosophy of "out of sight, out of mind" continues to dominate waste disposal attitudes, resulting in practices that contravene established waste management regulations and policies to maintain a clean environment (Yoada et al. 2014).

In the Global South, waste management practices often involve non-formal and unsustainable practices with human and environmental health impacts (Fig. 6.1). Current disposal practices involve various methods, including landfilling, incineration, recycling, and composting. However, the effectiveness of these methods is often compromised by inadequate infrastructure and public awareness. For instance, in many urban areas, waste collection systems are inefficient, accumulating waste in public spaces (Muiruri et al. 2020; Ololade et al. 2019). Biomedical waste management is particularly critical in healthcare settings, as improper handling can lead to severe health risks. Facilities implementing best practices, such as waste segregation and treatment, are committed to minimizing these risks (Nyametso 2023; Niyongabo et al. 2019).

Nevertheless, many healthcare facilities still struggle with proper disposal, often co-disposing medical waste with general waste, exacerbating public health concerns

Table 6.2 Waste generation and disposal processes

Aspect	Sources	Methods of waste generation	Current disposal practices	Impacts of improper disposal	References
Municipal Solid Waste (MSW)	Waste from households, businesses, and institutions	Food waste, packaging, yard waste	Landfilling, recycling, composting	Landfill overflow, soil, and water contamination	Abubakar et al. (2022)
Hazardous waste	Waste from diverse sources poses health and environmental risks	Chemicals, batteries, electronics	Hazardous landfills, incineration	Toxic exposure, air and water pollution	Fazzo et al. (2017)
Industrial waste	Waste from industrial manufacturing and processing	Scrap metal, chemicals, packaging	Recycling, incineration, landfilling	Environmental degradation, worker health risks	Siddiqua et al. (2022)
Construction and demolition waste	Waste from construction and demolition activities	Concrete, wood, asphalt	Recycling, landfilling, reuse	Habitat destruction, dust, and particulate pollution	Elshaboury et al. (2022)
Agricultural Waste	Waste from crop production and livestock	Crop residues, manure, pesticides	Composting, anaerobic digestion, landfilling	Soil contamination, water pollution from runoff	Burkholder et al. (2007)
E-Waste	Discarded electronics with hazardous materials	Computers, mobile phones, appliances	Recycling, specialized e-waste facilities	Heavy metal contamination, toxic chemical exposure	Pinto (2008)

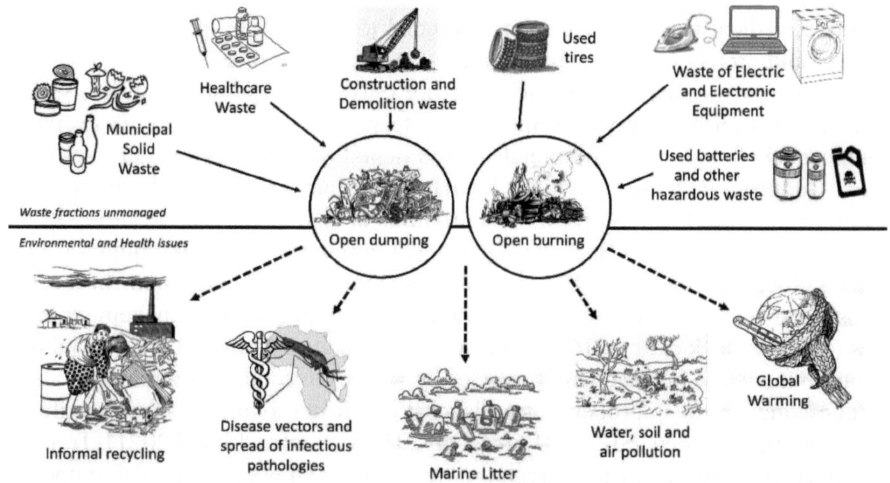

Fig. 6.1 Waste management practices in the global south and their environmental health impacts. *Source* Ferronato and Toretta (2019). http://creativecommons.org/licenses/by/4.0/

(Idowu et al. 2013). The impact of improper waste disposal on environmental health is profound. Research indicates that inadequate waste management contributes to air, soil, and water pollution, leading to the proliferation of vector-borne diseases such as cholera and malaria (Omang et al. 2021; Mohamed et al. 2023). Stagnant water from clogged drainage systems creates breeding grounds for mosquitoes, further exacerbating public health issues (Omang et al. 2021; Mohamed et al. 2023). Additionally, leaching hazardous substances from improperly managed landfills poses significant risks to groundwater quality, affecting human health and ecosystem integrity (Ololade et al. 2019). The consequences of these practices are particularly severe in low-income regions, where the lack of resources and infrastructure compounds the challenges of effective waste management (Aus der Beek et al. 2016).

The generation of e-waste presents another layer of complexity in waste disposal practices. The improper disposal of electronic waste can lead to environmental contamination by releasing toxic substances such as heavy metals and persistent organic pollutants (Nath et al. 2018). Awareness regarding the importance of recycling e-waste is still low in many communities, which hinders efforts to mitigate its environmental impact (Nath et al. 2018). Furthermore, while providing economic opportunities, the informal recycling sector often lacks the necessary safety measures, exposing workers and the surrounding community to hazardous materials (Nath et al. 2018). In the context of healthcare waste, the COVID-19 pandemic has exacerbated existing challenges. The surge in medical waste, particularly from personal protective equipment (PPE), has overwhelmed waste management systems in many countries (Singh et al. 2022; Husaini et al. 2023). The lack of adequate disposal facilities and the increased volume of waste have raised concerns about the potential for environmental contamination and public health crises (Singh et al. 2022; Husaini et al. 2023).

Effective medical waste management is crucial to prevent the spread of infectious diseases and protect environmental quality (Cheng et al. 2014).

Improper waste disposal practices are not only a local issue but also a global concern. The transboundary movement of waste, particularly hazardous waste, poses significant challenges to international environmental governance (Aus der Beek et al. 2016). Countries with lax regulations often become dumping grounds for waste from developed nations, leading to severe environmental degradation and health risks for local populations (Aus der Beek et al. 2016). Addressing these issues requires a coordinated global response, emphasizing stricter regulations and enforcement mechanisms to prevent illegal dumping and promote sustainable waste management practices (Aus der Beek et al. 2016). Public awareness and education play a crucial role in improving waste management practices. Initiatives to educate communities about the consequences of improper waste disposal can foster a culture of responsibility and environmental stewardship (Ouma et al. 2021; Akmal and Jamil 2021). Training programs focusing on waste segregation, recycling, and composting can empower individuals and businesses to adopt sustainable practices (Ouma 2021; Akmal and Jamil 2021).

Moreover, engaging local communities in waste management initiatives can enhance participation and compliance with waste disposal regulations (Ouma et al. 2021; Akmal and Jamil 2021). The role of technology in waste management cannot be overstated. Innovations such as waste-to-energy technologies, advanced recycling methods, and smart waste management systems offer promising solutions to the challenges of waste disposal (Tessema and Fantahun 2023; Hidayat et al. 2023). These technologies can enhance the efficiency of waste collection and processing efficiency, reduce waste's environmental impact, and promote resource recovery (Tessema and Fantahun 2023; Hidayat et al. 2023). However, successfully implementing such technologies requires investment, infrastructure development, and public acceptance (Tessema and Fantahun 2023; Hidayat et al. 2023).

6.4 Health Impacts of Hazardous Waste and Pollution

The health impacts of hazardous waste and pollution are multifaceted, affecting both short-term and long-term human health outcomes (Table 6.3). Unsafe waste, which includes various toxic materials generated from industrial, medical, and municipal sources, poses significant risks to human health through direct exposure and environmental contamination. Studies have shown that individuals living near hazardous waste sites are at a higher risk for various health issues, including respiratory diseases, neurological disorders, and even cancer (Fazzo et al. 2017). The systematic review conducted by Fazzo et al. (2017) highlighted the correlation between proximity to hazardous waste and adverse health outcomes, indicating that populations exposed to such environments experience a range of diseases that can manifest acutely and chronically.

Table 6.3 Health impacts of hazardous waste and pollution

Impact area	Sources	Short-term health Effects	Long-term health effects	Impact on environmental quality	References
Air quality	Atmospheric pollution from hazardous waste	Respiratory issues and eye irritation	Chronic respiratory diseases, cardiovascular issues	Reduced air quality, higher greenhouse gas emissions	Manisalidis et al. (2020), Ogwu and Izah (2024)
Water quality	Contamination of water bodies by waste runoff	Gastrointestinal issues and skin rashes	Heavy metal exposure leads to kidney damage, cancer	Polluted drinking water, loss of aquatic biodiversity	Iyiola et al. (2024), Izah et al. 2024a, b, c, d
Soil quality	Soil degradation due to hazardous waste accumulation	Acute poisoning and skin irritation	Food safety risks, long-term soil fertility loss	Decreased soil health and bioaccumulation in crops	Steffan et al. (2018)
Community health	Health impacts on communities near waste sites	Increased hospital visits and allergic reactions	Higher chronic disease rates and increased mortality	Declining living conditions and reduced life quality	Izah et al. (2022)
Vulnerable populations	Effects on sensitive groups (children, elderly, etc.)	Frequent respiratory infections, allergies	Developmental disorders, compromised immune systems	Higher pollutant exposure, environmental injustice	Mathiarasan and Huls (2021)

The short-term health effects of exposure to hazardous waste can include acute respiratory issues, skin irritations, and gastrointestinal problems, often resulting from direct contact with toxic substances or inhalation of contaminated air (Ibrahim et al. 2021; Izah et al. 2024e, f, g, h). For instance, the illegal dumping of toxic waste has been linked to immediate health crises, particularly among vulnerable populations such as children, who are more susceptible to the harmful effects of environmental toxins (Ibrahim et al. 2021). In contrast, long-term exposure to hazardous waste can lead to chronic health conditions, including developmental disorders, reproductive issues, and increased mortality rates due to cancers associated with specific hazardous materials (Fazzo et al. 2017; Ibrahim et al. 2021). The cumulative effects of these exposures underscore the urgent need for effective dangerous waste management strategies to mitigate health risks. The impact of hazardous waste extends beyond immediate health concerns, significantly affecting air, water, and soil quality. Air pollution from dangerous waste incineration or improper disposal can release particulate matter and toxic gases, contributing to respiratory diseases and cardiovascular problems (Sharma et al. 2013). Baxter et al. (2017) emphasized the relationship between air quality and health outcomes, noting that cities with higher levels of particulate matter (PM2.5) have increased mortality rates associated with respiratory and cardiovascular diseases. Furthermore, hazardous waste can contaminate water sources through leachate, the liquid that drains or 'leaches' from a landfill, carrying harmful chemicals into groundwater and surface water systems (Pandey et al. 2017). This contamination poses significant risks to human health, as polluted water can lead to various waterborne diseases and long-term health issues related to chemical exposure.

Hazardous waste also adversely affects soil quality, as toxic substances can persist in the environment, leading to bioaccumulation in the food chain (Pandey et al. 2017). Contaminated soil can affect agricultural productivity and food safety, posing risks to human health through consuming contaminated crops (Pandey et al. 2017). Moreover, the degradation of soil quality can have broader ecological impacts, disrupting local ecosystems and reducing biodiversity (Hong et al. 2022). The interconnectedness of air, water, and soil pollution highlights the need for comprehensive environmental health assessments that consider the cumulative effects of hazardous waste on human health and the environment. Managing hazardous waste is crucial for protecting public health and the environment. Effective waste management systems can significantly reduce the risks associated with hazardous waste, as demonstrated by various case studies (Aja et al. 2016; Mmereki et al. 2016). For example, countries that have implemented stringent regulations and advanced waste management technologies have seen improvements in environmental quality and public health outcomes (Aja et al. 2016). However, many developing countries still struggle with inadequate waste management practices, leading to increased health risks for their populations (Mmereki et al. 2016; NAVARRO et al. 2022). The relationship between industrial development and hazardous waste generation underscores the need for sustainable practices that prioritize human health and environmental protection. In addition to regulatory measures, community awareness, and education play a vital role in hazardous waste management. Public understanding of the risks associated with hazardous waste can drive demand for better management practices and encourage community involvement in monitoring local waste disposal activities (Baby and Mathew 2020; SEBASTIAN 2023). Programs to educate the public about proper waste disposal and the dangers of hazardous materials can empower communities to advocate for safer practices and hold polluters accountable (Baby and Mathew 2020).

Furthermore, engaging local populations in waste management initiatives can foster a sense of ownership and responsibility towards environmental health, ultimately leading to improved public health outcomes. The health impacts of hazardous waste and pollution are not uniformly distributed, often disproportionately affecting marginalized communities. Environmental justice issues arise when low-income and minority populations bear the brunt of dangerous waste exposure due to their proximity to industrial sites and landfills (Basu and Chakraborty 2016). This inequity highlights the need for targeted interventions that address the specific vulnerabilities of these communities, ensuring that all populations have equal protection from the health risks associated with hazardous waste (Basu and Chakraborty 2016). Policymakers must prioritize environmental justice in managing hazardous waste, recognizing that equitable access to clean air, water, and soil is a fundamental human right.

6.5 Regulatory Frameworks and Policies for Waste Management

The regulatory frameworks and policies governing hazardous waste management are critical for public health and environmental protection (Table 6.4). These frameworks can be categorized into international and national regulations, providing guidelines for safely handling, treating, and disposing of hazardous waste. Internationally, agreements such as the Basel Convention aim to control the transboundary movement of hazardous waste and ensure its environmentally sound management. The Convention emphasizes the need for countries to develop national legislation that aligns with its principles, thereby promoting a cohesive approach to hazardous waste management across borders (Fikru 2012). On the other hand, national regulations vary significantly in their scope and effectiveness, often reflecting a country's economic and infrastructural capabilities. For instance, in developing nations, regulatory frameworks may exist on paper but lack the necessary enforcement mechanisms, leading to widespread illegal dumping and inadequate waste treatment practices (Ferronato and Torretta 2019).

Various factors, including inadequate infrastructure, lack of public awareness, and insufficient funding for waste management initiatives, often challenge the effectiveness of these regulatory frameworks. In many developing countries, the absence of essential waste collection services and treatment facilities exacerbates the challenges of hazardous waste management (Subedi et al. 2023). For example, in Nepal,

Table 6.4 Regulatory Frameworks and Policies with Focus on Waste Management

Regulatory framework	Focus	Key regulations	Effectiveness	Challenges
International regulations	Global agreements for hazardous waste management	Basel, Stockholm, and Minamata Conventions	Facilitates cooperation and sets standards	Compliance varies across countries
Basel convention	Controls hazardous waste movement and disposal globally	Ratified by 187 countries, it focuses on waste reduction	Raises awareness and improves reporting	Enforcement and resource limitations in developing countries
Stockholm convention	Targets elimination of persistent organic pollutants (POPs)	Focuses on restricting specific hazardous chemicals	Promotes safer alternatives and protects health	Implementation and monitoring inconsistencies
National regulations	Country-specific hazardous waste management laws	RCRA (USA), Waste Management Act (U.K.)	Provides national waste management framework	Inconsistent enforcement and regulatory overlap
Resource Conservation and Recovery Act (RCRA)	U.S. law for solid and hazardous waste disposal	Sets standards for waste treatment and disposal	Reduces waste and encourages recycling	Monitoring and compliance resource constraints
Waste management policies	Local guidelines for sustainable waste management	National Waste Management Strategy (varies by country)	Encourages local solutions and community involvement	Public awareness and funding limitations
Effectiveness of policies	Assesses regulatory impact on waste reduction	Decreases in waste and pollution levels measure it	Some policies show positive health and environmental effects	Limited data for comprehensive assessment

the informal waste management sector operates without regulation, posing significant risks to public health and the environment (Subedi et al. 2023). Similarly, Nigeria's lack of a comprehensive national healthcare waste management policy has been identified as a barrier to effective hospital waste management practices (Ezirim and Agbo 2018). These challenges highlight the need for governments to invest in waste management infrastructure and develop policies that promote public–private partnerships, which can enhance the capacity for effective waste management (Ogutu et al. 2021). Implementing waste management policies effectively requires a multifaceted approach that involves various stakeholders, including government agencies, waste generators, and the community. The involvement of all stakeholders is crucial for developing and delivering effective hazardous waste management programs (Mmereki et al. 2016). For instance, a decentralized approach to waste management has been proposed in Lebanon to empower local authorities and improve governance in waste management practices (Massoud et al. 2019). However, implementing such policies often faces hurdles, such as political resistance, lack of technical expertise, and insufficient funding (Massoud et al. 2019). Moreover, the complexity of hazardous waste streams, which can include various materials from different sources, necessitates tailored approaches that consider local contexts and capabilities (Muzanni et al. 2022).

The challenges in implementing waste management policies are further compounded by the socioeconomic conditions prevalent in many regions. In Ghana, for example, despite governmental initiatives to empower local authorities in waste management, challenges such as inadequate infrastructure and insufficient operational funds persist (Lissah et al. 2021). This scenario is echoed in other developing countries where financial resources hinder the establishment of effective waste management systems (Ferronato and Torretta 2019). Furthermore, the informal sector often plays a significant role in waste management in these regions. Still, its unregulated nature can lead to unsafe practices that endanger public health and the environment (Subedi et al. 2023). Therefore, addressing these challenges requires a comprehensive understanding of the socioeconomic landscape and developing policies adaptable to local realities. In addition to infrastructural and financial challenges, public awareness and community engagement are critical components of successful waste management policies. Programs that promote community participation in waste management, such as waste bank initiatives, have shown promise in improving waste segregation and recycling rates (Khotimah et al. 2023). However, sustaining community involvement can be challenging, particularly in areas lacking education about the importance of proper waste management (Khotimah et al. 2023). Educational campaigns informing the public about hazardous waste's health and environmental impacts can foster greater community engagement and support for waste management initiatives (Asih et al. 2022). Moreover, integrating waste management education into school curricula can cultivate a culture of environmental stewardship among future generations (Muzanni et al. 2022).

The role of technology in enhancing hazardous waste management cannot be understated. Innovations in waste treatment technologies, such as advanced incineration methods and recycling processes, can significantly improve the efficiency and effectiveness of waste management systems (Ji et al. 2016). For instance, the adoption of life-cycle management concepts in hazardous waste management can help enterprises in the oil and gas sector to assess better and manage their waste streams (Wang et al. 2023). However, implementing such technologies often requires significant investment and technical expertise, which may be lacking in many developing countries (Mmereki et al. 2014). Therefore, fostering partnerships between governments, private sector entities, and research institutions can facilitate the transfer of technology and knowledge necessary for improving waste management practices. The global context of hazardous waste management is also influenced by transboundary waste movements, which can complicate national regulatory efforts. The Basel Convention provides a framework for regulating the international movement of hazardous waste, but enforcement remains a challenge, particularly in regions with weak regulatory systems (Fikru 2012). The movement of hazardous waste across borders can lead to environmental injustices, where wealthier nations export their waste to poorer countries, exacerbating existing ecological and health issues (Fikru 2012). Addressing these challenges requires international cooperation and a commitment to ensuring that all countries adhere to environmentally sound waste management principles.

6.6 Sustainable Waste Management Practices

Sustainable waste management practices are increasingly recognized as essential for addressing the growing environmental challenges of waste generation. The integration of innovative technologies plays a vital role in enhancing waste reduction and recycling efforts. For instance, Abera (2023) discussed various recycling systems, including source separation and on-site processing, which can significantly improve recycling procedures and contribute to sustainable development goals by reducing landfill utilization and greenhouse gas emissions. Table 6.5 provides an overview of sustainable waste management practices. Moreover, digital tracking technologies can optimize recycling, ensuring that materials are efficiently recovered and reused, thereby minimizing waste generation (Abera 2023). As highlighted by Swanson and Ferrari (2023), the importance of community-based interventions further emphasizes the need for long-term behavioral changes in recycling practices to ensure Sustainability.

The concept of a circular economy is integral to effective waste management, particularly in the context of hazardous waste. In a circular economy, the focus shifts from merely managing waste to creating systems prioritizing reuse, recycling, and recovery of materials (Moyo et al. 2022). This approach reduces waste and mitigates the environmental impacts associated with hazardous materials. Grabowski (2021) elaborates on how circular economy principles can transform waste management by closing the loop on resource use, thereby promoting sustainable practices that minimize waste generation. The challenges of implementing such systems are

Table 6.5 Sustainable waste management practices

Practice	Focus	Innovative technologies	Circular economy approaches	Benefits	References
Waste reduction	Minimizing waste at the source	Lean manufacturing, and eco-design	Product life extension, resource recovery	Lowers resource use and disposal costs	Gonzalez-Sanchez et al. (2023)
Recycling technologies	Advanced material recovery from waste	Automated sorting, pyrolysis, and bioremediation	Closed-loop recycling, upcycling	Conserves resources, and reduces landfill dependency	Kang et al. (2023)
Composting	Converting organic waste into soil amendments	In-vessel composting and anaerobic digestion	Composting services, community composting	It cuts methane emissions and enriches soil health	Policastro and Cesaro (2022)
Waste-to-Energy (WtE)	Transforming waste into energy	Incineration and gasification	Energy recovery, biogas production	Reduces waste volume and generates renewable energy	Kalair et al. (2021), Nizzy and Ogwu (2024)
Circular economy models	Systems minimizing waste via reuse and recycling	Design for disassembly, product-as-a-service	Take-back schemes, material exchange	Promotes Sustainability and reduces environmental impact	Milios (2018)
Smart waste management	Tech-driven waste collection optimization	IoT sensors, smart bins, and data analytics	Real-time tracking, predictive analytics	Boosts efficiency, and lowers operational costs	Addas et al. (2024)
Community-based initiatives	Local programs for sustainable waste practices	Educational campaigns and local recycling	Recycling cooperatives, repair cafes	Engages communities, and raises recycling rates	Abubakar et al. (2022)

significant, particularly in regions with less developed waste management infrastructures, as noted by Mihajlov et al. (Mihajlov et al. 2021). However, the potential benefits of adopting circular economy strategies are substantial, including reduced environmental pollution and enhanced resource efficiency. Innovative technologies are critical in advancing recycling and waste management practices. For instance, the integration of blockchain technology has emerged as a transformative force in waste management, offering secure and transparent transaction records that can facilitate recycling processes (Bułkowska et al. 2023). This technology can enhance accountability and traceability in waste management systems, improving efficiency and Sustainability (Akram et al. 2021a, b). Furthermore, the application of machine learning and the Internet of Things (IoT) in waste management can optimize recycling operations by providing real-time data and analytics, leading to better decision-making and resource allocation (Tanuja et al. 2023). These technological advancements not only improve the quality of recycled products but also contribute to significant reductions in energy consumption and waste generation (Tanuja et al. 2023).

The valorization of waste materials, mainly plastic waste, is another area where innovative technologies can significantly impact. Tang emphasizes the importance of incorporating plastic waste into construction materials as a sustainable solution to the ongoing plastic pollution crisis (Tang 2022). This approach reduces the volume of plastic waste and creates new economic opportunities within the construction industry. Additionally, enzymatic biocatalysis has been identified as an eco-friendly alternative for treating and recycling plastics, offering a promising avenue for waste valorization while minimizing environmental impacts (Zhu et al. 2022). Developing such innovative recycling technologies is crucial for addressing the challenges posed by plastic waste and achieving sustainability goals. In hazardous waste management, implementing effective policies and technologies is essential for minimizing risks to human health and the environment. Hennebert highlights the importance of waste characterization data in guiding the long-term management of hazardous waste, emphasizing a risk-based approach (Hennebert 2022). This approach establishes permissible concentration limits for hazardous materials, ensuring that waste is managed safely and effectively.

Furthermore, the integration of innovative recycling techniques for hazardous waste can significantly reduce environmental risks while promoting resource recovery (Abdelbasir et al. 2018). The need for comprehensive strategies that involve both technological innovation and regulatory frameworks is paramount for achieving sustainable hazardous waste management. The role of entrepreneurship in waste management cannot be overlooked, particularly in developing regions. As Dorcas et al. (2021) noted, innovative entrepreneurs in the waste recycling sector can drive economic growth and create employment opportunities while addressing waste challenges. By fostering a culture of innovation and resource optimization, these entrepreneurs can transform waste management practices and contribute to the development of sustainable communities. The promotion of sustainable business models that support the adoption of recycling technologies is essential for enhancing the effectiveness of waste management systems (Cui et al. 2022). This approach addresses waste challenges and contributes to broader economic and environmental goals. The integration of sustainable practices in waste management is also reflected in the concept of zero-waste manufacturing. Awogbemi et al. (2022) discussed various techniques for achieving zero waste through resource recycling, highlighting the benefits and challenges associated with these practices. Zero-waste manufacturing promotes a circular economy and minimizes environmental impacts by focusing on converting waste materials into valuable products. The adoption of such practices is crucial for achieving Sustainability in manufacturing processes and reducing the overall ecological footprint of industries.

6.7 Addressing Key Challenges in Waste Management

Addressing the key challenges in waste management requires a strategic approach that involves technological, economic, and social barriers as enumerated in Fig. 6.2. The increasing volume of municipal solid waste (MSW) generated globally, projected to reach 3.4 billion tons annually by 2050, highlights the urgency of effective waste management strategies (Cha et al. 2023). Integrating advanced technologies, such as Radio-Frequency Identification (RFID) and Big Data, has been recognized as vital in enhancing waste management systems by improving tracking, sorting, and recycling processes (Nielsen et al. 2010; Akram et al. 2021a, b). However, the successful implementation of these technologies is often hindered by economic constraints and a lack of institutional capacity, particularly in developing regions (Wismer and Gomez 2010; Shah et al. 2023). Technological barriers in waste management often stem from insufficient infrastructure and the high costs associated with adopting innovative solutions. For instance, while promising, deploying smart technologies in waste management requires significant investment in hardware and software systems (Ratnasabapathy et al. 2019).

Furthermore, the complexity of integrating these technologies into existing systems can pose challenges, particularly in regions where traditional waste management practices are deeply entrenched (Szpilko et al. 2023). Studies have shown that the effective use of technology, such as machine learning models for predicting waste generation rates, can significantly improve waste management outcomes (Cha et al. 2023). Still, these advancements are often underutilized due to waste management personnel lacking technical expertise and training (Bang and Andersen 2022).

Economic barriers also play a critical role in shaping waste management practices. The financial Sustainability of waste management systems is often compromised by inadequate funding and investment, particularly in low- and middle-income countries (Shah et al. 2023). The reliance on traditional waste disposal methods, such as landfilling, is environmentally detrimental and economically inefficient in the long run (Malinauskaitė et al. 2017). Transitioning to a circular economy model, which emphasizes recycling and resource recovery, requires substantial upfront investment but can yield significant economic benefits over time, including job

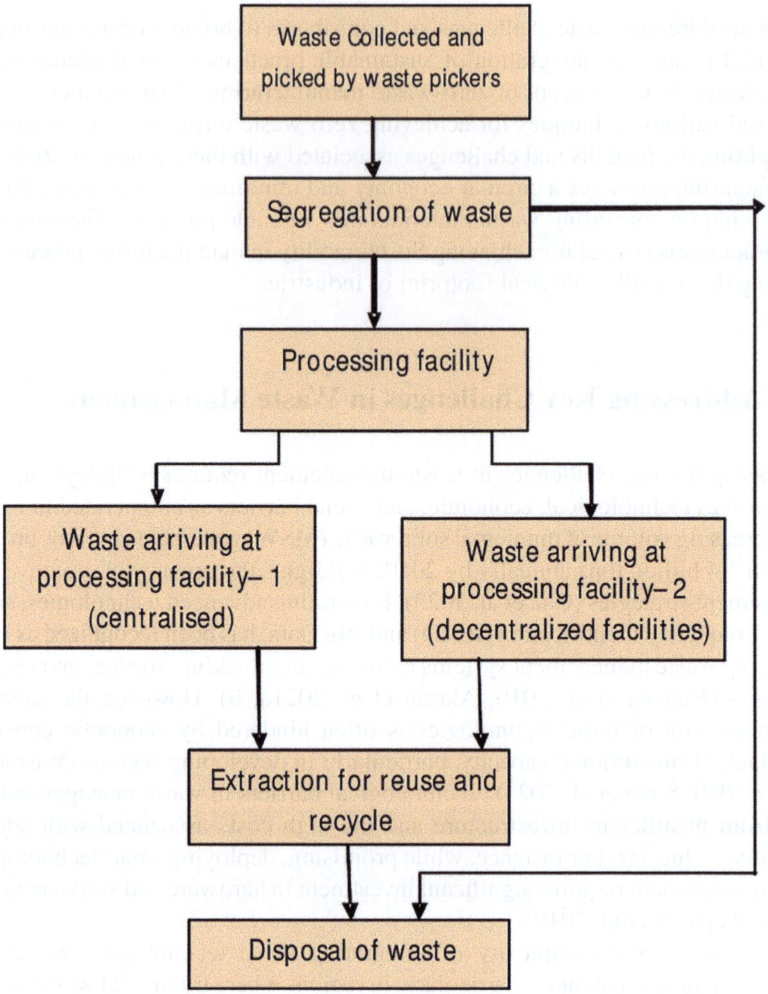

Fig. 6.2 Workflow for sustainable waste management. *Source* Gayathri et al. (2017)

creation and reduced environmental impact (Sundar et al. 2023). Moreover, the establishment of public–private partnerships can facilitate the mobilization of resources and expertise necessary for implementing innovative waste management solutions (Rodzi et al. 2023). Social barriers, including public perception and community engagement, are equally significant in waste management. The "Not In My Backyard" (NIMBY) phenomenon often leads to local opposition against waste management facilities, particularly those associated with waste-to-energy (WtE) processes (Malinauskaitė et al. 2017). Effective communication and community involvement are essential to overcoming these social barriers. Engaging residents in the waste management process through educational campaigns and participatory decision-making can foster a sense of ownership and responsibility towards waste management practices (Szpilko et al. 2023). Furthermore, understanding the attitudes and behaviors of specific demographics, such as college students, can inform targeted strategies to enhance participation in recycling and waste separation initiatives (Qu et al. 2023).

To address these multifaceted challenges, a comprehensive set of strategies is necessary. Firstly, enhancing institutional capacity at the local government level is crucial for effective waste management (Wismer and Gomez 2010). This includes training personnel, improving operational efficiency, and fostering collaboration among various stakeholders, including government agencies, private sector actors, and community organizations. Secondly, leveraging technological advancements can streamline waste management processes and improve data collection and analysis, leading to more informed decision-making (Ratnasabapathy et al. 2019; Akram et al. 2021a, b). For example, image-based analysis methods can enhance understanding of waste composition, essential for developing effective recycling strategies (Wagland et al. 2012).

Moreover, promoting economic incentives for waste reduction and recycling can drive behavioral change among consumers and businesses alike. Implementing policies that encourage waste minimization, such as pay-as-you-throw schemes or deposit return systems, can motivate individuals to adopt more sustainable waste practices (Rodzi et al. 2023). Fostering a corporate social responsibility (CSR) culture within businesses can lead to more sustainable supply chain practices, further contributing to waste reduction efforts (Nassani et al. 2023). The integration of green technologies into supply chains has shown promise in enhancing zero waste management initiatives, indicating a positive relationship between technological innovation and waste management outcomes (Nassani et al. 2023). Public awareness and education campaigns are also vital for addressing social barriers to effective waste management. By increasing knowledge about the environmental impacts of waste and the benefits of recycling, communities can be empowered to act (Szpilko et al. 2023). Educational programs targeting schools and universities can cultivate a culture of Sustainability among younger generations, ensuring long-term commitment to waste management practices (Qu et al. 2023). Furthermore, utilizing social media and digital platforms can enhance outreach efforts, making information about waste management more accessible to diverse audiences (Ratnasabapathy et al. 2019).

6.8 Conclusion

Effective waste management is critical for safeguarding both public health and the environment. Hazardous waste and pollution pose significant risks, harming air, water, soil quality, and human health. As urbanization and industrialization continue to expand, waste generation increases, necessitating robust and innovative solutions to mitigate these impacts. Proper classification, management, and disposal of hazardous waste are vital to reducing health risks, preventing environmental degradation, and ensuring sustainable development. A multifaceted approach that includes stringent regulatory frameworks, sustainable practices, and community engagement is essential to address the complex challenges posed by hazardous waste. Looking forward, innovations in hazardous waste management will play a crucial role in overcoming existing barriers. Technologies such as A.I. and predictive modeling can enhance waste tracking and disposal efficiency, while circular economy approaches emphasize waste reduction, recycling, and resource recovery. These strategies contribute to minimizing environmental impact and promoting Sustainability. Future waste management policies must focus on integrating these innovations into practice, with policymakers ensuring that regulatory frameworks support technological adoption, ecological justice, and equitable access to clean environments.

Moreover, stakeholders—policymakers, industries, communities, and researchers—are vital in shaping the future of waste management. Policymakers must enforce stringent regulations and provide incentives for sustainable waste practices, while industries should prioritize cleaner production methods and adopt waste-reduction technologies. Community engagement is equally important in raising awareness and fostering a culture of responsible waste management.

References

Abdelbasir SM, Hassan SS, Kamel AH, El-Nasr RS (2018) Status of electronic waste recycling techniques: a review. Environ Sci Pollut Res 25:16533–16547. https://doi.org/10.1007/s11356-018-2136-6

Abera Y (2023) Optimizing construction waste recycling: strategies, technologies, and environmental impacts. https://doi.org/10.21203/rs.3.rs-3497942/v1

Abubakar IR, Maniruzzaman KM, Dano UL, AlShihri FS, AlShammari MS, Ahmed SMS, Al-Gehlani WAG, Alrawaf TI (2022) Environmental sustainability impacts of solid waste management practices in the global South. Int J Environ Res Public Health 19(19):12717. https://doi.org/10.3390/ijerph191912717

Addas A, Khan MN, Naseer F (2024) Waste management 2.0 leveraging internet of things for an efficient and eco-friendly smart city solution. Plos one 19(7):e0307608. https://doi.org/10.1371/journal.pone.0307608

Ahmad S, Jamil F, Iqbal N, Kim D (2020) Optimal route recommendation for waste carrier vehicles for efficient waste collection: a step towards sustainable cities. IEEE Access 8:77875–77887. https://doi.org/10.1109/access.2020.2988173

Aja OC, Al-Kayiem HH, Zewge MG, Joo MS (2016) Overview of hazardous waste management status in Malaysia. Manag Hazardous Wastes (October). https://doi.org/10.5772/63682

Akmal T, Jamil F (2021) Testing the role of waste management and environmental quality on health indicators using structural equation modeling in Pakistan. Int J Environ Res Public Health 18(8):4193. https://doi.org/10.3390/ijerph18084193

Akpan VE, Olukanni DO (2020) Hazardous waste management: an African overview. Recycling 5(3):15. https://doi.org/10.3390/recycling5030015

Akram SV, Singh R, Gehlot A, Rashid M, AlGhamdi AS, Alshamrani SS, Prashar D (2021a) Role of wireless aided technologies in the solid waste management: a comprehensive review. Sustainability 13(23):13104

Akram SV, Alshamrani SS, Singh R, Rashid M, Gehlot A, AlGhamdi AS, Prashar D (2021b) Blockchain enabled automatic reward system in solid waste management. Secur Commun Netw 2021(1):6952121. https://doi.org/10.1155/2021/6952121

Asih AMS, Trapsilawati F, Sopha BM, Normasari NME (2022) Waste bank program for households as a means of processing inorganic waste. Jurnal Pengabdian kepada Masyarakat (Indonesian Journal of Community Engagement) 8(4):177–183. https://doi.org/10.22146/jpkm.73409

Aus der Beek T, Weber FA, Bergmann A, Hickmann S, Ebert I, Hein A, Küster A (2016) Pharmaceuticals in the environment—Global occurrences and perspectives. Environ Toxicol Chem 35(4):823–835

Awogbemi O, Kallon DVV, Bello KA (2022) Resource recycling with the aim of achieving zero-waste manufacturing. Sustainability 14(8):4503. https://doi.org/10.3390/su14084503

Baby A, Mathew S (2020) A correlative study to assess the knowledge and practice of housewives regarding household waste management in selected rural community at Mangalore with a view to provide an information pamphlet. Indian J Public Health Res Dev 11(3):466–470

Bang S, Andersen BS (2022) Utilising artificial intelligence in construction site waste reduction. J Eng, Project, Prod Manag 12(3):239–249. https://doi.org/10.32738/JEPPM-2022-0022

Basu P, Chakraborty J (2016) Environmental justice implications of industrial hazardous waste generation in India: a national scale analysis. Environ Res Lett 11(12):125001. https://doi.org/10.1088/1748-9326/11/12/125001

Baxter LK, Crooks JL, Sacks JD (2017) Influence of exposure differences on city-to-city heterogeneity in PM 2.5-mortality associations in U.S. cities. Environ Health 16:1–8. https://doi.org/10.1186/s12940-016-0208-y

Bedada TL, Eshete TB, Gebre SG, Dera FA, Sima WG, Negassi TY, Maheder RF, Teklu S, Awoke K, Feto TK, Tullu KD (2019) Virological quality of urban rivers and hospitals wastewaters in Addis Ababa, Ethiopia. Open Microbiol J 13(1). https://doi.org/10.2174/1874285801913010164

Budiman RCP, Suharto DG, Wahyunengseh RD (2023, November) Study of the implementation of household hazardous and toxic waste management policy in the province of the special region of Yogyakarta. In: IOP conference series: earth and environmental science, vol 1275, No 1, p 012037. IOP Publishing. https://doi.org/10.1088/1755-1315/1275/1/012037

Bułkowska K, Zielińska M, Bułkowski M (2023) Implementation of blockchain technology in waste management. Energies 16(23):7742. https://doi.org/10.3390/en16237742

Burkholder J, Libra B, Weyer P, Heathcote S, Kolpin D, Thorne PS, Wichman M (2007) Impacts of waste from concentrated animal feeding operations on water quality. Environ Health Perspect 115(2):308–312. https://doi.org/10.1289/ehp.8839

Cha GW, Hong WH, Kim YC (2023) Performance improvement of machine learning model using autoencoder to predict demolition waste generation rate. Sustainability 15(4):3691. https://doi.org/10.3390/su15043691

Chen M, Ogunseitan OA (2021) Zero E-waste: regulatory impediments and blockchain imperatives. Front Environ Sci Eng 15:1–10. https://doi.org/10.1007/s11783-021-1402-x

Cui Y, Cao Y, Ji Y, Chang IS, Wu J (2022) Determinant factors and business strategy in a sustainable business model: an explorative analysis for the promotion of solid waste recycling technologies. Bus Strateg Environ 31(5):2533–2545. https://doi.org/10.1002/bse.3042

Derso S, Taye G, Getachew T, Defar A, Teklie H, Amenu K, Gelibo T, Bekele A (2018) Biomedical waste disposal systems of health facilities in Ethiopia. Environ Health Eng Manag J 5(1):29–37. https://doi.org/10.15171/ehem.2018.05

Dorcas KD, Celestin BN, Yunfei S (2021) Entrepreneurs traits/characteristics and innovation performance of waste recycling start-ups in Ghana: an application of the upper echelons theory among SEED award winners. Sustainability 13(11):5794. https://doi.org/10.3390/su13115794

Elshaboury N, Al-Sakkaf A, Mohammed Abdelkader E, Alfalah G (2022) Construction and demolition waste management research: a science mapping analysis. Int J Environ Res Public Health 19(8):4496. https://doi.org/10.3390/ijerph19084496

Emmett EA (2021) Asbestos in high-risk communities: public health implications. Int J Environ Res Public Health 18(4):1579. https://doi.org/10.3390/ijerph18041579
Ezirim I, Agbo F (2018) Role of national policy in improving health care waste management in Nigeria. J Health Pollut 8(19):180913. https://doi.org/10.5696/2156-9614-8.19.180913
Fazzo L, Minichilli F, Santoro M, Ceccarini A, Della Seta M, Bianchi F, Comba P, Martuzzi M (2017) Hazardous waste and health impact: a systematic review of the scientific literature. Environ Health 16:1–11. https://doi.org/10.1186/s12940-017-0311-8
Ferronato N, Torretta V (2019) Waste mismanagement in developing countries: a review of global issues. Int J Environ Res Public Health 16(6):1060. https://doi.org/10.3390/ijerph16061060
Fikru MG (2012) Trans-boundary movement of hazardous waste: evidence from a new micro data in the European Union. Rev Eur Stud 4:3. https://doi.org/10.5539/res.v4n1p3
Folkers C, Gunter LP (2022) Radioactive releases from the nuclear power sector and implications for child health. BMJ Paediatrics Open 6(1). https://doi.org/10.1136/bmjpo-2021-001326
Frempong VE (2013) Municipal solid waste management in Tarkwa area council (TAC), Ghana. Int J Sustain Dev Plan 8(4):563–575. https://doi.org/10.2495/sdp-v8-n4-563-575
González-Sánchez R, Alonso-Muñoz S, Medina-Salgado MS (2023) Circularity in waste management: a research proposal to achieve the 2030 Agenda. Oper Manag Res 16(3):1520–1540. https://doi.org/10.1007/s12063-023-00373-0
Grabowski G (2021) Circular economy in the waste management sector. In: Sustainability and sustainable development, pp 243–250. https://doi.org/10.18559/978-83-8211-074-6/iv5
Greenstone M, Hanna R (2014) Environmental regulations, air and water pollution, and infant mortality in India. Am Econ Rev 104(10):3038–3072. https://doi.org/10.1257/aer.104.10.3038
Guerrero LA, Valverde VR, Maas G (2012) Waste management systems' impact on health and environment in developing countries. Linnaeus Eco-Tech. https://doi.org/10.15626/eco-tech.2012.019
Gutberlet J, Uddin SMN (2017) Household waste and health risks affecting waste pickers and the environment in low-and middle-income countries. Int J Occup Environ Health 23(4):299–310. https://doi.org/10.1080/10773525.2018.1484996
Hennebert P (2022) Risk management of hazardous solid wastes by hazardous property including mercury-containing wastes. Detritus (20):78–89. https://doi.org/10.31025/2611-4135/2022.15212
Hidayat T, Utama NA, Aini LN, Fikri M, Rachma AF, Bahar SNFR, Hasanah NAU, Annisa D (2023) The reduction of household waste using black soldier fly using stack bucket system. In: Proceeding international conference of technology on community and environmental development, vol 1, No 1, pp 207–216. https://doi.org/10.18196/ictced.v1i1.26
Hong SY, Yoon CW, Yoon YS, Kang JH, Jeon TW (2022) A study on safety management plan for recycling of medium-contact wastes via ecotoxicity assessment. Environ Health Insights 16:11786302221111872. https://doi.org/10.1177/11786302221111872
Husaini DC, Bernardez V, Zetina N, Mphuthi DD (2023) Healthcare industry waste and public health: a systematic review. Arab Gulf J Sci Res. https://doi.org/10.1108/agjsr-01-2023-0026
Ibrahim MF, Hod R, Toha HR, Mohammed Nawi A, Idris IB, Mohd Yusoff H, Sahani M (2021) The impacts of illegal toxic waste dumping on children's health: a review and case study from Pasir Gudang, Malaysia. Int J Environ Res Public Health 18(5):2221. https://doi.org/10.3390/ijerph18052221
Idowu I, Alo B, Atherton W, Al Khaddar R (2013) Profile of medical waste management in two healthcare facilities in Lagos, Nigeria: a case study. Waste Manag Res 31(5):494–501. https://doi.org/10.1177/0734242x13479429
Ikhajiagbe B, Ogwu MC (2020) Hazard quotient, microbial diversity and plant composition of spent crude oil polluted-soil. Beni-Suef Univ J Basic Appl Sci 9(26). http://doi.org/10.1186/s43088-020-00052-0

Iyiola AO, Kolawole AS, Setufe SB, Bilikoni J, Ofori E, Ogwu MC (2024) Fish as a sustainable biomonitoring tool in aquatic environments. In: Izah SC, Ogwu MC, Hamidifar H (eds) Biomonitoring of Pollutants in the Global South. Springer, Singapore. https://doi.org/10.1007/978-981-97-1658-6_12

Izah SC, Aigberua AO, Ogwu MC (2022) Trace element composition of Gallus gallus domesticus eggs and health risks associated with their consumption in Port Harcourt, Nigeria. J Food Saf Hyg 8(3):202–222

Izah SC, Stanley HO, Richard G, Sawyer WE, Uwaeme OR (2024a) Surface water quality: a statistical perspective on the efficacy of environmental and human health assessment tools. Water, Air, Soil Pollut 235(3). https://doi.org/10.1007/s11270-024-06965-1

Izah SC, Stanley HO, Richard G, Sawyer WE, Uwaeme OR (2024b) Environmental health risks of trace elements in sediment using multivariate approaches and contamination indices. Int J Environ Sci Technol. https://doi.org/10.1007/s13762-024-05974-1

Izah SC, Sylva L, Hait M (2024c) Cronbach's Alpha: a cornerstone in ensuring reliability and validity in environmental health assessment. E.S. Energy Environ 23:1057. https://doi.org/10.30919/esee1057

Izah SC, Richard G, Stanley HO, Sawyer WE, Ogwu MC, Uwaeme OR (2024d) Prospects and application of multivariate and reliability analyses to one health risk assessments of toxic elements. Toxicol Environ Health Sci 16(2):127–134

Izah SC, Ogwu MC, Shahsavani A, Kermani M, Hasanzadeh V (2024e) Understanding the one health implications of air pollution. In: Izah SC, Ogwu MC, Shahsavani A (eds). Air pollutants in the context of one health: fundamentals, sources and impacts. The Handbook in Environmental Chemistry. Springer, Berlin, Heidelberg. https://link.springer.com/chapter/10.1007/698_2024_1129

Izah SC, Sylva L, Ogwu MC, Shahsavani A, Bazzazpour S, Rahmatinia (2024f). Modeling and statistical approaches for air pollution analysis. In: Izah SC, Ogwu MC, Shahsavani A (eds). Air pollutants in the context of one health: fundamentals, sources and impacts. The Handbook in Environmental Chemistry. Springer. https://link.springer.com/chapter/10.1007/698_2024_1138

Izah SC, Sawyer WE, Ogwu MC, Shahsavani A, Kermani M, Hasanzadeh V (2024g). Air Pollution Principles, research and innovation. In: Izah SC, Ogwu MC, Shahsavani A (eds) Air pollutants in the context of one health: fundamentals, sources and impacts. The Handbook in Environmental Chemistry. Springer. https://link.springer.com/chapter/10.1007/698_2024_1131

Izah SC, Abdelkhalek ST, Ogwu MC, Hamidifar H (2024h) Overview and perspectives of biomonitoring in the global south. In: Izah SC, Ogwu MC, Hamidifar H (eds) Biomonitoring of pollutants in the global south. Springer, Singapore. https://doi.org/10.1007/978-981-97-1658-6_1

Jacob S, Nithianandam S, Rastogi S, Sakhuja S, Alankar SNSL (2021) Handling and treatment strategies of biomedical wastes and biosolids contaminated with SARS-CoV-2 in waste environment. In: Environmental and health management of novel coronavirus disease (COVID-19). Academic Press, pp 207–232. https://doi.org/10.1016/B978-0-323-85780-2.00012-3

Jamin N, Mahmood NZ (2015) Scheduled waste management in Malaysia: an overview. Adv Mater Res 1113:841–846. https://doi.org/10.4028/www.scientific.net/amr.1113.841

Ji L, Lu S, Yang J, Du C, Chen Z, Buekens A, Yan J (2016) Municipal solid waste incineration in China and the issue of acidification: a review. Waste Manag Res 34(4):280–297. https://doi.org/10.1177/0734242x16633776

Kalair AR, Seyedmahmoudian M, Stojcevski A, Abas N, Khan N (2021) Waste to energy conversion for a sustainable future. Heliyon, 7(10). https://doi.org/10.1016/j.heliyon.2021.e08155

Kang Z, Huang Z, Peng Q, Shi Z, Xiao H, Yin R, Fu G, Zhao J (2023) Recycling technologies, policies, prospects, and challenges for spent batteries. Iscience. 26(11):108072. https://doi.org/10.1016/j.isci.2023.108072

Khotimah SK, Nasruddin N, Santi HA, Padly A, An-Nafi AR (2023) Community empowerment through the 3R movement and the establishment of a waste bank in Wirittasi Village. J Transnat Univ Stud 1(11):919–927. https://doi.org/10.58631/jtus.v1i11.70

Kumar A, Thakur AK, Gaurav GK, Klemeš JJ, Sandhwar VK, Pant KK, Kumar R (2023) A critical review on sustainable hazardous waste management strategies: a step towards a circular economy. Environ Sci Pollut Res 30(48):105030–105055. https://doi.org/10.1007/s11356-023-29511-8

Laborde A, Tomasina F, Bianchi F, Bruné MN, Buka I, Comba P, Corra L, Cori L, Duffert CM, Harari R, Iavarone I, McDiarmid MA, Gray KA, Sly PD, Soares A, Suk WA, Landrigan PJ (2015) Children's health in Latin America: the influence of environmental exposures. Environ Health Perspect 123(3):201–209. https://doi.org/10.1289/ehp.1408292

Lema G, Mesfun MG, Eshete A, Abdeta G (2019) Assessment of status of solid waste management in Asella town, Ethiopia. BMC Public Health 19:1261. https://doi.org/10.1186/s12889-019-7551-1

Li SJ, Sun B, Hou DX, Jin WJ, Ji Y (2021) Does industrial agglomeration or foreign direct investment matter for environment pollution of public health? Evidence from China. Front Public Health 9:711033. https://doi.org/10.3389/fpubh.2021.711033

Lissah SY, Ayanore MA, Krugu JK, Aberese-Ako M, Ruiter RA (2021) Managing urban solid waste in Ghana: perspectives and experiences of municipal waste company managers and supervisors in an urban municipality. PLoS ONE 16(3):e0248392. https://doi.org/10.1371/journal.pone.0248392

Malinauskaite J, Jouhara H, Czajczyńska D, Stanchev P, Katsou E, Rostkowski P, Thorne RJ, Colon J, Ponsá S, Al-Mansour L, Anguilano R, Krzyżyńska IC, Vlasopoulos LA, Spencer N (2017) Municipal solid waste management and waste-to-energy in the context of a circular economy and energy recycling in Europe. Energy 141:2013–2044. https://doi.org/10.1016/j.energy.2017.11.128

Manisalidis I, Stavropoulou E, Stavropoulos A, Bezirtzoglou E (2020) Environmental and health impacts of air pollution: a review. Front Public Health 8:14. https://doi.org/10.3389/fpubh.2020.00014

Massoud MA, Mokbel M, Alawieh S, Yassin N (2019) Towards improved governance for sustainable solid waste management in Lebanon: centralised vs decentralised approaches. Waste Manag Res 37(7):686–697. https://doi.org/10.1177/0734242x19836705

Mathiarasan S, Hüls A (2021) Impact of environmental injustice on children's health—interaction between air pollution and socioeconomic status. Int J Environ Res Public Health 18(2):795. https://doi.org/10.3390/ijerph18020795

Mihajlov A, Mladenovic A, Jovanovic F (2021) Country in transition (Serbia) case: circular economy starts from waste management. Environ Res Technol 4(1):83–88

Milios L (2018) Advancing to a circular economy: three essential ingredients for a comprehensive policy mix. Sustain Sci 13(3):861–878. https://doi.org/10.1007/s11625-017-0502-9

Mmereki D, Baldwin A, Hong L, Li B (2016) The management of hazardous waste in developing countries. Manag Hazardous Wastes 6(8):39–50. https://doi.org/10.5772/63055

Mmereki D, Li B, Li'ao W (2014) Waste electrical and electronic equipment management in Botswana: prospects and challenges. J Air Waste Manag Assoc 65(1):11–26. https://doi.org/10.1080/10962247.2014.892544

Mohamed DA, Yousif MEA, Abdalla AI (2023) Environmental and health effects of solid waste dumpsites in Wadajir District, Mogadishu, Somalia. World J Adv Res Rev 19(3):621–631. https://doi.org/10.30574/wjarr.2023.19.3.1821

Ažman Momirski L, Mušič B, Cotič B (2021) Urban strategies enabling industrial and urban symbiosis: the case of Slovenia. Sustainability 13(9):4616. https://doi.org/10.3390/su13094616

Moyo T, Sadan Z, Lotter A, Petersen J (2022) Barriers to recycling e-waste within a changing legal environment in South Africa. South Afr J Sci 118(SPE):1–8. https://doi.org/10.17159/sajs.2022/12564

Muiruri JM, Wahome R, Karatu K (2020) Study of residents' attitude and knowledge on management of solid waste in Eastleigh, Nairobi, Kenya. J Environ Protect 11(10):779. https://doi.org/10.4236/jep.2020.1110048

Muzanni A, Lestari F, Chalid M, Wardani WK, Satyawardhani SA, Kristanto GA, Zulys A (2022) Multi-sectoral partnership for waste management evaluation and awards recognition in higher education. Int J Sustain Dev Plann 17(4):1205–1213. https://doi.org/10.18280/ijsdp.170419

Nassani AA, Hussain H, Condrea E, Grigorescu A, Yousaf Z, Haffar M (2023) Zero waste management: investigation of green technology, the green supply chain, and the moderating role of CSR intentions. Sustainability 15(5):4169. https://doi.org/10.3390/su15054169

Nath B, Kumari R, Gupta V, Vaswani ND, Lekhwani S (2018) A community-based study on e-waste disposal in Srinagar, Uttarakhand: assessment of awareness and practices. Int J Commun Med Public Health 5(8):3429. https://doi.org/10.18203/2394-6040.ijcmph20183075

Navarro PAC, Osorio JEA, Pinto JCC (2022) Network structures in hazardous waste management systems: advances, characteristics and challenges. Waste Manag Environ Impact XI 257:161. https://doi.org/10.2495/wmei220141

Ngwuluka NC, Ochekpe NA, Odumosu PO (2011) An assessment of pharmaceutical waste management in some Nigerian pharmaceutical industries. Afr J Biotech 10(54):11259–11264. https://doi.org/10.5897/ajb10.1194

Nielsen I, Lim M, Nielsen P (2010) Optimizing supply chain waste management through the use of RFID technology. In: 2010 IEEE international conference on RFID-technology and applications. IEEE, pp 296–301. https://doi.org/10.1109/rfid-ta.2010.5529921

Niyongabo E, Jang YC, Kang D, Sung K (2019) Current treatment and disposal practices for medical wastes in Bujumbura, Burundi. Environ Eng Res 24(2):211–219. https://doi.org/10.4491/eer.2018.095

Nizzy AM, Ogwu MC (2024) Valorization of cassava processing by-products into biofuel for a sustainable environment. In: Ogwu, MC, Izah SC, Cunha Alves AA, Babu SC (eds) Plant Biology, Sustainability and Climate Change, Sustainable Cassava. Academic Press. pp 291–309. https://doi.org/10.1016/B978-0-443-21747-0.00010-2

Nkuh YR, Namondo EC, Kah TB, Nteimaah LN, Ayuk AK (2023) Community perceptions on health and environmental implications of poor management of solid waste materials around local dump sites in the Buea municipality, Cameroon. Am J Educ Technol 2(3):136–146. https://doi.org/10.54536/ajet.v2i3.1896

Nyametso JK, Enchantee NSE (2023) The effectiveness of biomedical wastes handling and disposal in Accra: the case of the Korle-Bu Teaching and the La General Hospitals. Journal ISSN 2766:2276. https://doi.org/10.37871/jbres1727

Ogutu FA, Kimata DM, Kweyu RM (2021) Partnerships for sustainable cities as options for improving solid waste management in Nairobi city. Waste Manag Res 39(1):25–31. https://doi.org/10.1177/0734242x20967735

Ogwu MC, Izah SC (eds) (2024) Sustainable strategies for air pollution mitigation: development, economics, and technologies. Springer Cham. https://doi.org/10.1007/978-3-031-74157-9

Ololade OO, Mavimbela S, Oke SA, Makhadi R (2019) Impact of leachate from northern landfill site in Bloemfontein on water and soil quality: implications for water and food security. Sustainability 11(15):4238. https://doi.org/10.3390/su11154238

Kwun Omang DI, John GE, Inah SA, Bisong JO (2021) Public health implication of solid waste generated by households in Bekwarra Local Government area. Afr Health Sci 21(3):1467–1473. https://doi.org/10.4314/ahs.v21i3.58

Ouma B, Wandayi OM, Muthama JN, George A, Villacampa M (2021) Knowledge, attitudes and practices synthesis of waste management among horticultural processing MSMES in Kenya. East Afr J Sci, Technol Innov 2. https://doi.org/10.37425/eajsti.v2i.348

Padmanabhan KK, Barik D (2019) Health hazards of medical waste and its disposal. In: Energy from toxic organic waste for heat and power generation. Woodhead Publishing, pp 99–118. https://doi.org/10.1016/B978-0-08-102528-4.00008-0

Pandey RK, Mishra KP, Tiwari RP (2017) Characterization of municipal solid waste and evaluation of subsurface water quality in the vicinity of dumping sites. Int J Res Appl Sci Eng Technol V(X):425–431. https://doi.org/10.22214/ijraset.2017.10063

Parvez SM, Jahan F, Brune MN, Gorman JF, Rahman MJ, Carpenter D, Islam Z, Rahman M, Aich N, Knibbs LD, Sly PD (2021) Health consequences of exposure to e-waste: an updated systematic review. The Lancet. Planet Health 5(12):e905–e920. https://doi.org/10.1016/S2542-5196(21)00263-1

Pathak VM, Verma VK, Rawat BS, Kaur B, Babu N, Sharma A, Dewali S, Yadav M, Kumari R, Singh S, Mohapatra A, Pandey V, Rana N, Cunill JM (2022) Current status of pesticide effects on environment, human health and it's eco-friendly management as bioremediation: a comprehensive review. Front Microbiol 13:962619. https://doi.org/10.3389/fmicb.2022.962619

Pinto VN (2008) E-waste hazard: the impending challenge. Indian J Occup Environ Med 12(2):65–70. https://doi.org/10.4103/0019-5278.43263

Policastro G, Cesaro A (2022) Composting of organic solid waste of municipal origin: the role of research in enhancing its sustainability. Int J Environ Res Public Health 20(1):312. https://doi.org/10.3390/ijerph20010312

Purchase CK, Al Zulayq DM, O'Brien BT, Kowalewski MJ, Berenjian A, Tarighaleslami AH, Seifan M (2021) Circular economy of construction and demolition waste: a literature review on lessons, challenges, and benefits. Materials 15(1):76. https://doi.org/10.3390/ma15010076

Qu D, Shevchenko T, Shams Esfandabadi Z, Ranjbari M (2023) College students' attitude towards waste separation and recovery on campus. Sustainability 15(2):1620. https://doi.org/10.3390/su15021620

Ratnasabapathy S, Perera S, Alashwal A (2019) A review of smart technology usage in construction and demolition waste management. https://doi.org/10.31705/wcs.2019.5

Rodzi ZM, Hazri AN, Azri NASM, Rhmdan NDFS, Zaharudin ZA, Uttunggadewa S (2023) Uncovering obstacles to household waste recycling in Seremban, Malaysia through decision-making trial and evaluation laboratory (DEMATEL) analysis. Sci Technol Indonesia 8(3):422–428. https://doi.org/10.26554/sti.2023.8.3.422-428

Sebastian S (2023) A descriptive study to assess the knowledge on biomedical waste management among health team member in selected hospital, Mangalore. World J Biol Pharm Health Sci 13(2):309–314. https://doi.org/10.30574/wjbphs.2023.13.2.0100

Shah WUH, Yasmeen R, Sarfraz M, Ivascu L (2023) The repercussions of economic growth, industrialization, foreign direct investment, and technology on municipal solid waste: evidence from OECD economies. Sustainability 15(1):836. https://doi.org/10.3390/su15010836

Sharma R, Sharma M, Sharma R, Sharma V (2013) The impact of incinerators on human health and environment. Rev Environ Health 28(1):67–72. https://doi.org/10.1515/reveh-2012-0035

Siddiqua A, Hahladakis JN, Al-Attiya WAK (2022) An overview of the environmental pollution and health effects associated with waste landfilling and open dumping. Environ Sci Pollut Res 29(39):58514–58536. https://doi.org/10.1007/s11356-022-21578-z

Singh N, Ogunseitan OA, Tang Y (2022) Medical waste: current challenges and future opportunities for sustainable management. Crit Rev Environ Sci Technol 52(11):2000–2022. https://doi.org/10.1080/10643389.2021.1885325

Šomplák R, Kropáč J, Pluskal J, Pavlas M, Urbánek B, Vítková P (2022) A multi-commodity mathematical modelling approach—hazardous waste treatment infrastructure planning in the Czech Republic. Sustainability 14(6):3536. https://doi.org/10.3390/su14063536

Steffan JJ, Brevik EC, Burgess LC, Cerdà A (2018) The effect of soil on human health: an overview. Eur J Soil Sci 69(1):159–171. https://doi.org/10.1111/ejss.12451

Subedi M, Pandey S, Khanal A (2023) Integrated solid waste management for the circular economy: challenges and opportunities for Nepal. J Multidiscip Res Adv 1(1):21–26. https://doi.org/10.3126/jomra.v1i1.55100

Sundar D, Mathiyazhagan K, Agarwal V, Janardhanan M, Appolloni A (2023) From linear to a circular economy in the e-waste management sector: experience from the transition barriers in the United Kingdom. Bus Strateg Environ 32(7):4282–4298. https://doi.org/10.1002/bse.3365

Swanson HL, Ferrari JR (2023) Predictors of e-waste: considerations for community psychology prevention and intervention. J Community Psychol 51(5):2001–2009. https://doi.org/10.1002/jcop.22986

Szpilko D, de la Torre Gallegos A, Jimenez Naharro F, Rzepka A, Remiszewska A (2023) Waste management in the smart city: current practices and future directions. Resources 12(10):115. https://doi.org/10.3390/resources12100115

Tang KHD (2022) Valorization of plastic waste through incorporation into construction materials. Civ Sustain Urban Eng 2(2):96–109. https://doi.org/10.53623/csue.v2i2.141

Tang L, Li K, Jia P (2020) Impact of environmental regulations on environmental quality and public health in China: empirical analysis with panel data approach. Sustainability 12(2):623. https://doi.org/10.3390/su12020623

Tanuja GG, Srimani PS, Rao ALN, Sharma S, Mashkour MS, Singh N (2023) Innovative technologies for sustainable recycling and re-manufacturing of materials and components. In: E3S web of conferences, vol 430, p 01130. EDP Sciences. https://doi.org/10.1051/e3sconf/202343001130

Tchounwou PB, Yedjou CG, Patlolla AK, Sutton DJ (2012) Heavy metal toxicity and the environment. Experientia Suppl 2012(101):133–164. https://doi.org/10.1007/978-3-7643-8340-4_6

Tessema I, Fantahun Z (2023) Production and characterization of bricks from bottom ash and textile sludge using plastic waste as binding agent. J Eng 2023(1):7607677. https://doi.org/10.1155/2023/7607677

Wagland ST, Veltre F, Longhurst PJ (2012) Development of an image-based analysis method to determine the physical composition of a mixed waste material. Waste Manag 32(2):245–248. https://doi.org/10.1016/j.wasman.2011.09.019

Wang R, Xu Q, He C, Liu X, Feng Z, Zhang L, Gao J (2023) Analysis of hazardous waste management elements in oil and gas enterprises based on the life-cycle management concept. Sustainability 15(6):5504. https://doi.org/10.3390/su15065504

Wu J, Pu Y (2020) Air pollution, general government public-health expenditures and income inequality: empirical analysis based on the spatial Durbin model. PLoS ONE 15(10):e0240053. https://doi.org/10.1371/journal.pone.0240053

Yang N, Liu Z, Lin Y, Yang Y (2023) Does environmental regulation improve public health? Evidence from China's two control zones policy. Front Public Health 11:1059890. https://doi.org/10.3389/fpubh.2023.1059890

Yoada RM, Chirawurah D, Adongo PB (2014) Domestic waste disposal practice and perceptions of private sector waste management in urban Accra. BMC Public Health 14:1–10. https://doi.org/10.1186/1471-2458-14-697

Zangina AS, Ali AF (2021) An overview of hazardous waste management in Nigeria. Dutse J Pure Appl Sci 7:214–221. https://doi.org/10.4314/dujopas.v7i4b.22

Zhu B, Wang D, Wei N (2022) Enzyme discovery and engineering for sustainable plastic recycling. Trends Biotechnol 40(1):22–37. https://doi.org/10.1016/j.tibtech.2021.02.008

Part II
Technological Innovations for Environmental Quality Improvement

Chapter 7
Clean Energy Solutions and Public Health: Innovations in Renewable Energy for a Healthier Environment

Sylvester Chibueze Izah, Matthew Chidozie Ogwu, and Milan Hait

Abstract The association of clean energy, environmental sustainability, and public health presents a compelling case for transitioning from fossil fuels to renewable energy sources. Fossil fuel dependency contributes significantly to air and water pollution, exacerbating public health issues, including respiratory and cardiovascular diseases. Transitioning to renewable energy, such as solar, wind, and hydropower, offers a solution to these challenges by reducing environmental pollutants and improving health outcomes. Innovations in clean energy, including advanced solar panels, energy storage systems, and smart grids, are transforming the energy landscape and promoting sustainability. These technologies help mitigate climate change, enhance community resilience, create economic opportunities, and reduce health disparities. A robust approach to clean energy adoption must integrate sustainability principles into public health initiatives, ensure equity in energy access, and involve proactive community engagement in energy planning. Collaboration across sectors, including government, academia, and industry, fosters innovation and drives the widespread adoption of clean energy solutions. Finally, clean energy goes beyond technological necessity but is a moral imperative to safeguard public health, promote environmental stewardship, and ensure a healthier future for future generations.

S. Chibueze Izah
Department of Community Medicine, Faculty of Clinical Sciences, Bayelsa Medical University, Yenagoa, Bayelsa State, Nigeria

Department of Microbiology, Faculty of Science, Bayelsa Medical University, Yenagoa, Bayelsa State, Nigeria

S. Chibueze Izah
e-mail: Sylvester.izah@bmu.edu.ng

M. C. Ogwu
Goodnight Family Department of Sustainable Development, Appalachian State University, 212 Living Learning Center, 305 Bodenheimer Drive, Boone, NC 28608, USA
e-mail: ogwumc@appstate.edu

M. Hait (✉)
Department of Chemistry, Dr. C. V. Raman University, Bilaspur, Chhattisgarh, India
e-mail: haitmilan@gmail.com

S. Chibueze Izah and M. Chidozie Ogwu (eds.), *Innovative Approaches in Environmental Health Management*, Environmental Science and Engineering,
https://doi.org/10.1007/978-3-031-81966-7_7

Keywords Renewable energy · Public health · Environmental sustainability · Clean energy technologies · Air pollution · Community resilience · Sustainable development

7.1 Introduction

The intricate relationship between energy, environment, and public health is a critical area of study that has garnered increasing attention in recent years. The interdependencies among these domains highlight the necessity of transitioning towards sustainable energy systems to mitigate environmental degradation and enhance public health outcomes. The reliance on fossil fuels has been linked to adverse health effects, including respiratory diseases and cardiovascular issues, primarily due to air pollution generated from combustion processes. The World Health Organization has recognized air pollution as a leading environmental risk factor for health, contributing to millions of premature deaths annually (Guidry et al. 2020). This highlights the urgent need for a paradigm shift towards cleaner energy sources to alleviate these health burdens while promoting environmental sustainability.

The importance of clean energy in reducing environmental pollution cannot be overstated. Renewable energy sources such as solar, wind, and biogas offer sustainable alternatives to fossil fuels, notorious for their high carbon emissions and detrimental environmental impacts (D'Amico et al. 2020). Transitioning to these renewable sources stabilizes energy prices and enhances energy security, particularly in urban settings with high energy demand (Fig. 7.1; D'Amico et al. 2020).

Furthermore, integrating clean energy technologies into public health strategies can significantly reduce healthcare costs associated with pollution-related illnesses. For instance, Malik et al. emphasize that energy efficiency measures can decrease energy use by over 40%, thereby reducing the carbon footprint of healthcare systems (Malik et al. 2018). This reduction in energy consumption directly correlates with improved air quality and, consequently, better health outcomes for populations exposed to harmful pollutants. Moreover, the relationship between energy consumption and public health extends beyond immediate health impacts to involve broader social determinants of health. The reliance on bioenergy, particularly in developing regions, has adversely affected health and biodiversity due to unsustainable harvesting practices (Qureshi et al. 2016). This highlights the need for regulatory policies that promote sustainable energy practices while safeguarding public health. The Gulf Cooperation Council (GCC) countries have been identified as significant contributors to carbon emissions, necessitating a shift towards more sustainable energy consumption patterns to mitigate pollution-related health risks (Qureshi et al. 2016). Implementing such policies can foster a healthier environment, ultimately improving public health outcomes. The sustainability of public health interventions is also critical in addressing the nexus of energy, environment, and health. Research indicates that the sustainability of health programs is often hampered by a lack of consistent definitions and frameworks, which complicates the assessment of

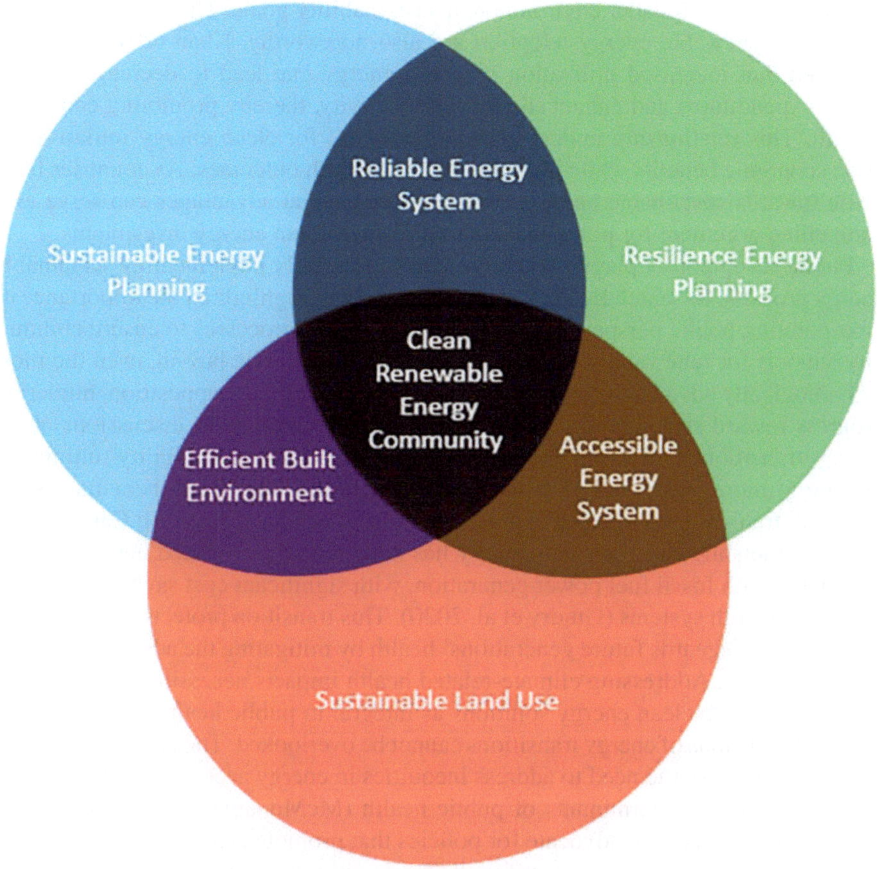

Fig. 7.1 Framework from transitioning to sustainable clean energy. *Source* Wehbi (2024). https://creativecommons.org/licenses/by/4.0/

their long-term impacts (Hailemariam et al. 2019). Schell et al. (2013) proposed a comprehensive framework for evaluating the sustainability of public health programs, emphasizing the importance of political support, funding stability, and community engagement. By integrating sustainability principles into public health initiatives, stakeholders can enhance the resilience of health systems against environmental challenges, thereby improving health outcomes in the long run. Public meals and their role in promoting health and sustainability further illustrate the interconnectedness of these domains. The consumption patterns influenced by public meals can significantly impact health outcomes and environmental sustainability (Höijer et al. 2020). Public health initiatives can reduce food systems' ecological footprint by prioritizing sustainable food sourcing and preparation methods while promoting healthier dietary choices among populations. This dual approach addresses immediate health concerns and contributes to broader environmental sustainability goals. The economic implications of renewable energy adoption are also noteworthy. Khan (2019) research indicated that increased utilization of green energy can lead to decreased public health expenditures and enhanced labor productivity, thereby promoting economic growth. This relationship underscores the potential for clean energy initiatives to drive economic benefits while improving public health outcomes. As countries transition towards sustainable energy systems, the financial advantages can serve as a compelling argument for policymakers to prioritize clean energy investments.

Public acceptability plays a crucial role in successfully implementing sustainable energy projects. Research by Perlaviciute et al. (2018) highlighted the importance of incorporating public perspectives into energy planning processes to ensure community support for renewable energy initiatives. Without public buy-in, even the most technologically advanced energy projects may face significant opposition, hindering progress toward sustainability goals. Engaging communities in discussions about energy transitions can foster a sense of ownership and responsibility, ultimately leading to more successful and sustainable outcomes. The health benefits associated with transitioning to clean energy are well-documented. In North Carolina, for instance, the shift towards clean energy has been linked to reduced health impacts associated with fossil fuel power generation, with significant cost savings projected for public health systems (Guidry et al. 2020). This transition protects current populations and safeguards future generations' health by mitigating the adverse effects of climate change. Addressing climate-related health impacts necessitates a concerted effort to promote clean energy solutions as integral to public health strategies. The ethical dimensions of energy transitions cannot be overlooked. The concept of energy justice emphasizes the need to address inequities in energy access and affordability, which are critical determinants of public health (McMonagle et al. 2021). Public health professionals can advocate for policies that promote clean energy and ensure equitable access to these resources for all communities by prioritizing energy justice. This holistic approach to energy, environment, and health underscores the necessity of integrating ethical considerations into energy decision-making processes to achieve sustainable and just outcomes.

The chapter focuses on the connection between renewable energy and public health, examining the harmful effects of traditional energy sources on health and how clean energy mitigates these issues. It highlights innovative approaches to integrating clean energy with environmental health goals, emphasizing air and water quality improvements. Additionally, the chapter explores cutting-edge renewable technologies, successful case studies, and policy strategies that promote the adoption of renewable energy for sustainable health outcomes.

7.2 Linking Renewable Energy to Environmental and Public Health

The relationship between traditional energy sources and public health is a critical area of study, particularly as the world grapples with the consequences of air pollution and its associated health impacts. Traditional energy sources, particularly fossil fuels, have been linked to significant public health issues, including respiratory diseases and cardiovascular problems (Table 7.1). According to Smith et al. (2013), the Global Burden of Disease study in 2010 attributed approximately 3.5 million premature deaths to household air pollution (HAP) from solid fuels and another 3.1 million to outdoor air pollution, much of which is a direct result of energy production from fossil fuels. The combustion of coal, oil, and natural gas releases a plethora of harmful pollutants, including particulate matter (PM), nitrogen oxides (NOx), and sulfur dioxide (SO_2), which can exacerbate respiratory conditions such as asthma and chronic obstructive pulmonary disease (COPD) (Hendryx et al. 2020). Furthermore, the health impacts of air pollution are not evenly distributed; marginalized communities often bear a disproportionate burden due to their proximity to industrial sites and highways, which are significant sources of air pollution (Giannadaki et al. 2016).

The transition to renewable energy sources such as solar, wind, and hydropower presents a viable solution to mitigate these public health issues. Renewable energy technologies are characterized by low or zero emissions during operation, significantly reducing the air pollutants associated with traditional energy sources. For instance, Wiser et al. (2016) highlighted that achieving high penetrations of solar energy can lead to substantial reductions in air pollution emissions, thereby improving public health outcomes. The environmental benefits of renewable energy extend beyond just reduced emissions; they also decrease the incidence of diseases linked to air quality degradation. Silva et al. (2021) further emphasized that increasing the share of renewable energy in the energy mix can lead to essential health co-benefits, including reduced hospital admissions and healthcare costs associated with pollution-related illnesses.

Moreover, the temporal dynamics of renewable energy generation can further enhance public health benefits. Siler-Evans et al. (2013) noted that wind energy generation often peaks when there is low demand, which can displace coal-fired power generation that typically operates during these periods. This displacement is crucial because coal plants are among the largest sources of harmful emissions. Similarly, solar energy generation peaks during midday when electricity demand is high, allowing for a reduction in reliance on fossil fuel sources that are more polluting (Siler-Evans et al. 2013). By strategically integrating renewable energy into the grid, optimizing health benefits while meeting energy demands is possible. In addition to direct health benefits, the adoption of renewable energy can also contribute to broader societal resilience and economic stability. Anderson et al. (2018) discussed how renewable energy resources can enhance community resilience by diversifying energy sources and reducing dependence on fossil fuels, subject to volatile market fluctuations. This diversification stabilizes energy prices and fosters job creation in

Table 7.1 Linking renewable energy and public health

Impact of traditional energy sources on public health	Type of renewable energy	Renewable energy solution
Air pollution (e.g., particulate matter, NOx, SOx)	Solar energy	Solar power generates electricity without combustion, eliminating sulfur oxides (SOx) and nitrogen oxides (NOx), reducing smog, and improving air quality
Respiratory diseases (e.g., asthma, bronchitis)	Wind energy	Wind energy produces zero emissions during operation, reducing the incidence of respiratory illnesses linked to fossil fuel air pollutants
Cardiovascular diseases	Hydropower	Hydropower reduces reliance on fossil fuels, lowering the risk of cardiovascular diseases linked to poor air quality
Greenhouse gas emissions (CO_2)	Geothermal energy	Geothermal energy has minimal carbon emissions compared to coal or natural gas, significantly reducing the contribution to climate change and associated health impacts
Indoor air pollution (from biomass and coal in low-income households)	Biomass energy (modern biomass)	Modern biomass systems burn cleanly and efficiently, reducing indoor air pollution that leads to respiratory and cardiovascular diseases
Water pollution (from coal mining and oil spills)	Solar, wind, hydropower	Solar, wind, and hydropower have minimal water usage and do not contaminate water sources, reducing waterborne diseases and improving public health
Health impacts from climate change (e.g., heatwaves, floods, disease spread)	All renewable sources	Renewable energy sources reduce greenhouse gas emissions, mitigating climate change impacts on health (e.g., heat stress and vector-borne diseases)

the renewable sector, which can positively impact communities, particularly in areas that have historically relied on fossil fuel industries (Nahhas 2023). The economic benefits of transitioning to renewable energy are complemented by the potential for improved public health outcomes, as healthier populations are often more productive and incur lower healthcare costs. However, it is essential to recognize that the transition to renewable energy has challenges. Implementing renewable energy technologies must be carefully managed to avoid unintended consequences, such as the potential health impacts of producing and disposing solar panels and wind turbines

(Behrani et al. 2023). For instance, the manufacturing processes for photovoltaic cells can involve hazardous materials that pose occupational health risks if not managed properly (Behrani et al. 2023). Therefore, while the shift towards renewable energy benefits public health, it necessitates a comprehensive approach that includes stringent regulations and best practices to ensure that the entire lifecycle of renewable technologies is environmentally friendly and safe for workers.

Furthermore, integrating renewable energy into existing energy systems requires robust policy frameworks and community engagement. Zulu et al. emphasize the importance of public perception and trust in adopting renewable energy solutions, suggesting that enhancing awareness of the benefits of solar energy can lead to greater acceptance and implementation (Zulu et al. 2021). Policymakers must prioritize education and outreach to ensure that communities understand the health and environmental benefits of renewable energy, which can facilitate smoother transitions and more excellent public support for renewable initiatives.

7.3 Innovative Approaches in Environmental Health Management

Innovative approaches in environmental health management are increasingly recognized as critical for integrating clean energy initiatives with public health goals. The intersection of ecological innovation and public health is essential for addressing the pressing challenges of climate change and pollution. A comprehensive strategy aligning clean energy technologies with public health objectives can significantly improve air and water quality, enhancing overall community health outcomes. For instance, Huang and Li (2018) emphasized the importance of developing an environmental innovation strategy incorporating ecological protection into business operations, which can subsequently positively influence public health by reducing environmental hazards. Table 7.2 shows an overview of the Innovative Approaches in Environmental Health Management.

The integration of clean energy technologies has demonstrated tangible benefits for air quality. Zhang et al. (2019) highlighted that promoting clean fuels in residential sectors has led to considerable improvements in PM2.5 air quality in China, with a national reduction of annual mean PM2.5 concentrations by 2.2 $\mu g/m^3$ in 2017. This reduction is crucial, as PM2.5 significantly contributes to respiratory diseases and other health issues. Furthermore, as discussed by Shi et al. (2021), the co-benefits of clean air actions indicated that reductions in PM2.5 levels during China's clean air initiatives have been linked to a decrease in chronic health effects, suggesting that effective air pollution control policies can save lives. Such evidence underscores the necessity of integrating clean energy solutions into public health strategies to mitigate the adverse health impacts of air pollution.

Moreover, the role of innovative educational approaches in environmental health management cannot be overlooked. Maipas et al. (2021) argued that ecological health

Table 7.2 Innovative approaches in environmental health management

Process	Focus	Examples of clean energy technologies improving air and water quality	References
Renewable energy adoption	I am transitioning from fossil fuels to clean energy sources to reduce air pollution and greenhouse gas emissions	Solar power reduces air pollutants like sulfur dioxide (SO_2) and nitrogen oxides (NOx), improving air quality and reducing respiratory illnesses	Gyimah et al. (2024)
Energy efficiency improvements	Enhancing energy systems' efficiency to lower energy consumption and reduce harmful emissions	Energy-efficient lighting and appliances reduce the need for power generation from coal or oil, contributing to improved air quality	Gillingham et al. (2021)
Electrification of transportation	Shifting from gasoline and diesel-powered vehicles to electric vehicles (EVs) to reduce emissions	EVs eliminate tailpipe emissions, which lowers urban air pollution and decreases risks of asthma and other respiratory diseases	Carey (2023)
Water conservation through clean energy	Renewable energy technologies require minimal water usage compared to traditional power plants	Solar and wind energy use minimal water, reducing water contamination from cooling processes and preventing heavy metal runoff	Jones and Olsson (2017)
Clean cooking initiatives	Implementing modern biomass and electric stoves to replace traditional wood-burning stoves, which cause indoor air pollution	Improved cookstoves reduce indoor air pollution, lowering the risk of respiratory diseases like chronic obstructive pulmonary disease	Phillip et al. (2023)
Hydropower for clean water	Utilizing hydropower systems to generate energy while managing water resources sustainably	Small-scale hydropower plants generate clean energy and help manage water quality by regulating river flows and reducing pollution	Moran et al. (2018)

education is evolving to address the threats posed by climate change and urban air pollution, thereby fostering a more informed public that can advocate for sustainable practices. This educational shift is vital for empowering communities to engage in health-promoting behaviors and supporting policies prioritizing clean energy and environmental sustainability. By incorporating innovative educational interventions, public health professionals can enhance community resilience against environmental health threats. In addition to improving air quality, clean energy technologies also play a vital role in improving water quality. As Soewarno et al. (2019) noted, green innovation strategies emphasize reducing waste and pollution prevention, directly benefiting water resources. For instance, adopting sustainable industry practices can decrease effluent discharges, protecting aquatic ecosystems and public health. Furthermore, the collaborative efforts between various stakeholders in environmental management, as highlighted by Green et al. (2012), can facilitate the development of comprehensive strategies that simultaneously address air and water quality issues.

The concept of the exposome, introduced by Juárez et al. (2014), provided a framework for understanding the cumulative impact of environmental exposures on health. This approach is particularly relevant in integrating clean energy with public health goals, as it allows for a more nuanced understanding of how various environmental factors, including air and water quality, influence health outcomes. By employing tools such as health impact assessments and participatory geographic information systems, public health researchers can better assess the implications of clean energy initiatives on community health and identify areas for improvement. The innovative use of technology in environmental health management also holds promise for enhancing the effectiveness of clean energy initiatives. For example, using energy recovery ventilators, as discussed by Huizing et al. (2015), can significantly improve indoor air quality while optimizing energy efficiency in buildings. This dual benefit improves health outcomes and aligns with broader sustainability goals. By leveraging such technologies, public health and environmental management can work synergistically to create healthier living environments.

Furthermore, the importance of stakeholder engagement in promoting environmental sustainability cannot be overstated. As Kimbara et al. (2012) noted, fostering collaboration among various stakeholders, including government agencies, businesses, and community organizations, is essential to successfully implement environmental innovations. This collaborative approach can enhance the diffusion of clean energy technologies and ensure that public health considerations are integrated into environmental policies. By creating a shared vision for sustainable development, stakeholders can collectively address the challenges posed by pollution and climate change. The impact of clean energy technologies on health disparities is another critical area of focus. Ferrante et al. (2022) emphasized the need to address environmental determinants of health, particularly in vulnerable populations. By integrating clean energy initiatives with public health strategies, it is possible to reduce exposure to harmful pollutants and improve health equity. This approach aligns with the principles of One Health, which advocate for a holistic understanding of health that encompasses human, animal, and environmental factors (Veras-Estévez and Chapman 2023; Izah et al. 2023, 2024a, b, c, d). By adopting this integrated perspective, public health professionals can develop more effective interventions addressing health disparities' root causes.

7.4 Technological Innovations in Clean Energy

Technological innovations in clean energy have become crucial in addressing the global energy crisis and mitigating climate change (Table 7.3). Among these innovations, advanced solar panels, energy storage systems, and intelligent grids are transformative technologies (Izah et al. 2024e). Advanced solar panels, particularly those utilizing bifacial and perovskite materials, have significantly improved efficiency and cost-effectiveness. Bifacial solar panels can capture sunlight from both sides, increasing energy yield, while perovskite solar cells promise lower production costs and higher efficiencies compared to traditional silicon-based cells (Wang et al. 2019; Sherif et al. 2022). These advancements enhance the viability of solar energy and contribute to the overall reduction of carbon emissions, aligning with global sustainability goals.

Energy storage technologies are crucial for managing the intermittency of renewable energy sources such as solar and wind. Various methods, including lithium-ion batteries, compressed air energy storage (CAES), and thermal energy storage, have been developed to store excess energy generated during peak production times for use during periods of high demand (Yao et al. 2019; Wei et al. 2023). Lithium-ion batteries, widely used in electric vehicles and grid storage, have significantly increased energy density and lifecycle (Sun et al. 2022a, b). CAES utilizes compressed air to store energy, which can be released to generate electricity when needed, thus providing a reliable solution for energy supply fluctuations (Zhang 2023; Llamas et al. 2020). These energy storage innovations enhance the reliability of renewable energy systems and facilitate the transition towards a more sustainable energy infrastructure. Smart grids represent another significant technological advancement in clean energy. By integrating digital technology into the electricity grid, smart grids enhance the efficiency and reliability of energy distribution. They enable real-time monitoring and management of energy flow, allowing for better integration of renewable energy sources and improved demand response capabilities (Sârbu and Sebarchievici 2018; Chen and Xie 2022). Implementing smart grids can reduce energy losses, lower operational costs, and increased resilience against outages. Furthermore, they support the deployment of electric vehicles and distributed energy resources, which are essential components of a sustainable energy future (Yao 2023a). The synergy between advanced solar technologies, energy storage systems, and smart grids creates a robust framework for achieving energy sustainability.

Studies of successful clean energy projects provide valuable insights into the health impacts of these technologies. For instance, deploying solar energy systems in urban areas has been linked to improved air quality and public health outcomes. A study conducted in California demonstrated that increased solar energy adoption significantly reduced air pollutants, decreasing respiratory illnesses among residents (Pitso 2019). Similarly, projects incorporating energy storage solutions have been shown to enhance grid reliability, thereby reducing the frequency and duration of power outages that disproportionately affect vulnerable populations (Detka and

Table 7.3 Technological innovations in clean energy addressing

Technology	Description	Example of a successful clean energy project	Health impact	References
Advanced solar panels	Next-generation photovoltaic panels with higher efficiency and better energy conversion	Topaz solar farm in California, USA: A large-scale solar farm generating over 500 MW of clean energy	Reduction in air pollution-related illnesses such as asthma and respiratory diseases due to decreased reliance on fossil fuels	Battersby (2023)
Energy storage (Battery Technology)	Advanced batteries store excess renewable energy, enabling a stable and reliable supply	Hornsdale Power Reserve in Australia: World's most extensive lithium-ion battery system	Ensures continuous clean energy supply, reducing the need for diesel generators contributing to air pollution	Trahey et al. (2020), Shaikk et al. (2024)
Smart grids	Electrical grid systems that integrate renewable energy sources improve efficiency and manage demand	Smart Grid City project in Boulder, Colorado, USA: A city-wide smart grid implementation	Improved energy efficiency reduces harmful emissions and leads to cleaner air, promoting better public health	Ohanu et al. (2024)
Offshore wind turbines	Advanced turbines are installed offshore to capture higher wind speeds and generate more electricity	Hornsea one offshore wind farm in the UK: The largest offshore wind farm generating 1.2 GW	Reducing greenhouse gas emissions lowers global warming-related health impacts, such as heat-related illnesses	Akhtar et al. (2024)
Concentrated Solar Power (CSP)	It uses mirrors to focus sunlight, producing thermal energy that can be stored and used to generate electricity	Noor solar complex in Morocco: One of the largest CSP plants, powering over one million homes	Cleaner air due to reduced fossil fuel use leads to fewer respiratory diseases and improved cardiovascular health	ESFC Investment Group (N.D.)

(continued)

Table 7.3 (continued)

Technology	Description	Example of a successful clean energy project	Health impact	References
Floating solar farms	Solar panels are installed on water bodies to maximize land use and increase efficiency	Anhui floating solar farm in China: The world's largest floating solar farm on a former coal-mining site	Reduced air pollution and water contamination, leading to improved respiratory and cardiovascular health outcomes	Pouran et al. (2022)
Hydrogen fuel cells	Produces electricity by combining hydrogen and oxygen, emitting only water vapor as a by-product	Tokyo Hydrogen Project in Japan: Hydrogen fuel cells used in homes and vehicles	Zero harmful emissions, leading to a significant decrease in urban air pollution and related respiratory problems	Le et al. (2023)

Górecki 2023). These health benefits underscore the importance of investing in clean energy technologies for environmental sustainability and public health improvement.

Moreover, integrating renewable energy technologies has been associated with economic benefits, particularly job creation and local economic development. The growth of the solar and wind industries has generated millions of jobs worldwide, ranging from manufacturing to installation and maintenance (Yao 2023b; Tabrizian 2019). In regions where clean energy projects have been implemented, local economies have experienced revitalization through increased investment and infrastructure development. For example, deploying wind farms in rural areas has provided new revenue streams for farmers and landowners, contributing to economic resilience (Salkuti 2018). These financial advantages further reinforce the case for transitioning to clean energy technologies.

The role of government policies in promoting clean energy innovations cannot be overstated. Effective policies, such as tax incentives, subsidies, and renewable energy mandates, have been instrumental in driving the adoption of advanced technologies (Habib and Sou 2018; Wang et al. 2022). Countries implementing comprehensive clean energy policies have seen significant increases in renewable energy capacity and reductions in greenhouse gas emissions. For instance, Germany's Energiewende policy has successfully transformed its energy landscape, leading to a substantial increase in renewable energy generation and decreased reliance on fossil fuels (Torres et al. 2020). Such policy frameworks are essential for fostering innovation and ensuring the long-term sustainability of clean energy technologies. In addition to government support, collaboration between academia, industry, and research institutions is critical for advancing clean energy technologies. Joint research initiatives and public–private partnerships have facilitated the development and commercializing of innovative solutions, such as next-generation batteries and smart grid technologies (Wu 2023; Sun et al. 2022a, b). These collaborations enhance technological advancements and promote knowledge sharing and capacity building within the clean energy sector.

Furthermore, international cooperation on clean energy research and development can accelerate the deployment of innovative solutions globally, addressing the urgent need for sustainable energy systems. The future of clean energy technologies is promising, with ongoing research focused on enhancing efficiency, reducing costs, and expanding applications. Emerging technologies, such as hydrogen fuel cells and advanced biofuels, hold great potential for diversifying the clean energy portfolio (Li et al. 2021; Tabrizian 2019). Additionally, advancements in artificial intelligence and machine learning are being integrated into energy management systems, optimizing energy consumption and improving grid stability (Liu et al. 2019). As these technologies evolve, they will be crucial in achieving global climate goals and transitioning to a low-carbon economy.

7.5 Strategies for Sustainable Energy and Health

The intersection of sustainable energy and public health is increasingly recognized as a critical area for policy development and community initiatives (Table 7.4). Strategies that promote renewable energy address climate change and have profound implications for public health. The transition to renewable energy sources, such as solar, wind, and biomass, can significantly reduce air pollution, contributing to respiratory diseases and other health issues. Research indicates that urban areas that adopt renewable energy technologies can experience a marked decrease in health-related costs associated with air quality deterioration (Kusch-Brandt 2019; Wall et al. 2021). Furthermore, integrating renewable energy into urban planning can lead to healthier living environments, enhancing overall public health outcomes (Kusch-Brandt 2019; Kut and Pietrucha-Urbanik 2023).

Policy strategies that effectively promote renewable energy and public health must consider a multifaceted approach. This includes the establishment of regulatory frameworks that incentivize the adoption of clean energy technologies and implementing educational programs that raise awareness about the health benefits of renewable energy (Kut and Pietrucha-Urbanik 2023; Sharma 2023). For instance, tax credits and subsidies for renewable energy installations can encourage individuals and businesses to transition away from fossil fuels, thereby reducing greenhouse gas emissions and improving air quality (Sharma 2023). Additionally, policies that support research and development in renewable energy technologies can foster innovation and lead to more efficient energy solutions that are both economically viable and environmentally friendly (Chang and Lo 2022; Kumar and Majid 2020). Community initiatives play a vital role in the adoption of cleaner energy sources. Local governments and community organizations can facilitate the transition to renewable

Table 7.4 Strategies for sustainable energy and health

Strategy	Description	Example of community or global initiative	Health impact	References
Clean energy subsidies and incentives	Governments provide financial incentives (e.g., tax credits and rebates) for renewable energy adoption	U.S. Clean Energy Tax Credit—Provides credits for solar, wind, and energy-efficient home improvements	Reduced reliance on fossil fuels decreases air pollution, leading to lower rates of respiratory diseases	Bölük and Kaplan (2022)
Carbon pricing and emissions trading	Policies that set a price on carbon emissions or establish cap-and-trade systems to limit greenhouse gases	European Union Emissions Trading System (EU ETS)—Caps carbon emissions and allows trading	Reduces greenhouse gas emissions, mitigating climate change and associated health risks (e.g., heat stress)	Bayer and Aklin (2020)
Energy efficiency standards	Regulations requiring products, buildings, and vehicles to meet energy efficiency standards	Energy Star Program(U.S.)—Sets energy efficiency benchmarks for appliances and buildings	Lower energy consumption reduces harmful emissions, improves air quality, and decreases cardiovascular diseases	Gillingham et al. (2021)
Clean cooking initiatives	Programs aimed at transitioning households from solid fuels (e.g., wood, coal) to cleaner cooking technologies	Global Alliance for Clean Cookstoves—Promotes adopting clean cookstoves in developing countries	Reduces indoor air pollution, preventing respiratory illnesses such as chronic obstructive pulmonary disease	Agbokey et al. (2019)
Renewable energy mandates	Government policies require a certain percentage of energy to come from renewable sources	Renewable Portfolio Standards (RPS) in various U.S. states—Mandate a percentage of renewable energy	Promotes cleaner air by reducing coal and natural gas emissions, improving public health	Zhou et al. (2024)

(continued)

Table 7.4 (continued)

Strategy	Description	Example of community or global initiative	Health impact	References
Sustainable transport policies	Policies that promote electric vehicles, public transportation, and active transport (walking, cycling)	Netherlands' electric vehicle promotion—Incentives for electric vehicles and cycling infrastructure	Reduced vehicular emissions result in lower urban air pollution, lowering asthma and heart disease risks	Winkler et al. (2023)
International climate agreements	Global agreements that commit countries to reducing emissions and increasing renewable energy use	Paris Agreement—Global commitment to limiting temperature rise and promoting renewable energy	Mitigates climate change, which reduces health impacts like heat-related illnesses and vector-borne diseases	Maslin et al. (2023)
Community solar programs	Programs that allow multiple households or businesses to share the benefits of a solar energy system	New York State's Community Solar Program—Provides solar energy access to low-income communities	Cleaner air leads to reduced respiratory problems, particularly in underprivileged communities	U.S. DoE (2023)
Health and energy education campaigns	Initiatives that raise awareness about the health benefits of renewable energy and energy conservation practices	UN Sustainable Energy for All (SEforALL) promotes global energy access and sustainability	Improved public understanding of energy-health linkages leads to greater adoption of clean energy solutions	Sharma et al. (2024)

energy through grassroots campaigns that promote energy efficiency and sustainability (Gasimli et al. 2022; Haidi and Cheddadi 2022). For example, community solar projects allow residents to invest in solar energy collectively, making renewable energy more accessible to those who may not have the means to install solar panels on their properties (Sharma 2023; Sait et al. 2018). Such initiatives empower communities and create local jobs in the renewable energy sector, contributing to economic resilience while enhancing public health through cleaner air and reduced emissions (Kumar and Majid 2020).

Globally, initiatives aimed at cleaner energy adoption are gaining momentum, particularly in developing countries where energy access remains a significant challenge. The United Nations' Sustainable Energy for All initiative emphasizes the importance of providing universal access to modern energy services, crucial for improving health outcomes in underserved populations (Bensch et al. 2015). By investing in renewable energy infrastructure, these countries can leapfrog traditional fossil fuel-based energy systems, thereby reducing their carbon footprint and improving public health simultaneously (Kumar and Majid 2020; Haidi and Cheddadi 2022). Moreover, international collaborations and funding mechanisms can support deploying renewable energy technologies in regions that are most vulnerable to the impacts of climate change (Gasimli et al. 2022; Hoang et al. 2021). The role of education in promoting renewable energy adoption cannot be overstated. Educational programs that inform the public about the health benefits of renewable energy and the dangers of fossil fuel consumption can drive behavioral change (Piran et al. 2023). For instance, initiatives that educate communities about the health risks associated with indoor air pollution from traditional cooking methods can lead to increased adoption of improved cooking stoves and cleaner fuels (Bensch et al. 2015; Jagger and Jumbe 2016). Furthermore, integrating renewable energy education into school curricula can foster a culture of sustainability among future generations, ensuring long-term commitment to clean energy solutions (Piran et al. 2023).

Innovative governance models are essential for the successful implementation of renewable energy policies. Collaborative governance that involves multiple stakeholders, including government agencies, non-governmental organizations, and the private sector, can enhance the effectiveness of renewable energy initiatives (Kaartemo and Gonzalez-Perez 2020). This approach allows for pooling resources and expertise, leading to more comprehensive and inclusive energy strategies prioritizing environmental sustainability and public health (Kaartemo and Gonzalez-Perez 2020). For example, partnerships between local governments and health organizations can facilitate the development of policies addressing energy and health challenges, creating synergies that benefit communities (Kusch-Brandt 2019; Kut and Pietrucha-Urbanik 2023). The economic implications of transitioning to renewable energy are significant. Studies have shown that investments in renewable energy can lead to job creation and economic growth, particularly in regions that have historically relied on fossil fuels (Kumar and Majid 2020). The renewable energy sector is labor-intensive, meaning it can provide employment opportunities for a wide range of skill levels, thereby contributing to economic stability (Kumar and Majid 2020).

Additionally, reducing healthcare costs associated with cleaner air and improved public health can offset the initial investments in renewable energy infrastructure, making it a financially sound strategy for governments and communities (Kusch-Brandt 2019; Piran et al. 2023). Challenges remain in the widespread adoption of renewable energy technologies. Barriers such as high initial costs, lack of access to financing, and insufficient infrastructure can hinder progress (Sharma 2023; Haidi and Cheddadi 2022). Policymakers must address these challenges by creating supportive environments that facilitate investment in renewable energy (Sharma 2023; Lee et al. 2020). This includes developing financing mechanisms that lower the cost of entry for individuals and businesses and investing in infrastructure that supports renewable energy generation and distribution (Lee et al. 2020). Moreover, addressing social and cultural factors influencing energy choices is crucial for overcoming resistance to change and fostering a more sustainable energy future (Makki and Mosly 2020).

7.6 Conclusion

Clean energy solutions are integral to fostering a healthier and more sustainable environment. The transition from fossil fuels to renewable energy mitigates air and water pollution and enhances public health by reducing the incidence of respiratory and cardiovascular diseases. Renewable energy sources such as solar, wind, and geothermal have shown significant potential in minimizing environmental degradation while promoting community well-being and economic resilience. Innovations in clean energy technologies, including advancements in energy storage and smart grids, are transforming how energy is produced and consumed, making it more accessible and efficient. These technological breakthroughs and supportive policies are crucial for achieving long-term sustainability goals. The future of renewable energy is promising, with ongoing research and development poised to drive further innovations. As clean energy solutions become more efficient and cost-effective, their adoption is expected to expand, benefiting environmental sustainability and public health.

Moreover, integrating clean energy into public health initiatives offers opportunities to address health disparities by providing cleaner air, reducing pollution-related illnesses, and fostering equitable energy access for marginalized communities. However, ensuring these innovations are accessible requires collaboration between governments, industries, and local communities. Public education, policy reforms, and community engagement are critical components for ensuring that the benefits of clean energy reach all segments of society. Converging renewable energy and public health initiatives ultimately leads to a healthier, more resilient future. By prioritizing sustainability and equity in energy planning, stakeholders can contribute to an environment where people and the planet thrive together. The transition to clean energy involves more than environmental necessity; it is also a vital public health intervention, offering a brighter, healthier future for future generations.

References

Agbokey F, Dwommoh R, Tawiah T, Ae-Ngibise KA, Mujtaba MN, Carrion D, Ali Abdulai M, Afari-Asiedu S, Owusu-Agyei S, Asante KP, Jack DW (2019) Determining the enablers and barriers for the adoption of clean cookstoves in the middle belt of Ghana—A qualitative study. Int J Environ Res Public Health 16(7):1207. https://doi.org/10.3390/ijerph16071207

Akhtar N, Geyer B, Schrum C (2024) Larger wind turbines as a solution to reduce environmental impacts. Sci Rep 14(1):6608. https://doi.org/10.1038/s41598-024-56731-w

Anderson K, Laws ND, Marr S, Lisell L, Jimenez T, Case T, Li X, Lohmann D, Cutler D (2018) Quantifying and monetizing renewable energy resiliency. Sustainability 10(4):933. https://doi.org/10.3390/su10040933

Battersby S (2023) How to expand solar power without using precious land. Proc Natl Acad Sci USA 120(9):e2301355120. https://doi.org/10.1073/pnas.2301355120

Bayer P, Aklin M (2020) The European Union emissions trading system reduced CO_2 emissions despite low prices. Proc Natl Acad Sci USA 117(16):8804–8812. https://doi.org/10.1073/pnas.1918128117

Behrani P, Isha ASN, Salleh R, Al Mekhlafi ABA (2023) Determination of occupational health and safety risks in solar energy. KnE Soc Sci 142–150. https://doi.org/10.18502/kss.v8i20.14599

Bensch G, Grimm M, Peters J (2015) Why do households forego high returns from technology adoption? Evidence from improved cooking stoves in Burkina Faso. J Econ Behav Organ 116:187–205. https://doi.org/10.1016/j.jebo.2015.04.023

Bölük G, Kaplan R (2022) Effectiveness of renewable energy incentives on sustainability: evidence from dynamic panel data analysis for the EU countries and Turkey. Environ Sci Pollut Res Int 29(18):26613–26630. https://doi.org/10.1007/s11356-021-17801-y

Carey J (2023) The other benefit of electric vehicles. Proc Natl Acad Sci USA 120(3):e2220923120. https://doi.org/10.1073/pnas.2220923120

Chang CH, Lo SF (2022) Impact analysis of a national and corporate carbon emission reduction target on renewable electricity use: a review. Energies 15(5):1794. https://doi.org/10.3390/en15051794

Chen S, Xie J (2022) Application prospect analysis of molten salt energy storage technology. Highlights Sci Eng Technol 26, 46–51. https://doi.org/10.54097/hset.v26i.3641

D'Amico G, L'Abbate P, Liao W, Yigitcanlar T, Ioppolo G (2020) Understanding sensor cities: insights from technology giant company driven smart urbanism practices. Sensors 20(16):4391. https://doi.org/10.3390/s20164391

Detka K, Górecki K (2023) Selected technologies of electrochemical energy storage—a review. Energies 16(13):5034. https://doi.org/10.3390/en16135034

ESFC Investment Group (n.d.) Noor Ouarzazate: the world's largest concentrated solar power plant (CSP) built in Morocco. ESFC Investment Group. https://esfccompany.com/en/articles/solar-energy/noor-ouarzazate-the-world-s-largest-concentrated-solar-power-plant-csp-built-in-morocco/. Accessed 14 Oct 2024

Ferrante G, Fasola S, Cilluffo G, Piacentini G, Viegi G, Grutta S (2022) Addressing exposome: an innovative approach to environmental determinants in pediatric respiratory health. Front Public Health 10. https://doi.org/10.3389/fpubh.2022.871140

Gasimli O, Haq IU, Munir S, Khalid MH, Gamage SKN, Khan A, Ishtiaq M (2022) Globalization and sustainable development: empirical evidence from CIS countries. Sustainability 14(22):14684. https://doi.org/10.3390/su142214684

Giannadaki D, Lelieveld J, Pozzer A (2016) Implementing the us air quality standard for pm2.5 worldwide can prevent millions of premature deaths per year. Environ Health 15(1). https://doi.org/10.1186/s12940-016-0170-8

Gillingham KT, Huang P, Buehler C, Peccia J, Gentner D R. (2021) The climate and health benefits from intensive building energy efficiency improvements. Sci Adv 7(34):eabg0947. https://doi.org/10.1126/sciadv.abg0947

Green KW, Zelbst PJ, Bhadauria VS, Meacham J (2012) Do environmental collaboration and monitoring enhance organizational performance? Ind Manag Data Syst 112(2):186–205. https://doi.org/10.1108/02635571211204254

Guidry VT, Thie L, Money EB (2020) Sidebar: health benefits of north Carolina's transition to clean energy. North Carolina Med J 81(5):334–335. https://doi.org/10.18043/ncm.81.5.334

Gyimah J, Batasuma S, Yao X, Wauk G (2024) The adoption of renewable energy towards environmental sustainability: evidence from Partial Least Square Structural Equation Modelling (PLS-SEM). PLoS ONE 19(4):e0299727. https://doi.org/10.1371/journal.pone.0299727

Habib A, Sou C (2018) Analytical review of the trends and present situation of large-scale sustainable energy storage technology. Eur J Sustain Dev Res 2(3). https://doi.org/10.20897/ejosdr/86200

Haidi T, Cheddadi B (2022) Wind energy integration in Africa: development, impacts and barriers. Int J Electr Comput Eng (IJECE) 12(5):4614. https://doi.org/10.11591/ijece.v12i5.pp4614-4622

Hailemariam M, Bustos T, Montgomery B, Barajas R, Evans LB, Drahota A (2019) Evidence-based intervention sustainability strategies: a systematic review. Implement Sci 14(1). https://doi.org/10.1186/s13012-019-0910-6

Hendryx M, Zullig KJ, Luo J (2020) Impacts of coal use on health. Annu Rev Public Health 41(1):397–415. https://doi.org/10.1146/annurev-publhealth-040119-094104

Hoang AT, Nižetić S, Olcer AI, Ong HC, Chen WH, Chong CT, Thomas S, Bandh SA, Nguyen XP (2021) Impacts of covid-19 pandemic on the global energy system and the shift progress to renewable energy: opportunities, challenges, and policy implications. Energy Policy 154:112322. https://doi.org/10.1016/j.enpol.2021.112322

Höijer K, Lindö C, Mustafa A, Nyberg M, Olsson V, Rothenberg E, Sepp H, Wendin K (2020) Health and sustainability in public meals—an explorative review. Int J Environ Res Public Health 17(2):621. https://doi.org/10.3390/ijerph17020621

Huang JW, Li YH (2018) How resource alignment moderates the relationship between environmental innovation strategy and green innovation performance. J Bus Ind Market 33(3):316–324. https://doi.org/10.1108/jbim-10-2016-0253

Huizing R, Chen H, Wong F (2015) Contaminant transport in membrane-based energy recovery ventilators. Sci Technol Built Environ 21(1):54–66. https://doi.org/10.1080/10789669.2014.969171

Izah SC, Richard G, Stanley HO, Sawyer WE, Ogwu MC, Uwaeme OR (2023) Integrating the one health approach and statistical analysis for sustainable aquatic ecosystem management and trace metal contamination mitigation. ES Food Agroforest 14:1012. https://doi.org/10.30919/esfaf1012

Izah SC, Richard G, Stanley HO, Sawyer WE, Ogwu MC, Uwaeme OR (2024a) Potential applications of linear regression models in studying the relationship between fish and contaminants in their environment: one health perspective. Juniper Online J Public Health 8(4):555743. https://doi.org/10.19080/JOJPH.2024.08.555743

Izah SC, Richard G, Stanley HO, Sawyer WE, Ogwu MC, Uwaeme OR (2024b) Prospects and application of multivariate and reliability analyses to one health risk assessments of toxic elements. Toxicol Environ Heal Sci 16(2):127–134

Izah SC, Ogwu MC, Shahsavani A, Kermani M, Hasanzadeh V (2024c) Understanding the one health implications of air pollution. In: Izah SC, Ogwu MC, Shahsavani A (eds) Air pollutants in the context of one health: fundamentals, sources and impacts. The Handbook in Environmental Chemistry. Springer, Berlin, Heidelberg. https://link.springer.com/chapter/10.1007/698_2024_1129

Izah SC, Ogwu MC, Etim NG, Shahsavani A, Namvar Z (2024d). Short-term health effects of air pollution. In: Izah SC, Ogwu MC, Shahsavani A (eds) Air pollutants in the context of one health: fundamentals, sources and impacts. The Handbook in Environmental Chemistry. Springer, Berlin, Heidelberg.

Izah SC, Amabie T, Aigberua AO (2024e) Technological advancements for sustainable clean air. In Ogwu MC, Izah SC (eds) Sustainable strategies for air pollution mitigation: development,

economics and technologies. Part of the Bok Series: The Handbook in Environmental Chemistry, Springer, Berlin, Heidelberg. https://link.springer.com/chapter/10.1007/698_2024_1120
Jagger P, Jumbe C (2016) Stoves or sugar? willingness to adopt improved cookstoves in Malawi. Energy Policy 92:409–419. https://doi.org/10.1016/j.enpol.2016.02.034
Jones LE, Olsson G (2017) Solar photovoltaic and wind energy providing water. Global Chall (Hoboken, NJ) 1(5):1600022. https://doi.org/10.1002/gch2.201600022
Juarez PD, Matthews-Juarez P, Hood DB, Im W, Levine RS, Kilbourne BJ, Langston MA, Al-Hamdan MZ, Crosson WL, Estes MG et al (2014) The public health exposome: a population-based, exposure science approach to health disparities research. Int J Environ Res Public Health 11(12):12866–12895. https://doi.org/10.3390/ijerph111212866
Kaartemo V, Gonzalez-Perez MA (2020) Renewable energy in international business. Crit Perspect Int Bus 16(4):325–336. https://doi.org/10.1108/cpoib-08-2019-0062
Khan SAR (2019) The role of renewable energy, public health expenditure, logistics and environmental performance in economic growth: an evidence from structural equation modelling.. https://doi.org/10.20944/preprints201901.0102.v1
Kimbara T, Fujii H, Kaneko S (2012) Analysis of environmental patents in Japan from 1964 to 2008. Int Bus Res 6(1). https://doi.org/10.5539/ibr.v6n1p145
Kumar JCR, Majid MA (2020) Renewable energy for sustainable development in India: current status, future prospects, challenges, employment, and investment opportunities. Energy Sustain Soc 10(1). https://doi.org/10.1186/s13705-019-0232-1
Kusch-Brandt S (2019) Urban renewable energy on the upswing: a spotlight on renewable energy in cities in ren21's "renewables 2019 global status report." Resources 8(3):139. https://doi.org/10.3390/resources8030139
Kut P, Pietrucha-Urbanik K (2023) Bibliometric analysis of renewable energy research on the example of the two European countries: insights, challenges, and future prospects. Energies 17(1):176. https://doi.org/10.3390/en17010176
Le PA, Trung VD, Nguyen PL, Bac Phung TV, Natsuki J, Natsuki T (2023) The current status of hydrogen energy: an overview. RSC Adv 13(40):28262–28287. https://doi.org/10.1039/d3ra05158g
Lee SH, Lee W, Hyun JH, Bhang BG, Choi J, Ahn HK (2020) Capacity design and cost analysis of converged renewable energy resources by considering base load conditions in residential and industrial areas. Appl Sci 10(21):7822. https://doi.org/10.3390/app10217822
Li P, Lv H, Li Z, Meng X, Lin Z, Wang R, Li X (2021) The electrostatic attraction and catalytic effect enabled by ionic–covalent organic nanosheets on mxene for separator modification of lithium–sulfur batteries. Adv Mater 33(17). https://doi.org/10.1002/adma.202007803
Liu S, Wu S, Hu Y, Li H (2019) Comparative analysis of air and CO2 as working fluids for compressed and liquefied gas energy storage technologies. Energy Convers Manag 181:608–620. https://doi.org/10.1016/j.enconman.2018.12.031
Llamas B, Ortega M, Barthelemy G, Godos I, Acién FG (2020) Development of an efficient and sustainable energy storage system by hybridization of compressed air and biogas technologies (bio-cases). Energy Convers Manag 210:112695. https://doi.org/10.1016/j.enconman.2020.112695
Maipas S, Konstantinidou A, Lazaris AC, Kavantzas N (2021) Environmental health education: from museum specimens and math word problems to virtual and augmented reality. Environ Health Insights 15:117863022110663. https://doi.org/10.1177/11786302211066355
Makki A, Mosly I (2020) Factors affecting public willingness to adopt renewable energy technologies: an exploratory analysis. Sustainability 12(3):845. https://doi.org/10.3390/su12030845
Malik AA, Lenzen M, McAlister S, McGain F (2018) The carbon footprint of Australian health care. Lancet Planet Health 2(1):e27–e35. https://doi.org/10.1016/s2542-5196(17)30180-8
Maslin MA, Lang J, Harvey F (2023) A short history of the successes and failures of the international climate change negotiations. UCL Open Environ 5:e059. https://doi.org/10.14324/111.444/ucloe.000059

McMonagle R, Castellanos I, Patel S (2021) Advancing energy justice as a climate and public health solution. World Med Health Policy 13(1):146–153. https://doi.org/10.1002/wmh3.420

Moran EF, Lopez MC, Moore N, Müller N, Hyndman DW (2018) Sustainable hydropower in the 21st century. Proc Natl Acad Sci USA 115(47):11891–11898. https://doi.org/10.1073/pnas.1809426115

Nahhas AM (2023) Introductory chapter: overview of renewable energy. https://doi.org/10.5772/intechopen.112619

Ohanu CP, Rufai SA, Oluchi UC (2024) A comprehensive review of recent developments in smart grid through renewable energy resources integration. Heliyon 10(3):e25705. https://doi.org/10.1016/j.heliyon.2024.e25705

Perlaviciute G, Schuitema G, Devine-Wright P, Ram B (2018) At the heart of a sustainable energy transition: the public acceptability of energy projects. IEEE Power Energ Mag 16(1):49–55. https://doi.org/10.1109/mpe.2017.2759918

Phillip E, Langevin J, Davis M, Kumar N, Walsh A, Jumbe V, Clifford M, Conroy R, Stanistreet D (2023) Improved cookstoves to reduce household air pollution exposure in sub-Saharan Africa: a scoping review of intervention studies. PLoS ONE 18(4):e0284908. https://doi.org/10.1371/journal.pone.0284908

Piran M, Sharifi A, Safari MM (2023) Exploring the roles of education, renewable energy, and global warming on health expenditures. Sustainability 15(19):14352. https://doi.org/10.3390/su151914352

Pitso T (2019) Clean coal technology adaptability and r and d support for efficiency and sustainability. https://doi.org/10.5772/intechopen.81605

Pouran HM, Lopes MPC, Nogueira T, Branco DAC, Sheng Y (2022) Environmental and technical impacts of floating photovoltaic plants as an emerging clean energy technology. iScience 25(11):105253. https://doi.org/10.1016/j.isci.2022.105253

Qureshi MI, Rasli AM, Zaman K (2016) Energy crisis, greenhouse gas emissions and sectoral growth reforms: repairing the fabricated mosaic. J Clean Prod 112:3657–3666. https://doi.org/10.1016/j.jclepro.2015.08.017

Sait MA, Chigbu UE, Hamiduddin I, De Vries WT (2018) Renewable energy as an underutilised resource in cities: germany's 'energiewende' and lessons for post-brexit cities in the united kingdom. Resources 8(1):7. https://doi.org/10.3390/resources8010007

Salkuti SR (2018) Large-scale electricity storage technology options for smart grid. Int J Eng Technol 7(2):635. https://doi.org/10.14419/ijet.v7i2.11395

Sârbu I, Sebarchievici C (2018) A comprehensive review of thermal energy storage. Sustainability 10(1):191. https://doi.org/10.3390/su10010191

Schell SF, Luke DA, Schooley MW, Elliott MB, Herbers SH, Mueller NB, Bunger AC (2013) Public health program capacity for sustainability: a new framework. Implement Sci 8(1). https://doi.org/10.1186/1748-5908-8-15

Shaikh RA, Vowles DJ, Dinovitser A, AllisonA, Abbott D (2024) Robust capital cost optimization of generation and multitimescale storage requirements for a 100% renewable Australian electricity grid. PNAS nexus 3(4):127. https://doi.org/10.1093/pnasnexus/pgae127

Sharma L, Singh J, Dhiman R, Vargas Nunez DR, Ba AE, Joshi KJ, Modi B, Padhi A, Tandon JK, Seidel M (2024) Advancing solar energy for primary healthcare in developing nations: addressing current challenges and enabling progress through UNICEF and collaborative partnerships. Cureus 16(1):e51571. https://doi.org/10.7759/cureus.51571

Sharma V (2023) Overcoming barriers: strategies for accelerating adoption of renewable energy technologies for net zero goal. J Waste Manag Recycl Technol 1(1):1–3. https://doi.org/10.47363/jwmrt/2023(1)124

Sherif M, Ibrahiem DM, El-Aasar KM (2022) Investigating the potential role of innovation and clean energy in mitigating the ecological footprint in n11 countries. Environ Sci Pollut Res 29(22):32813–32831. https://doi.org/10.1007/s11356-021-18477-0

Shi W, Bi J, Liu R, Liu M, Ma Z (2021) Decrease in the chronic health effects from pm2.5 during the 13th five-year plan in china: impacts of air pollution control policies. J Clean Prod 317:128433. https://doi.org/10.1016/j.jclepro.2021.128433

Siler-Evans K, Azevedo IL, Morgan MG, Apt J (2013) Regional variations in the health, environmental, and climate benefits of wind and solar generation. Proc Natl Acad Sci 110(29):11768–11773. https://doi.org/10.1073/pnas.1221978110

Silva S, Laranjeira E, Soares I (2021) Health benefits from renewable electricity sources: a review. Energies 14(20):6678. https://doi.org/10.3390/en14206678

Smith KR, Frumkin H, Balakrishnan K, Butler CD, Chafe ZA, Fairlie I, Kinney P, Kjellstrom T, Mauzerall DL, McKone TE et al (2013) Energy and human health. Annu Rev Public Health 34(1):159–188. https://doi.org/10.1146/annurev-publhealth-031912-114404

Soewarno N, Tjahjadi B, Fithrianti F (2019) Green innovation strategy and green innovation: the roles of green organizational identity and environmental organizational legitimacy. Manag Decis 57(11):3061–3078. https://doi.org/10.1108/md-05-2018-0563

Sun H, Xu H, Teng X, Qian T (2022) Research status and development of air-liquid energy storage technology. In International conference on mechanisms and robotics (ICMAR 2022), vol 12331. SPIE, pp 318–323. https://doi.org/10.1117/12.2652594

Sun H, Xu H, Teng X, Liu X, Zhu J (2022, February) Characteristic analysis of compressed air energy storage system based on intermediate cooling. J Phys: Conf Series, vol 2205, No 1. IOP Publishing, p 012002. https://doi.org/10.1088/1742-6596/2205/1/012002

Tabrizian S (2019) Technological innovation to achieve sustainable development—renewable energy technologies diffusion in developing countries. Sustain Dev 27(3):537–544. https://doi.org/10.1002/sd.1918

Torres J, Blanco M, Lafoz M, Navarro G, Nájera J, Santos-Herrán M (2020) Dimensioning methodology of energy storage systems for power smoothing in a wave energy conversion plant considering efficiency maps and filtering control techniques. Energies 13(13):3380. https://doi.org/10.3390/en13133380

Trahey L, Brushett FR, Balsara NP, Ceder G, Cheng L, Chiang YM, Hahn NT, Ingram BJ, Minteer SD, Moore JS, Mueller KT, Nazar LF, Persson KA, Siegel DJ, Xu K, Zavadil KR, Srinivasan V, Crabtree GW (2020) Energy storage emerging: a perspective from the Joint Center for Energy Storage Research. Proc Natl Acad Sci USA 117(23):12550–12557. https://doi.org/10.1073/pnas.1821672117

U.S. DoE (Department of Energy) (2023) DOE expands programs to increase access to solar energy for disadvantaged communities. Office of Energy Efficiency & Renewable Energy. https://www.energy.gov/eere/articles/doe-expands-programs-increase-access-solar-energy-disadvantaged-communities. Accessed 14 Oct 2024

Veras-Estévez BA, Chapman HJ (2023) Incorporating one health in health professions curriculum in the Dominican republic. Innovare Revista De Ciencia Y Tecnología 12(3):122–123. https://doi.org/10.5377/innovare.v12i3.17171

Wall WP, Khalid B, Urbański M, Kot M (2021) Factors influencing consumer's adoption of renewable energy. Energies 14(17):5420. https://doi.org/10.3390/en14175420

Wang P, Yang X, Zhang Q (2022) Physical energy storage technologies: basic principles, parameters and applications. Highlights Sci Eng Technol 3, 73–84. https://doi.org/10.54097/hset.v3i.695

Wang Q, Tan Z, De G, Pu L, Wu J (2019) Research on promotion incentive policy and mechanism simulation model of energy storage technology. Energy Sci Eng 7(6):3147–3159. https://doi.org/10.1002/ese3.485

Wehbi H (2024) Powering the future: an integrated framework for clean renewable energy transition. Sustainability 16:5594. https://doi.org/10.3390/su16135594

Wei P, Abid M, Adun H, Kemena Awoh D, Cai D, Zaini JH, Bamisile O (2023) Progress in energy storage technologies and methods for renewable energy systems application. Appl Sci 13(9):5626. https://doi.org/10.3390/app13095626

Winkler L, Pearce D, Nelson J, Babacan O (2023) The effect of sustainable mobility transition policies on cumulative urban transport emissions and energy demand. Nat Commun 14(1):2357. https://doi.org/10.1038/s41467-023-37728-x

Wiser R, Mai T, Millstein D, Macknick J, Carpenter A, Cohen S, Cole W, Frew B, Heath G (2016) On the path to sunshot. The environmental and public health benefits of achieving high solar penetrations in the United States (No. NREL/TP-6A20-65628; LBNL-1004373). National Renewable Energy Lab.(NREL), Golden, CO (United States)

Wu T (2023) Frontiers of energy storage technologies. Int J Energy Res 2023:1–14. https://doi.org/10.1155/2023/7138669

Yao J, Hu Y, Liu S, Li H (2019) An experimental study on compressed carbon dioxide energy storage. Destech Trans Environ Energy Earth Sci (ICEEE). https://doi.org/10.12783/dteees/iceee2019/31825

Yao P (2023a) Development trends and challenges of energy storage technology in coal-fired power plants. Smart Syst Green Energy 5(1):11–16. https://doi.org/10.23977/ssge.2023.050102

Yao Y (2023b) The effect of spherical filled particles on the energy storage density of composite materials. J Phys: Conf Ser 2578(1):012022. https://doi.org/10.1088/1742-6596/2578/1/012022

Zhang Q, Zheng Y, Tong D, Shao M, Wang S, Zhang Y, Xu X, Wang J, He H, Liu W et al (2019) Drivers of improved pm 2.5 air quality in China from 2013 to 2017. Proc Natl Acad Sci 116(49):24463–24469. https://doi.org/10.1073/pnas.1907956116

Zhang Y (2023) Diverse promising electrode materials for advanced sodium-ion batteries. Highlights Sci Eng Technol 46:298–305. https://doi.org/10.54097/hset.v46i.7717

Zhou S, Solomon BD, Brown MA (2024) The spillover effect of mandatory renewable portfolio standards. Proc Natl Acad Sci USA 121(25):e2313193121. https://doi.org/10.1073/pnas.2313193121

Zulu S, Zulu E, Chabala M (2021) Factors influencing households' intention to adopt solar energy solutions in zambia: insights from the theory of planned behaviour. Smart Sustain Built Environ 11(4):951–971. https://doi.org/10.1108/sasbe-01-2021-0008

Chapter 8
Smart Technologies in Environmental Monitoring: Enhancing Real-Time Data for Health Management

Sylvester Chibueze Izah

Abstract Integrating smart technologies into environmental monitoring represents a transformative approach to managing health risks associated with environmental changes. This chapter explores various intelligent technologies, such as the Internet of Things (IoT), remote sensing, wearable health sensors, and data analytics, which provide real-time data crucial for enhancing health response systems. These technologies enable continuous monitoring of environmental parameters—air quality, water quality, temperature, and more—allowing for rapid identification of health hazards and enabling timely interventions. Additionally, the chapter examines how advances in artificial intelligence (AI) are revolutionizing data analytics by supporting predictive models that can forecast disease outbreaks, pollution events, and other environmental health risks. By leveraging AI-driven models, public health officials and environmental agencies can proactively address potential threats, ensuring a quicker and more targeted response to emerging health concerns. The role of AI in data analytics is also discussed, highlighting how AI-driven predictive models can support disease surveillance and anticipate environmental health risks. Examples of smart technology applications in health management illustrate the potential for these innovations to enhance public health responses. However, implementing smart environmental monitoring systems is challenging, as issues such as data privacy, infrastructure costs, and data reliability remain significant barriers. The chapter concludes by promoting a proactive, data-driven approach to environmental health management. To overcome current smart technologies challenges in environmental health monitoring, there is a need for an integrated approach that brings together technology, health, and environmental management.

S. Chibueze Izah (✉)
Department of Community Medicine, Faculty of Clinical Sciences, Bayelsa Medical University, Yenagoa, Bayelsa State, Nigeria
e-mail: sylvester.izah@bmu.edu.ng

Department of Microbiology, Faculty of Science, Bayelsa Medical University, Yenagoa, Bayelsa State, Nigeria

S. Chibueze Izah and M. Chidozie Ogwu (eds.), *Innovative Approaches in Environmental Health Management*, Environmental Science and Engineering,
https://doi.org/10.1007/978-3-031-81966-7_8

Keywords Smart technologies · Environmental monitoring · Data analytics · Health response systems · Public health · Real-time data · Wearable health sensors · Predictive analytics

8.1 Introduction

Integrating smart technologies in environmental monitoring significantly advances our ability to collect, analyze, and respond to environmental data in real time. These technologies enhance our understanding of ecological conditions and play a crucial role in health management by providing timely information that can inform public health interventions. These technologies leverage advancements in the Internet of Things (IoT), artificial intelligence (AI), and data analytics to enhance our understanding of environmental conditions and their impacts on public health. Innovative technologies involve a range of devices and systems that utilize sensors, connectivity, and data processing capabilities to monitor and manage environmental conditions. These technologies include IoT Sensors for ecological data collection; remote sensing technologies like satellite imagery and aerial drones, which provide valuable data on land use, vegetation cover, and climate patterns and data analytics and machine learning employed to analyze the vast amounts of data generated by smart technologies (Haji and Sallow 2021; Ngwenya et al. 2023). These tools can identify patterns, predict trends, and inform decision-making processes related to environmental management and public health. By leveraging big data, stakeholders can gain insights into the relationships between environmental factors and health outcomes (Zhe et al. 2019). Real-time data collection is critical for effective environmental health management.

Monitoring environmental conditions continuously allows for rapid identification of environmental hazards, such as air or water pollution events (Jacob et al. 2024a, b, c, d). For instance, smart air quality monitoring systems can alert authorities and the public to hazardous pollution levels, facilitating timely interventions to protect public health. This proactive approach can mitigate health risks associated with exposure to environmental contaminants (Grishin et al. 2021). Access to real-time data empowers policymakers and public health officials to make informed decisions regarding environmental management and health interventions. For example, data on water quality can guide decisions about water treatment and distribution, ensuring communities have access to safe drinking water (Dwevedi et al. 2018). By integrating real-time data into decision-making processes, stakeholders can enhance the effectiveness of public health strategies. Real-time environmental monitoring can strengthen community engagement by providing residents with information about local environmental conditions. Mobile applications and online platforms can disseminate real-time air and water quality data, empowering individuals to make informed choices about their health and well-being. This transparency fosters a sense of community involvement in environmental stewardship (Zhang and Huang 2019).

The Internet of Things (IoT) has revolutionized environmental monitoring by enabling the deployment of interconnected sensors that collect data on various environmental parameters. These sensors can monitor air quality, water quality, temperature, humidity, and other critical factors influencing human health (Zhang and Huang 2019). Edge computing further enhances this capability by allowing data processing to occur closer to the source of data collection, reducing latency and enabling real-time analysis (Zhang and Huang 2019). IoT and edge computing combination facilitates rapid responses to environmental changes and potential health risks. Wireless Sensor Networks (WSNs) are distributed sensor nodes that communicate wirelessly to monitor ecological conditions over large areas. These networks can be deployed in various settings, including urban areas, agricultural fields, and natural ecosystems, to monitor environmental health indicators (Lazarescu 2013) continuously. The data collected by WSNs can be used to assess pollution levels, track biodiversity changes, and monitor conservation efforts' effectiveness, ultimately contributing to public health management. Advances in satellite and aerial remote sensing technologies allow for the monitoring of environmental changes on a global scale. These technologies can provide valuable data on land use changes, deforestation, and climate change impacts, which are critical for understanding the broader context of environmental health. By integrating remote sensing data with ground-based monitoring, researchers can comprehensively understand ecological conditions and their implications for human health (Haahtela 2019).

These smart technologies are revolutionizing real-time air quality monitoring, providing critical information about pollutants affecting respiratory health. For example, drones equipped with air quality sensors allow for collecting data on pollutants at different altitudes and locations, facilitating a more nuanced understanding of air quality dynamics (Ranganathan et al. 2023). This information can inform public health advisories and interventions, particularly in urban areas where air pollution is a significant concern. Smart environmental monitoring systems can also assess water quality in real-time. Technologies like wireless sensor networks can continuously monitor water bodies' parameters like pH, turbidity, and contaminant levels (Mingfei et al. 2010). This data is essential for ensuring safe drinking water and preventing waterborne diseases. By providing timely information on water quality, these systems can help public health officials respond quickly to contamination events and protect community health. Integrating environmental monitoring with health surveillance systems can enhance our understanding of the relationships between environmental factors and health outcomes. For instance, the Stell-ARI study focuses on integrated surveillance for acute respiratory infections (ARIs), combining clinical and ecological data to identify risk factors and inform prevention strategies (Carducci et al. 2023). This holistic approach allows for a more comprehensive understanding of how environmental conditions influence health, enabling targeted interventions.

Bioremediation is a practical application of smart technologies in environmental monitoring that significantly impacts human health. This process involves using microorganisms to degrade pollutants in contaminated environments, restoring ecosystem health and reducing health risks associated with pollution. The effectiveness of bioremediation relies on understanding the microbial communities involved in the degradation of pollutants. Smart technologies can facilitate real-time monitoring of these microbial populations, providing insights into their dynamics and effectiveness in breaking down contaminants. Researchers can assess the diversity and functionality of microbial communities involved in bioremediation efforts by employing advanced techniques such as DNA sequencing and metagenomics (Hait et al. 2024b). Practical bioremediation can promote public health by reducing exposure to harmful pollutants. For example, the successful cleanup of oil spills through bioremediation restores the affected ecosystems and protects human health by minimizing the risk of exposure to toxic substances. By monitoring the progress of bioremediation efforts and assessing the resulting improvements in environmental quality, public health officials can better understand the health impacts of pollution and the benefits of remediation. Engaging local communities in bioremediation projects can enhance the effectiveness of these initiatives while promoting public health awareness. Community-led bioremediation efforts empower residents to restore their environments actively, fostering a sense of ownership and responsibility for local ecosystems. By combining scientific expertise with local knowledge, these projects can achieve more sustainable outcomes and improve community health.

This chapter explores smart technologies' role in environmental monitoring and how real-time data contributes to effective health management. It will cover various smart technologies—such as IoT sensors, remote sensing, and wearable devices—that enable the continuous monitoring of air, water, and soil quality, providing critical insights for public health interventions. By integrating data analytics and AI, these technologies enhance our ability to interpret complex environmental data, anticipate health risks, and support decision-making processes. Through case studies, the chapter will illustrate real-world applications of these technologies, demonstrating their impact on public health in diverse settings. Additionally, the chapter will address the challenges of adopting smart environmental monitoring systems, including data privacy concerns, infrastructure costs, and issues related to data reliability. It will also explore emerging trends, such as advancements in sensor technology and AI integration, and emphasize the importance of sustainable and collaborative approaches for impactful environmental health management. Ultimately, this chapter aims to provide policymakers, health professionals, and technology developers with insights and recommendations for leveraging smart technologies to support environmental health policies and build resilient health management systems.

8.2 Some Types of Smart Environmental Monitoring Technologies

Integrating smart technologies into environmental monitoring has revolutionized collecting, analyzing, and responding to environmental data (Izah et al. 2024). These technologies facilitate real-time tracking of various ecological parameters, which are crucial for managing public health and environmental risks. The different types of smart environmental monitoring technologies include IoT sensors, remote sensing and satellite technologies, and wearable and mobile health sensors. These smart technologies are transforming our ability to assess and respond to environmental conditions that impact human health.

Internet of Things (IoT) Sensors

IoT sensors are vital in modern environmental monitoring, providing continuous data collection and analysis capabilities. These sensors are embedded in various environments to monitor air quality, water quality, soil health, and agricultural conditions. IoT sensors are extensively used for air quality monitoring, providing real-time data on pollutants such as particulate matter (PM), nitrogen dioxide (NO2), sulfur dioxide (SO2), and ozone (O3). These sensors can be deployed in urban areas, industrial sites, and near traffic hotspots to assess air quality and identify pollution sources (Jo et al. 2020). For example, a study demonstrated the effectiveness of an IoT-based air quality monitoring system in subway tunnels, where real-time data on PM concentrations were collected and analyzed, allowing for timely interventions to improve air quality (Jo et al. 2020). The data generated by these sensors can be integrated into public health systems to inform communities about air quality levels and associated health risks. This information is particularly vital for vulnerable populations, such as children and the elderly, who may be more susceptible to the adverse effects of air pollution (Grishin et al. 2021).

IoT sensors are also employed in water quality monitoring, enabling the continuous assessment of parameters such as pH, turbidity, dissolved oxygen, and contaminant levels. These sensors can be installed in rivers, lakes, and water treatment facilities to detect pollutants and ensure safe drinking water (Zhang and Huang 2019). For instance, a smart water quality monitoring system can provide real-time alerts when contaminant levels exceed safe thresholds, allowing immediate action to protect public health. The integration of IoT technology with cloud computing facilitates data sharing and analysis, enabling stakeholders to monitor water quality trends over time and make informed decisions regarding water management (Haji and Sallow 2021). This capability is crucial for preventing waterborne diseases and ensuring access to clean water. IoT sensors are critical in monitoring soil health and optimizing resource use in agriculture. These sensors can measure soil moisture, temperature, and nutrient levels, providing farmers with valuable insights into crop conditions (Dai, 2021). By utilizing real-time data, farmers can make informed decisions about irrigation, fertilization, and pest management, ultimately enhancing crop yields and sustainability. For example, a study highlighted IoT-based platforms for precision agriculture, where sensors collected data on soil parameters to optimize irrigation practices (Jo et al. 2020). This approach improves agricultural productivity, reduces water consumption, and minimizes environmental impacts.

Remote Sensing and Satellite Technologies

Remote sensing technologies, including satellite imagery and aerial drones, provide valuable data for environmental monitoring on a larger scale. These technologies enable assessing land use changes, habitat loss, and climate impacts. Satellite imagery is a powerful tool for monitoring environmental changes over time. It allows researchers to observe large areas and assess changes in land cover, vegetation health, and urban development (Dwevedi et al. 2018). For instance, satellite data can track deforestation rates in tropical regions, providing insights into the impacts of land use changes on biodiversity and ecosystem services (Kencono and Ahsany 2023). The ability to analyze historical satellite imagery enables researchers to identify trends and patterns in environmental change, informing conservation efforts and policy decisions. This data is precious for understanding the long-term impacts of climate change on ecosystems and human health. Drones equipped with sensors and cameras are increasingly used for environmental monitoring, providing high-resolution data on specific areas. Drones can be deployed to assess habitat conditions, monitor wildlife populations, and evaluate the effectiveness of conservation efforts (Baviskar et al. 2024). Their ability to access remote or difficult-to-reach areas makes them a valuable tool for researchers and conservationists. For example, drones have been used to monitor coral reef health and assess the impacts of climate change on marine ecosystems. By capturing detailed imagery and data, drones can provide insights into coral bleaching events and inform management strategies (Sagheer et al. 2020). A notable case study involves satellite monitoring to track deforestation and habitat loss in the Amazon rainforest. Researchers have utilized satellite imagery to assess changes in forest cover over time, identifying areas at risk of deforestation due to agricultural expansion and logging activities (Bibri 2021). This data has been instrumental in informing conservation policies and enforcement efforts to protect critical habitats. Integrating satellite monitoring with ground-based data collection enhances the understanding of the drivers of deforestation and their implications for biodiversity and human health. By providing real-time data on habitat loss, stakeholders can take proactive measures to mitigate environmental impacts and protect ecosystems (Ibaseta et al. 2018).

Wearable and Mobile Health Sensors

Wearable and mobile health sensors are emerging as valuable tools for monitoring individual exposure to environmental hazards and assessing health outcomes. These technologies enable real-time personal health monitoring, providing insights into how environmental conditions affect health. Wearable sensors can monitor personal exposure to air pollutants and UV radiation, providing individuals with real-time data on their ecological exposure (Pamudji 2023). For instance, smart wristbands equipped with air quality sensors can alert users to high pollution levels, enabling them to take precautions to protect their health. This capability is significant for individuals with respiratory conditions or heightened sensitivity to environmental pollutants. Mobile applications that integrate data from wearable sensors can provide users with personalized health recommendations based on exposure levels. This information empowers individuals to make informed decisions about their activities and environments, ultimately promoting better health outcomes (Majumder et al. 2017). In high-risk environments, such as industrial settings or areas prone to natural disasters, wearable sensors can simultaneously monitor health indicators and environmental conditions. For example, workers in hazardous environments can use wearable devices to track their exposure to toxic substances while monitoring vital signs such as heart rate and temperature (Fathy and Ali 2023). This dual monitoring capability enhances workplace safety and enables timely interventions in case of health emergencies. Several examples illustrate the application of wearable health technology in field settings. For instance, researchers have developed wearable devices that monitor air quality exposure among agricultural workers, providing insights into the health impacts of pesticide use (Hajjaji et al. 2021). Similarly, wearable sensors have been employed in disaster response scenarios to monitor the health of first responders and assess their exposure to hazardous conditions (Roşca et al. 2021).

8.3 Role of Data Analytics and Artificial Intelligence

Integrating data analytics and artificial intelligence (AI) into environmental monitoring has transformed how we understand and manage environmental health risks (Hait et al. 2024a). By leveraging big data, machine learning, and geospatial analytics, stakeholders can make informed decisions that enhance public health outcomes and promote sustainable practices.

Data Collection and Processing

Big data refers to the vast volumes of structured and unstructured data generated from various sources, including IoT sensors, satellite imagery, and social media. In environmental monitoring, big data is crucial in providing comprehensive insights into environmental conditions and trends Hariri et al. (2019). Collecting and analyzing large datasets allows researchers and policymakers to identify patterns and correlations that may not be apparent from smaller datasets. For example, the integration of big data analytics in air quality monitoring enables the identification of pollution sources and trends over time. By analyzing data from multiple sensors across urban areas, researchers can assess the effectiveness of pollution control measures and develop targeted interventions to improve air quality (Wu 2023). Similarly, big data can be used to monitor water quality, track changes in biodiversity, and assess the impacts of climate change on ecosystems.

Data Storage and Management

While the potential of big data in environmental monitoring is significant, it also presents challenges related to data storage and management. The sheer volume of data generated requires robust infrastructure and systems to store, process, and analyze the information effectively (Yang et al. 2021). Traditional data management systems may struggle to handle the scale and complexity of big data, necessitating the development of new storage solutions and architectures. Moreover, ensuring data quality and consistency is critical for accurate analysis and decision-making. Data from various sources may vary in format, accuracy, and reliability, making it essential to establish standardized protocols for data collection and processing (Munim et al. 2020). Addressing these challenges is vital for maximizing the benefits of big data in environmental monitoring.

Machine Learning and Predictive Analytics

Machine learning, a subset of AI, involves the development of algorithms that enable computers to learn from data and make predictions or decisions without explicit programming. In environmental monitoring, machine learning models can predict pollution patterns based on historical data and real-time inputs (Bernatska et al. 2023). For instance, AI algorithms can analyze air quality data to forecast pollution levels, allowing authorities to proactively issue health advisories and implement mitigation measures. These predictive models can also be used to assess the impact of various factors on pollution levels, such as traffic patterns, weather conditions, and industrial activities. By identifying the key drivers of pollution, stakeholders can develop targeted strategies to reduce emissions and improve air quality (Ragimova et al. 2020).

Identifying Emerging Environmental Health Risks

AI and machine learning can also play a critical role in identifying emerging environmental health risks. AI algorithms can detect patterns indicating potential health threats by analyzing large datasets from various sources, including health records, environmental monitoring data, and social media (Abdullah & Fakieh, 2020). For example, machine learning models can analyze data on respiratory illnesses in conjunction with air quality data to identify correlations and predict outbreaks of pollution-related health issues. Furthermore, AI can enhance surveillance systems for infectious diseases by integrating environmental data with health data to identify potential links between environmental changes and disease outbreaks. This capability is precious in the context of zoonotic diseases, where environmental factors can influence the transmission dynamics of pathogens (Stepanov 2023).

Geospatial Analytics and Geographic Information Systems

Geospatial analytics and Geographic Information Systems (GIS) are powerful tools for visualizing and analyzing spatial data related to environmental hazards. By integrating data from various sources, including satellite imagery and ground-based monitoring, GIS can create detailed maps highlighting areas at risk of environmental degradation or pollution (Akanbi 2020). These maps can inform decision-making processes by identifying priority intervention and resource allocation areas. For instance, GIS can map air pollution hotspots in urban areas, enabling targeted air quality management strategies. Similarly, GIS can help visualize the distribution of water quality issues in rivers and lakes, guiding efforts to protect aquatic ecosystems and public health (Gul et al. 2018). Geospatial analytics can also predict disease outbreaks by analyzing the spatial relationships between environmental factors and health outcomes. For example, researchers can use GIS to assess the impact of climate variables, land use changes, and pollution levels on the spread of vector-borne diseases (Kumar et al. 2021). Public health officials can implement targeted interventions to mitigate outbreaks by identifying high-risk disease transmission areas. Moreover, spatial analysis can enhance the understanding of how environmental changes, such as deforestation or urbanization, influence the emergence of infectious diseases. By integrating environmental data with health data, stakeholders can develop predictive models that inform public health strategies and resource allocation.

Real-Time data visualization for decision-making

Visualizing real-time data is essential for effective decision-making in environmental health management. Advanced data visualization tools can present complex datasets in an accessible format, enabling stakeholders to quickly grasp trends and patterns (Gavela et al. 2019). For example, interactive dashboards can display real-time air quality data alongside health advisories, allowing individuals and communities to make informed decisions about outdoor activities. Furthermore, real-time data visualization can facilitate stakeholder collaboration by providing a shared platform for monitoring environmental conditions and health risks. By integrating data from various sources, these tools can enhance communication and coordination among public health officials, environmental agencies, and community organizations (Fatima et al. 2023) (Fig. 8.1).

8.4 Applications of Real-Time Environmental Data in Health Management

Integrating real-time environmental data into health management strategies is crucial for addressing the complex interplay between environmental conditions and public health. By leveraging advanced monitoring technologies, stakeholders can gain insights into air and water quality, climate impacts, and associated health risks.

Air Quality Monitoring and Respiratory Health

The Air Quality Index (AQI) is a widely used tool for communicating air quality levels to the public. It provides a standardized measure of air pollution based on concentrations of key pollutants, including particulate matter (PM), ozone (O3), nitrogen dioxide (NO2), and sulfur dioxide (SO2). Real-time monitoring of these pollutants allows for the timely issuance of public health alerts when air quality deteriorates Edagawa et al. (2015). For example, during periods of high pollution, local health authorities can issue advisories recommending that sensitive populations, such as children and individuals with respiratory conditions, limit outdoor activities. Studies have shown timely AQI alerts can significantly reduce health risks associated with

Fig. 8.1 Value of data analytics and AI to environmental monitoring

poor air quality, such as asthma exacerbations and respiratory infections (Zhao et al. 2022). Integrating real-time AQI data into public health systems enhances community awareness and promotes proactive health management. Urban areas are often hotspots for air pollution due to high traffic volumes, industrial emissions, and other anthropogenic activities. Real-time air quality monitoring systems equipped with IoT sensors can provide continuous data on pollutant levels across different locations in a city (Di Napoli et al. 2023). This data can identify pollution hotspots and inform targeted interventions, such as traffic management strategies or emissions controls. For instance, cities like Los Angeles have implemented comprehensive air quality monitoring networks that utilize real-time data to assess pollution levels and inform public health policies (Pathaburee 2022). By analyzing patterns in air quality data, urban planners can develop strategies to reduce emissions and improve overall air quality, ultimately benefiting public health. For indoor air quality monitoring using smart devices, Benammar et al. (2018) described a system that measures multiple air pollutants and environmental factors, uses a local gateway for reliable data processing and dissemination via an open-source IoT platform (Emoncms), and features a scalable modular architecture for integrating diverse sensing technologies and wireless networks, along with a backup mechanism for data continuity during internet outages (Fig. 8.2).

Water Quality Monitoring for Safe Drinking Water

Real-time water quality monitoring is essential for safe drinking water and public health protection. A typical smart water quality monitoring system integrates different components into the IoT device (Fig. 8.3; Ullo and Sinha 2020). IoT sensors can be deployed in water distribution systems to continuously monitor parameters such as pH, turbidity, and contaminant levels, including heavy metals and pathogens (Delgado-Gardea et al. 2017). This capability allows for the immediate detection of water quality issues, enabling swift responses to contamination events. For example, advanced sensors can detect the presence of pathogens like *E. coli* or heavy metals such as lead in real time, triggering alerts for water treatment facilities and public health officials (Besmer et al. 2017). This proactive approach is critical for preventing waterborne disease outbreaks and ensuring communities can access safe drinking water. An illustrative case study involves IoT technology monitoring water quality in regions prone to waterborne diseases. In a pilot project in rural India, researchers deployed IoT sensors to monitor water quality in real-time, collecting data on microbial contamination and chemical pollutants (Saab et al. 2017). The data was integrated into a centralized system that provided alerts to local health authorities when contamination levels exceeded safe thresholds. This real-time monitoring system enabled rapid responses to potential outbreaks, allowing health officials to implement preventive measures, such as distributing safe drinking water and conducting community awareness campaigns (Casson et al. 2023). The success of this project demonstrates the potential of IoT technology in enhancing public health through effective water quality monitoring.

Climate Monitoring and Health Impacts

Climate monitoring is essential for understanding the health impacts of environmental changes. Real-time data on temperature and weather patterns can inform public health strategies related to heat-related illnesses, vector-borne diseases, and respiratory conditions exacerbated by climate change (González-González et al. 2022). For example, monitoring temperature extremes can help health authorities issue heat advisories and implement cooling centers for vulnerable populations during heatwaves. Additionally, climate data can be integrated with health data to assess the relationships between climate variables and health outcomes. This integration allows for the identification of trends and patterns that can inform public health interventions (Hall et al. 2022). Urban areas are particularly vulnerable to heat-related health risks due to the urban heat island effect, where built environments absorb and

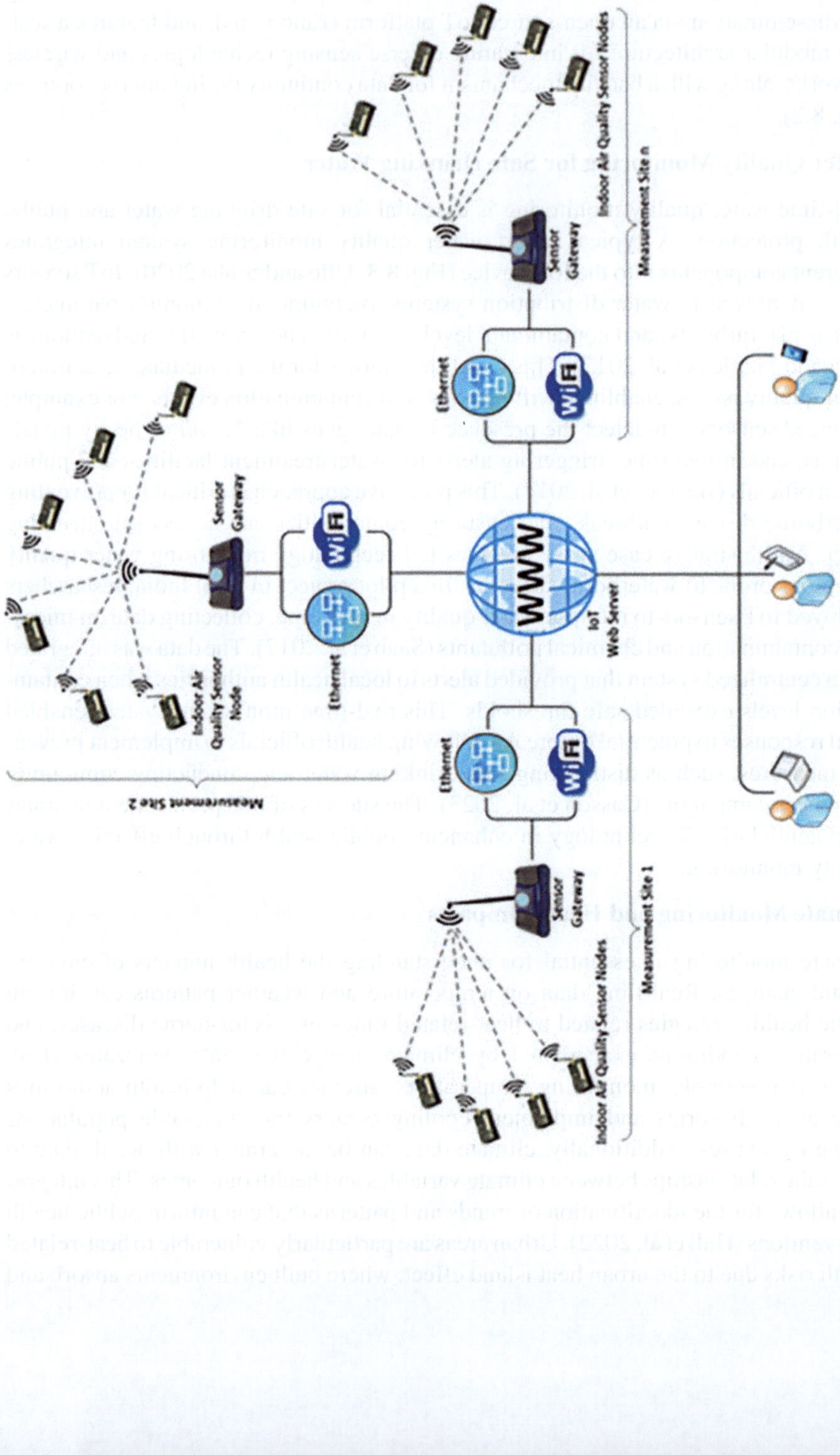

Fig. 8.2 A typical smart system for indoor air quality monitoring system. *Source* Benammar et al. (2018). http://creativecommons.org/licenses/by/4.0/

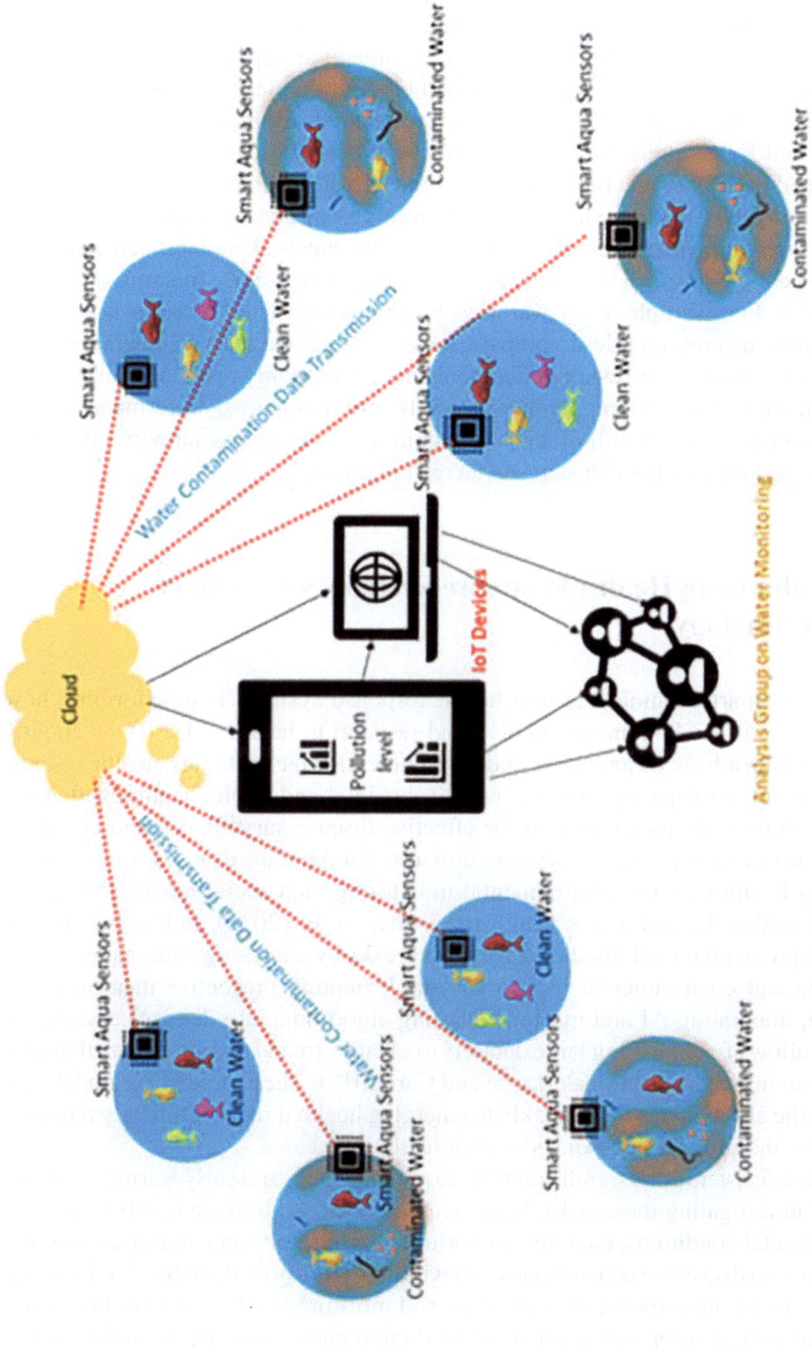

Fig. 8.3 Smart environmental monitoring system for water quality. *Source* Ullo and Sinha (2020). http://creativecommons.org/licenses/by/4.0/

retain heat. Real-time temperature monitoring systems can provide critical data to identify areas at risk of extreme heat (Abdallah et al. 2015). Public health officials can use this information to implement heat management strategies, such as opening cooling centers and providing resources for vulnerable populations. For instance, cities like Phoenix, Arizona, have developed heat action plans that utilize real-time temperature data to issue alerts and mobilize resources during extreme heat events (Tsitsifli and Kanakoudis 2022). These proactive measures can significantly reduce heat-related morbidity and mortality. Real-time climate monitoring is also crucial in early warning systems for extreme weather events like hurricanes, floods, and wildfires. By integrating meteorological data with health data, stakeholders can assess the potential health impacts of these events and implement preventive measures (Wasswa et al. 2019). For example, early warning systems can provide alerts for impending floods, allowing communities to prepare and evacuate if necessary. Integrating real-time data into emergency response plans enhances community resilience and reduces the health impacts of extreme weather events. By providing timely information, these systems enable public health officials to mobilize resources and support vulnerable populations during crises (Masters-Awatere et al. 2022).

8.5 Enhancing Health Response Systems with Smart Technology

Integrating smart technologies into health response systems is transforming how public health officials monitor, predict, and respond to health risks. By leveraging real-time data, advanced analytics, and communication technologies, health systems can enhance their capacity to manage environmental and public health challenges. Early warning systems are critical for effective disease surveillance and epidemic prediction (Table 8.1). These systems utilize real-time data from various sources, including health records, environmental monitoring, and social media, to identify potential outbreaks and assess health risks Chen et al. (2022). Public health officials can predict the likelihood of disease outbreaks by analyzing patterns in disease incidence and environmental factors and implementing preventive measures. For example, integrating AI and machine learning algorithms into disease surveillance systems allows for analyzing large datasets to identify trends and anomalies that may indicate an impending outbreak (Lurie and Carr 2018). These predictive models can enhance the ability to respond quickly to emerging health threats, ultimately reducing morbidity and mortality associated with infectious diseases.

Natural disasters pose significant risks to public health, and early warning systems are vital in mitigating these risks. Smart technologies can provide real-time data on environmental conditions, enabling authorities to issue alerts and mobilize resources in response to disasters such as floods and wildfires (Reilly et al. 2020). For instance, IoT sensors can monitor weather patterns, soil moisture levels, and river flow rates, providing critical information for flood prediction and management. In the case of

Table 8.1 Ways to enhance health response systems using smart technology, with examples of specific technologies, applications, and the health benefits they offer

Technology	Application	Health Benefits	Examples
Early Warning Systems	Real-time alerts for pollution, disease outbreaks, and extreme weather events	Reduces health risks by enabling timely preventive measures and rapid response	Air Quality Index alerts, epidemic outbreak warning systems, flood and storm alert systems
Mobile Health (mHealth) Applications	Health monitoring and data collection in remote areas	Increasing access to healthcare improves health tracking and supports rural health management	mHealth apps for remote diagnostics, SMS-based health information services
Remote Sensing and Satellite Imaging	Environmental surveillance and monitoring of changes in land use, deforestation, and habitat loss	Supports proactive health planning by monitoring environmental risks linked to health	NASA's MODIS for tracking fires and deforestation, satellite-based air quality monitoring
Internet of Things Sensors	Continuous monitoring of environmental factors (air, water, soil quality)	Provides real-time data for health protection, reducing exposure to pollutants	Smart city air quality networks and water quality monitoring in rivers and lakes
Wearable Health Sensors	Personal monitoring of air quality, UV exposure, and health indicators in high-risk areas	Enhances personal health management and reduces exposure to harmful environmental conditions	Wearable air quality monitors, UV exposure wristbands, and fitness trackers with environmental sensors
Artificial Intelligence and Predictive Analytics	Analyzing data for predicting disease outbreaks and pollution patterns	Increases preparedness for health crises improves disease management and response time	Predictive modeling for flu outbreaks and AI-based pollution forecasting
Geographic Information Systems	Mapping health hazards and tracking disease spread spatially	Facilitates targeted health interventions and resource allocation based on geographic data	GIS mapping of malaria outbreaks and spatial analysis for flood risk areas
Telemedicine and Virtual Health Assistance	Remote consultations and health services in emergencies	Expands healthcare access, especially in remote or disaster-affected areas	Telemedicine platforms, and virtual health consultations for remote locations
Drone Technology	Delivery of medical supplies and real-time environmental monitoring	Ensures timely delivery of medical resources, and enhances emergency response	Drones for medicine delivery in rural areas and drone-based air quality monitoring

wildfires, remote sensing technologies can detect changes in vegetation health and moisture levels, allowing for early identification of fire risks. By integrating these data sources into emergency response plans, public health officials can implement timely interventions to protect communities and reduce health impacts associated with natural disasters. Mobile health (mHealth) applications are increasingly used to address health challenges in rural and remote areas, particularly during environmental health crises. These applications enable individuals to access health information, monitor their health status, and communicate with healthcare providers from their mobile devices (Intrieri et al. 2017). For example, mHealth applications can provide real-time alerts about air quality or water contamination, empowering users to take preventive measures to protect their health. In remote communities, where access to healthcare services may be limited, mHealth solutions can bridge the gap by facilitating teleconsultations and remote monitoring. This capability is precious during public health emergencies, such as disease outbreaks or environmental disasters, where timely access to healthcare is critical (Lurie and Carr 2018).

AI-powered virtual health assistants can enhance health response systems by providing real-time risk assessments and health information. These systems can analyze data from various sources, including environmental monitoring, health records, and social media, to assess potential health risks and provide personalized recommendations (Lurie and Carr 2018). For example, a virtual health assistant could alert individuals to elevated air pollution levels and suggest actions to minimize exposure, such as staying indoors or using air purifiers. Moreover, virtual health assistants can facilitate communication between healthcare providers and patients, enabling remote consultations and follow-up care. This capability is critical during environmental health crises, where traditional healthcare delivery may be disrupted (Vlahu-Gjorgievska et al. 2019). The integration of real-time environmental data into public health policies is essential for developing evidence-based health guidelines. Public health officials can identify trends and correlations that inform health recommendations by analyzing air and water quality, climate conditions, and disease incidence (Brigo et al. 2022). For instance, real-time air quality data can guide the development of health advisories for vulnerable populations, such as children and individuals with respiratory conditions.

Additionally, environmental data can inform the establishment of regulatory standards for air and water quality. By utilizing real-time monitoring data, policymakers can assess the effectiveness of existing regulations and make necessary adjustments to protect public health (Donner et al. 2021). A notable case study involves real-time environmental data in policymaking for air and water quality standards. In California, the implementation of a comprehensive air quality monitoring network has allowed public health officials to assess pollution levels and their impacts on health outcomes (Marshall et al. 2022). This data-driven approach has informed the development of stringent air quality regulations to reduce emissions and improve public health.

Similarly, real-time water quality monitoring systems have been implemented in various regions to assess drinking water safety. By continuously monitoring contaminants and providing timely alerts, these systems enable public health officials to respond quickly to contamination events and protect community health. The integration of real-time data into policy decisions enhances the effectiveness of public health interventions and promotes a proactive approach to environmental health management.

8.6 Challenges in Implementing Smart Environmental Monitoring

Implementing smart environmental monitoring technologies presents numerous challenges that must be addressed to maximize their effectiveness and ensure the protection of public health. These challenges various data privacy and security concerns, infrastructure and resource limitations, and data accuracy and reliability issues.

Data Privacy and Security Concerns

As smart environmental monitoring systems collect vast amounts of data, including sensitive health and ecological information, ensuring data privacy is paramount. Integrating IoT devices in environmental monitoring raises significant concerns regarding protecting personal data, particularly when health information is involved. For instance, data collected from air quality sensors may be linked to individual health records, creating potential privacy risks if this information is not adequately protected. Moreover, sharing environmental data among various stakeholders, including government agencies, researchers, and private companies, necessitates robust data governance frameworks to ensure compliance with privacy regulations. Failure to protect sensitive data can lead to breaches of trust, legal repercussions, and adverse health outcomes for individuals whose data is compromised.

Cybersecurity Challenges in IoT Systems

The interconnected nature of IoT systems makes them vulnerable to cyberattacks, which can compromise the integrity and availability of environmental monitoring data. Cybersecurity threats, such as hacking and data breaches, pose significant risks to the functionality of smart environmental monitoring systems. For example, a cyberattack on a water quality monitoring system could disrupt data collection and lead to delayed responses to contamination events, ultimately jeopardizing public health. To mitigate these risks, it is essential to implement robust cybersecurity measures, including encryption, secure authentication protocols, and regular system updates. Additionally, educating stakeholders about cybersecurity best practices is crucial for ensuring the resilience of smart environmental monitoring systems against potential threats.

Infrastructure and Resource Limitations

Deploying smart environmental monitoring technologies often requires significant financial investment, which can hinder implementation, particularly in low-resource settings. The costs of purchasing, installing, and maintaining IoT sensors and other monitoring equipment can be prohibitive for many communities. This financial barrier can result in disparities in access to advanced monitoring technologies, particularly in developing regions with limited resources. Moreover, the ongoing costs associated with data storage, processing, and analysis can further strain budgets, making it challenging for organizations to sustain monitoring efforts over the long term. To address these challenges, innovative funding models, such as public–private partnerships and grants, may be necessary to support the implementation of smart environmental monitoring systems.

Barriers to Implementation in Developing Regions

In addition to financial constraints, developing regions may face infrastructural challenges that hinder the implementation of smart environmental monitoring technologies. Limited access to reliable internet connectivity, electricity, and technical expertise can impede the deployment and effectiveness of IoT systems. For instance, remote areas may lack the necessary infrastructure to support continuous data transmission from monitoring devices, limiting the ability to collect real-time data. Targeted investments in infrastructure development and capacity building are essential to overcome these barriers. Collaborations between governments, non-governmental organizations, and private sector stakeholders can help facilitate the implementation of smart technologies in underserved regions.

Data Accuracy and Reliability

Ensuring the accuracy and reliability of data collected by smart environmental monitoring devices is critical for effective health management. Regular calibration and maintenance of sensors are necessary to provide accurate measurements. However, the logistical challenges associated with maintaining a network of sensors can be significant, particularly in remote or hard-to-reach areas. Inaccurate data can lead to misguided health interventions and policies, ultimately compromising public health outcomes. Therefore, establishing standardized protocols for sensor calibration and maintenance is essential to ensure the reliability of monitoring systems. Real-time monitoring systems generate large volumes of data, and ensuring the quality of this data is paramount for informed decision-making. Factors such as sensor drift, environmental conditions, and interference from other sources can affect data quality. Implementing data validation techniques, such as cross-referencing data from multiple sensors and employing machine learning algorithms to detect anomalies, can enhance data quality and reliability. Moreover, developing robust data management systems that facilitate the integration of data from various sources can improve the overall accuracy of environmental monitoring efforts. By ensuring high-quality data, stakeholders can make informed decisions that address environmental health risks effectively.

8.7 Future Trends in Smart Environmental Monitoring

The future of smart environmental monitoring is poised to be shaped by advances in sensor technology, artificial intelligence (AI), and innovative applications that enhance our understanding of environmental health. As the challenges of climate change, pollution, and public health continue to escalate, the integration of these technologies will be crucial for effective monitoring and management. The continuous evolution of sensor technology is leading to the development of more accurate, reliable, and affordable sensors for environmental monitoring. Innovations in materials science and nanotechnology enable the creation of susceptible sensors capable of detecting a wide range of environmental parameters, including air and water quality indicators, soil health metrics, and even biological contaminants Chen et al. (2024). These advancements are particularly significant for low-resource settings, where cost-effective monitoring solutions are essential. The miniaturization of sensors and the reduction in manufacturing costs are making it feasible to deploy large networks of sensors for comprehensive environmental monitoring (Anwar et al. 2023). For example, low-cost air quality sensors can be distributed throughout urban areas to provide real-time data on pollution levels, empowering communities to act against air quality issues.

Integrating AI with environmental monitoring technologies enhances our ability to predict ecological health risks. Machine learning algorithms can analyze vast datasets from various sources, including sensor data, satellite imagery, and historical health records, to identify patterns and predict future environmental conditions (Danju 2023). For instance, AI models can forecast pollution levels based on meteorological data, traffic patterns, and industrial activities, allowing public health officials to issue timely alerts and implement mitigation strategies (Tedyyana et al. 2024). Additionally, AI can be used to model the potential impacts of climate change on biodiversity and public health, providing valuable insights for policymakers and conservationists. Innovations in bio-sensing technologies and environmental DNA (eDNA) monitoring are opening new avenues for environmental health assessment. Biosensors can detect specific biological indicators, such as pathogens or toxins, in environmental samples, providing real-time data on ecosystem health (Lee et al. 2022). eDNA monitoring, which involves analyzing genetic material shed by organisms into their environment, is becoming a powerful tool for assessing biodiversity and ecosystem health. This method allows for the detection of species presence and abundance without the need for direct observation, making it particularly useful for monitoring elusive or endangered species (Jafar et al. 2021). By integrating eDNA data with traditional monitoring methods, researchers can better understand ecosystem dynamics and their implications for public health.

Blockchain technology is emerging as a promising solution for enhancing data security and transparency in environmental monitoring systems. By providing a decentralized and immutable ledger for data storage, blockchain can ensure the integrity and authenticity of environmental data collected from various sources

(Setiawan 2024). This technology is particularly relevant for environmental monitoring, where data accuracy is critical for informed decision-making. For instance, blockchain can be used to track the provenance of environmental data, ensuring that it has not been tampered with or altered. Additionally, blockchain can facilitate secure data sharing among stakeholders, promoting collaboration and transparency in environmental health initiatives (Ferrández-Pastor et al. 2022). Public–private partnerships are increasingly recognized as vital for advancing smart environmental monitoring technologies. These collaborations can leverage the strengths of both sectors, combining public-sector resources and regulatory frameworks with private-sector innovation and expertise (Sanjeeva et al. 2023). For example, partnerships between government agencies and technology companies can facilitate developing and deploying advanced monitoring systems that address specific environmental health challenges. By pooling resources and expertise, PPPs can accelerate the adoption of smart technologies and enhance the effectiveness of environmental monitoring efforts (Xue et al. 2022). As environmental health issues transcend national borders, international collaboration is essential for effective monitoring and management. Collaborative efforts among countries, international organizations, and research institutions can enhance data sharing, standardize monitoring protocols, and promote best practices in environmental health monitoring (Park et al. 2023). For instance, initiatives such as the Global Environmental Monitoring System aim to provide a framework for international cooperation in monitoring water quality and other environmental parameters. By fostering collaboration and knowledge exchange, these initiatives can strengthen global efforts to address environmental health challenges and promote sustainable practices (Andrei 2024).

8.8 Conclusion

Smart technologies in environmental monitoring are transforming our ability to collect and analyze real-time data, significantly enhancing health management strategies. Integrating IoT, edge computing, and remote sensing technologies provides valuable insights into environmental conditions directly impacting human health. By applying these innovations in air and water quality monitoring and bioremediation efforts, we can better understand and address the health risks associated with environmental changes. As the challenges posed by pollution and climate change continue to grow, leveraging smart technologies will be essential for promoting public health and ensuring a sustainable future. The role of data analytics and artificial intelligence in environmental monitoring is transformative, enabling more effective management of environmental health risks. By leveraging big data, machine learning, and geospatial analytics, stakeholders can gain valuable insights into environmental conditions and their impacts on public health. As the challenges posed by environmental degradation and public health threats continue to grow, the integration of these technologies will be essential for promoting sustainable practices and safeguarding human health. The applications of real-time environmental data in health management are diverse and impactful. These technologies provide critical insights that inform public health strategies and interventions, from air quality monitoring to water quality assessments and climate monitoring. By leveraging real-time data, stakeholders can proactively address environmental health risks, enhance community resilience, and promote better health outcomes. As the challenges posed by environmental degradation and climate change continue to grow, integrating real-time monitoring technologies will be essential for safeguarding public health and ensuring a sustainable future. The enhancement of health response systems through smart technology is transforming the landscape of public health management. Early warning systems, remote health solutions, and real-time data integration into public health policies are all critical components of this transformation. By leveraging advanced technologies and data analytics, stakeholders can improve their capacity to monitor and respond to environmental health risks, ultimately promoting better health outcomes for communities. As the challenges posed by environmental degradation and public health threats continue to grow, integrating smart technologies will be essential for safeguarding public health and ensuring a sustainable future. Implementing smart environmental monitoring technologies presents several challenges that must be addressed to enhance their effectiveness in protecting public health. Data privacy and security concerns, infrastructure and resource limitations, and data accuracy and reliability issues are critical considerations for stakeholders. The future of smart environmental monitoring is characterized by advances in sensor technology, AI integration, and innovative applications that enhance our understanding of environmental health. By developing more accurate and affordable sensors, leveraging AI for predictive modeling, and exploring emerging technologies such as blockchain and eDNA monitoring, stakeholders can improve their capacity to monitor and manage environmental health risks.

References

Abdallah A, Shahrour I, Sadek M, Abdallah M (2015) Real-time detection of microbial intrusion in potable water using advanced devices. WIT Trans Built Environ 168:511–518. https://doi.org/10.2495/sd150451

Akanbi A (2020) Estemd: a distributed processing framework for environmental monitoring based on apache Kafka streaming engine. In: Proceedings of the 4th international conference on big data research, pp 18–25. (Nov 2020). https://doi.org/10.1145/3445945.3445949

Andrei N (2024) Marine applications: the future of autonomous maritime transportation and logistics. https://doi.org/10.5772/intechopen.1004275

Anwar A, Goyal SB, Jan T (2023) Blockchain-based clinical trials: a meta-model framework for enhancing security and transparency with a novel algorithm. Int J Technol 14(6):1380. https://doi.org/10.14716/ijtech.v14i6.6703

Baviskar HD, Aringale SS, Jagtap TV, Shende TS, Dashore P (2024). Iot based air pollution monitoring system. Int J Res Appl Sci Eng Technol 12(4):3050–3053. https://doi.org/10.22214/ijraset.2024.60601

Benammar M, Abdaoui A, Ahmad SHM, Touati F, Kadri A (2018) A modular IoT platform for real-time indoor air quality monitoring. Sensors 18(2):581. https://doi.org/10.3390/s18020581

Bernatska N, Dzhumelia E, Dyakiv V, Mitryasova O, Саламон I (2023). Web-based information and analytical monitoring system tools–online visualization and analysis of surface water quality of mining and chemical enterprises. Ecol Eng Environ Technol 24(3):99–108. https://doi.org/10.12912/27197050/159885

Besmer MD, Sigrist JA, Props R, Buysschaert B, Mao G, Boon N, Hammes F (2017) Laboratory-scale simulation and real-time tracking of a microbial contamination event and subsequent shock-chlorination in drinking water. Front Microbiol 8. https://doi.org/10.3389/fmicb.2017.01900

Bibri SE (2021) A novel model for data-driven smart sustainable cities of the future: the institutional transformations required for balancing and advancing the three goals of sustainability. Energy Inf 4(1). https://doi.org/10.1186/s42162-021-00138-8

Brigo E, Rintala A, Kossi O, Verwaest F, Vanhoof O, Feys P, Bonnechère B (2022) Using telehealth to guarantee the continuity of rehabilitation during the COVID-19 pandemic: a systematic review. Int J Environ Res Public Health 19(16):10325. https://doi.org/10.3390/ijerph191610325

Carducci A, Arzilli G, Atomsa NT, Lauretani G, Verani M, Pistelli F, Tavoschi L, Federigi I, Fornili M, Petri D et al (2023). Integrated environmental and clinical surveillance for the prevention of acute respiratory infections (aris) in indoor environments and vulnerable communities (stell-ari). https://doi.org/10.21203/rs.3.rs-3687565/v1

Casson N, Cameron L, Mauro I, Friesen-Hughes K, Rocque R (2023) Perceptions of the health impacts of climate change among Canadians. BMC Public Health 23(1). https://doi.org/10.1186/s12889-023-15105-z

Chen L, Xia C, Zhao Z, Fu H, Chen Y (2024) AI-Driven sensing technology. Sensors 24(10):2958. https://doi.org/10.3390/s24102958

Chen X, Xu L, Pan Z (2022) Design and preliminary realization of a screening and early warning health management system for populations at high risk for depression. Int J Environ Res Public Health 19(6):3599. https://doi.org/10.3390/ijerph19063599

Dai W (2021) Intelligent environment platform for industrial clusters based on cloud computing technology. Mob Inf Syst 2021:1–12. https://doi.org/10.1155/2021/2858357

Danju D (2023) Innovations in accounting: the use of blockchain technologies to increase transparency and security of financial transactions. Bull Toraighyrov Univ Econ Ser 4(2023):57–66. https://doi.org/10.48081/nuug8234

Delgado-Gardea MCE, Tamez-Guerra P, Gomez-Flores R, Mendieta-Mendoza A, Zavala-Díaz de la Serna FJ, Contreras-Cordero JF, Erosa-de la Vega G, Pérez-Recoder MC, Sánchez-Ramírez B, González-Horta C, Infante-Ramírez R (2017) Prevalence of rotavirus genogroup a and norovirus genogroup ii in bassaseachic falls national park surface waters in chihuahua, mexico. Int J Environ Res Public Health 14(5):482. https://doi.org/10.3390/ijerph14050482

Di Napoli C, Romanello M, Minor K, Chambers J, Dasgupta S, Escobar LE, Hang Y, Hänninen R, Liu Y, Lotto Batista M, Lowe R (2023) The role of global reanalyses in climate services for health: Insights from the Lancet Countdown. Meteorol Appl 30(2):e2122. https://doi.org/10.1002/met.2122

Donner CF, ZuWallack R, Nici L (2021) The role of telemedicine in extending and enhancing medical management of the patient with chronic obstructive pulmonary disease. Medicina 57(7):726. https://doi.org/10.3390/medicina57070726

Dwevedi R, Krishna V, Kumar A (2018) Environment and big data: role in smart cities of India. Resources 7(4):64. https://doi.org/10.3390/resources7040064

Edagawa A, Kimura A, Kawabuchi-Kurata T, Adachi S, Furuhata K, Miyamoto H (2015) Investigation of legionella contamination in bath water samples by culture, amoebic co-culture, and real-time quantitative PCR methods. Int J Environ Res Public Health 12(10):13118–13130. https://doi.org/10.3390/ijerph121013118

Fathy C, Ali HM (2023) A secure iot-based irrigation system for precision agriculture using the expeditious cipher. Sensors 23(4):2091. https://doi.org/10.3390/s23042091

Fatima SA, Nasim SF, Ahmed S (2023) Enhancing agricultural operations: big data analytics using distributed and parallel computing. Int J Emerg Eng Technol 2(2):1–7.https://doi.org/10.57041/v3jj9f69

Ferrández-Pastor FJ, Mora-Pascual J, Díaz-Lajara D (2022) Agricultural traceability model based on IoT and blockchain: application in industrial hemp production. J Ind Inf Integr 29:100381. https://doi.org/10.1016/j.jii.2022.100381

González-González RB, Flores-Contreras EA, Gonzalez-Gonzalez E, Torres Castillo NE, Parra-Saldivar R, Iqbal HM (2022) Biosensor constructs for the monitoring of persistent emerging pollutants in environmental matrices. Ind Eng Chem Res 62(11):4503–4520. https://doi.org/10.1021/acs.iecr.2c00421

Grishin OE, Nesterchuk OA, Popov SI, Tolochko AV, Levashova OV (2021) Smart city technologies in the context of ensuring environmental safety and political stability. In: E3s Web of conferences, vol 311, pp 07004. https://doi.org/10.1051/e3sconf/202131107004

Gul SE, Cody D, Kharchenko A, Martin S, Mintova S, Cassidy J, Naydenova I (2018) Ltl type nanozeolites utilized in surface photonics structures for environmental sensors. Microporous Mesoporous Mater 261:268–274. https://doi.org/10.1016/j.micromeso.2017.11.019

Haahtela T (2019) A biodiversity hypothesis. Allergy 74(8):1445–1456. https://doi.org/10.1111/all.13763

Hait M, Patel D, Izah SC (2024b) Molecular techniques and technologies in biomonitoring for environmental sustainability. In: Izah SC, Ogwu MC, Hamidifar H (eds) Biomonitoring of pollutants in the global south. Springer, Singapore, pp 605–637. https://doi.org/10.1007/978-981-97-1658-6_17

Hait M, Sahu P, Biswas S, Izah SC (2024a) Sustainable application of artificial intelligence in biomonitoring for environmental sustainability: challenges and prospects. In: Izah SC, Ogwu MC, Hamidifar H (eds) Biomonitoring of pollutants in the global south. Springer, Singapore, pp 747–778. https://doi.org/10.1007/978-981-97-1658-6_22

Haji SH, Sallow AB (2021) IoT for smart environment monitoring based on python: a review. Asian J Res Comput Sci 57–70. https://doi.org/10.9734/ajrcos/2021/v9i130215

Hajjaji Y, Boulila W, Farah IR, Romdhani I, Hussain A (2021). Big data and iot-based applications in smart environments: a systematic review. Comput Sci Rev 39:100318. https://doi.org/10.1016/j.cosrev.2020.100318

Hall R, Pal D, Ariya PA (2022) Novel dynamic technique, nano-dihm, for rapid detection of oil, heavy metals, and biological spills in aquatic systems. Anal Chem 94(32):11390–11400. https://doi.org/10.1021/acs.analchem.2c02396

Hariri RH, Fredericks EM, Bowers KM (2019) Uncertainty in big data analytics: survey, opportunities, and challenges. J Big Data 6(1). https://doi.org/10.1186/s40537-019-0206-3

Ibaseta D, Molleda J, Díez F, Granda JC (2018) Indoor air quality monitoring sensor for the web of things. In Proceedings, vol 2, No 23, p 1466. MDPI (Nov 2018). https://doi.org/10.3390/proceedings2231466

Intrieri E, Bardi F, Fanti R, Gigli G, Fidolini F, Casagli N, Costanzo S, Raffo A, Di Massa G, Capparelli G, Versace P (2017) Big data managing in a landslide early warning system: experience from a ground-based interferometric radar application. https://doi.org/10.5194/nhess-2017-178

Izah SC, Abdelkhalek ST, Ogwu MC, Hamidifar H (2024) Overview and perspectives of biomonitoring in the global south. In: Izah SC, Ogwu MC, Hamidifar H (eds) Biomonitoring of pollutants in the global south. Springer, Singapore, pp 3–36. https://doi.org/10.1007/978-981-97-1658-6_1

Jacob DE, Nelson IU, Okweche SI, Oko PA, Izah SC (2024a) Suitability of mammals indigenous to the global south as bioindicator species for assessing environmental health. In: Izah SC, Ogwu MC, Hamidifar H (eds) Biomonitoring of pollutants in the global south. Springer, Singapore, pp 451–484. https://doi.org/10.1007/978-981-97-1658-6_13

Jacob DE, Nelson IU, Efenakpo OD, Izah SC (2024b) Reptiles as environmental sentinels: exploring their significance. In: Izah SC, Ogwu MC, Hamidifar H (eds) Biomonitoring of pollutants in the global south. Springer, Singapore, pp 485–533. https://doi.org/10.1007/978-981-97-1658-6_14

Jacob DE, Nelson IU, Oluwafemi OJ, Izah SC, Ogwu MC (2024c) Environmental stewardship: safeguarding biodiversity in protected landscapes and recreational parks using biosecurity. In: Izah SC, Ogwu MC, Hamidifar H (eds) Biomonitoring of pollutants in the global south. Springer, Singapore, pp 779–834. https://doi.org/10.1007/978-981-97-1658-6_23

Jacob DE, Nelson IU, Izah SC, Ukpong E, Akpan UU, Ogwu MC (2024d) Bioindicators in recreational planning and development: balancing nature and human activities. In: Izah SC, Ogwu MC, Hamidifar H (eds) Biomonitoring of pollutants in the global south. Springer, Singapore, pp 835–878. https://doi.org/10.1007/978-981-97-1658-6_24

Jafar U, Aziz MJA, Shukur Z (2021) Blockchain for electronic voting system—review and open research challenges. Sensors 21(17):5874. https://doi.org/10.3390/s21175874

Jo JH, Jo B, Kim JH, Choi I (2020) Implementation of iot-based air quality monitoring system for investigating particulate matter (Pm10) in subway tunnels. Int J Environ Res Public Health 17(15):5429. https://doi.org/10.3390/ijerph17155429

Kencono DS, Ahsany MR (2023) The use of cctv model analysis for a smart environment in semarang city. In: E3s web of conferences, vol 440, pp 07006. https://doi.org/10.1051/e3sconf/202344007006

Kumar A, Madaan G, Sharma P, Kumar A (2021) Application of disruptive technologies on environmental health: an overview of artificial intelligence, blockchain and internet of things. Asia Pacific J Health Manag 16(4):251–259. https://doi.org/10.24083/apjhm.v16i4.1297

Lazarescu MT (2013) Design of a wsn platform for long-term environmental monitoring for iot applications. IEEE J Emerg Select Top Circuits Syst 3(1):45–54. https://doi.org/10.1109/jetcas.2013.2243032

Lee, D., Park, M., Baek, Y., Bae, B., Heo, J., & Lee, K. (2022). In-sensor image memorization and encoding via optical neurons for bio-stimulus domain reduction toward visual cognitive processing. Nat Commun 13(1). https://doi.org/10.1038/s41467-022-32790-3

Lurie N, Carr BG (2018) The role of telehealth in the medical response to disasters. JAMA Intern Med 178(6):745. https://doi.org/10.1001/jamainternmed.2018.1314

Majumder S, Aghayi E, Noferesti M, Memarzadeh-Tehran H, Mondal T, Pang Z, Deen MJ (2017) Smart homes for elderly healthcare—recent advances and research challenges. Sensors 17(11):2496. https://doi.org/10.3390/s17112496

Marshall CJ, El-Ansary D, Pranata A, Ganderton C, O'Donnell J, Takla A, Tran P, Wickramasinghe N, Tirosh O (2022) Validity and reliability of a novel smartphone tele-assessment solution for quantifying hip range of motion. Sensors 22(21):8154. https://doi.org/10.3390/s22218154

Masters-Awatere B, Howard D, Young P. (2022) Haumanu hauora: refining public health institution policy to include māori and climate change. https://doi.org/10.21203/rs.3.rs-1966481/v1

Mingfei Z, Daoliang L, Lianzhi W, Daokun M, Qisheng D (2010) Design and development of water quality monitoring system based on wireless sensor network in aquaculture. In: Computer and computing technologies in agriculture IV. CCTA 2010. IFIP advances in information and communication technology, vol. 347. https://doi.org/10.1007/978-3-642-18369-0_76

Munim ZH, Dushenko M, Jimenez VJ, Shakil MH, Imset M (2020) Big data and artificial intelligence in the maritime industry: a bibliometric review and future research directions. Marit Policy Manag 47(5):577–597. https://doi.org/10.1080/03088839.2020.1788731

Ngwenya S, Mashau NS, Mhlongo SE, Traoré AN (2023) A systematic review of the risk management frameworks for potentially toxic chemical elements. Toxicol Ind Health 39(12):679–686. https://doi.org/10.1177/07482337231209092

Pamudji AK(2023) Iot-driven environmental support system for smart cities. Sisforma 10(1):14–18. https://doi.org/10.24167/sisforma.v10i1.10209

Park SM, Hong S, Rachim VP, Baek JH (2023) Integrating computer vision and a strain sensor for human-machine interfaces with enhanced versatility and scalability. https://doi.org/10.21203/rs.3.rs-2518755/v1

Pathaburee P (2022) Taq man multiplex real-time pcr for identification of porcine and canine dna in halal food. J Halal Sci Res 2(1):1–7. https://doi.org/10.12928/jhsr.v2i1.6052
Ragimova NA, Abdullayev VH, Khalilov ME, Abuzarova VA, Askerov TK (2020) Analysis of tools for the description and analytical processing of enviromental data. EKSAKTA: J Sci Data Anal 129–132. https://doi.org/10.20885/eksakta.vol1.iss2.art6.
Ranganathan RH, Balusamy S, Partheeban P, Mani C, Sridhar M, Rajasekaran V (2023) Air quality monitoring and analysis for sustainable development of solid waste dump yards using smart drones and geospatial technology. Sustainability 15(18):13347. https://doi.org/10.3390/su151813347
Reilly SE, Zane KL, McCuddy WT, Soulliard ZA, Scarisbrick DM, Miller LE, Mahoney Iii JJ (2020) Mental health practitioners' immediate practical response during the covid-19 pandemic: observational questionnaire study. Jmir Mental Health 7(9):e21237. https://doi.org/10.2196/21237
Roşca MI, Nicolae C, Sanda E, Madan A (2021) Internet of things (IoT) and sustainability. In: Basiq international conference, pp 346–352. https://doi.org/10.24818/basiq/2021/07/044
Saab C, Farah E, Shahrour I, Chehade FH (2017). Field study of real-time water quality control. https://doi.org/10.2495/ws170231
Sagheer A, Mohammed M, Riad K, Alhajhoj M (2020) A cloud-based IoT platform for precision control of soilless greenhouse cultivation. Sensors 21(1):223. https://doi.org/10.3390/s21010223
Sanjeeva P, Sathwik MS, Prasad GS, Reddy GP, Sajwan V, Ganesh B (2023) Decentralized and automated online voting system using blockchain technology. In: E3s web of conferences, pp 430, 01046. https://doi.org/10.1051/e3sconf/202343001046
Setiawan A (2024) Blockchain-based management information systems: benefits and challenges. Esaprom 3(01):1–8. https://doi.org/10.58471/esaprom.v3i01.3783
Stepanov N (2023) Models for monitoring the implementation of environmental projects of enterprises in the region. In: E3s web of conferences, vol 458, pp 07015. https://doi.org/10.1051/e3sconf/202345807015
Tedyyana A, Ghazali O, Asnafi T, Purbo OW, Harun NZ, Riza F. (2024) Transforming the voting process integrating blockchain into e-voting for enhanced transparency and securiy. Telkomnika (Telecommunication Computing Electronics and Control) 22(2):311. https://doi.org/10.12928/telkomnika.v22i2.25758
Tsitsifli S, Kanakoudis V (2022) Identification of suitable locations in a small water supply network for the placement of water quality sensors based on different criteria under demand-driven conditions. Water 14(16):2504. https://doi.org/10.3390/w14162504
Ullo SL, Sinha GR (2020) Advances in smart environment monitoring systems using IoT and sensors. Sensors 20:3113. https://doi.org/10.3390/s20113113
Vlahu-Gjorgievska E, Nagapuri S, Win KT (2019) Tele-monitoring technology as a tool for monitoring and management of patients with congestive heart failure. Australas J Inf Syst 23. https://doi.org/10.3127/ajis.v23i0.1816
Wasswa J, Mladenov N, Pearce W (2019) Assessing the potential of fluorescence spectroscopy to monitor contaminants in source waters and water reuse systems. Environ Sci: Water Res Technol 5(2):370–382. https://doi.org/10.1039/c8ew00472b
Wu T (2023) Application and development of artificial intelligence in environmental health. Appl Comput Eng 7(1):35–40. https://doi.org/10.54254/2755-2721/7/20230335
Xue Z, Jin T, Xu S, Bai K, He Q, Zhang F, Cheng X, Ji Z, Pang W, Shen Z, Song H (2022) Assembly of complex 3d structures and electronics on curved surfaces. Sci Adv 8(32). https://doi.org/10.1126/sciadv.abm6922
Yang X, Xi W, Chen A, Wang C (2021) An environmental monitoring data sharing scheme based on attribute encryption in cloud-fog computing. PLoS ONE 16(9):e0258062. https://doi.org/10.1371/journal.pone.0258062
Zhang X, Huang Z (2019) Research on smart environmental protection iot application dased on edge computing. https://doi.org/10.2991/cnci-19.2019.76

Zhao J, Wang D, Zhang F, Pan J, Claesson P, Larsson R, Shi Y (2022) Self-powered, long-durable, and highly selective oil–solid triboelectric nanogenerator for energy harvesting and intelligent monitoring. Nano-Micro Lett 14(1) https://doi.org/10.1007/s40820-022-00903-8

Zhe W, Yonghong L, Bike Z, Jinghuan R, Fanxiao J (2019) Application of public health risk assessment methodology for typhoon disaster in China. Eur J Prev Med 7(1):11. https://doi.org/10.11648/j.ejpm.20190701.13

Chapter 9
Green Infrastructure and Environmental Health Security: The Role of Nature-Based Solutions in Improving Urban Health

Matthew Chidozie Ogwu, Enoch Akwasi Kosoe, Amarachi Chekosiba Alaka, and Jiajun Xu

Abstract Green infrastructure (GI) represents a critical approach to urban planning, integrating natural and semi-natural features into cities to address environmental and health challenges associated with rapid urbanization and climate change. This chapter discussed the multifaceted role of green infrastructure, focusing on various types, such as urban parks, green roofs, rain gardens, urban forests, and wetlands. These GI elements offer substantial environmental benefits, including improved air and water quality, climate regulation, and biodiversity conservation. Additionally, they contribute to urban resilience by mitigating the effects of urban heat islands, managing stormwater, and reducing flood risks. The chapter also examined the socio-economic advantages of GI, highlighting how it fosters recreation, community cohesion, and enhanced property values. GI has been linked to positive mental health outcomes by improving access to nature, such as reduced stress and enhanced psychological well-being, essential in densely populated urban areas. The

M. C. Ogwu
Goodnight Family Department of Sustainable Development, Appalachian State University, 212 Living Learning Center, 305 Bodenheimer Drive, Boone, NC 28608, USA
e-mail: ogwumc@appstate.edu

E. A. Kosoe (✉)
Department of Environment and Resource Studies, Faculty of Integrated Development Studies, Simon Diedong Dombo University of Business and Integrated Development Studies, Wa, P. O. Box 520, Wa, Upper West Region, Ghana
e-mail: ekosoe@ubids.edu.gh

A. C. Alaka
Department of Agricultural and Bio-Environmental Engineering, Imo State Polytechnic, Omuma, Imo State, Nigeria
e-mail: alaka.amarachi@imopoly.edu.ng

J. Xu
Institute for Advanced Studies, University of Malaya, 50603 Kuala Lumpur, Malaysia

S. Chibueze Izah and M. Chidozie Ogwu (eds.), *Innovative Approaches in Environmental Health Management*, Environmental Science and Engineering,
https://doi.org/10.1007/978-3-031-81966-7_9

chapter emphasizes the importance of nature-based solutions (NbS) as an effective strategy for urban health improvement, integrating GI to deliver ecosystem services that promote physical, mental, and social well-being. Case studies on NbS implementations demonstrated their potential to enhance urban health by reducing heat stress, filtering pollutants, and encouraging physical activity through accessible green spaces. However, challenges remain in implementing GI, including financial constraints, space limitations, and the need for sustained community engagement. Addressing these barriers requires innovative policies, cross-sector collaboration, and community-driven planning. This chapter provides a comprehensive analysis of the role of GI in fostering sustainable, healthy urban environments. It presents actionable insights for urban planners, policymakers, and community leaders to optimize NbS for urban health.

Keywords Urban health · Nature-based solutions · Environmental benefits · Climate resilience · Mental well-being · Urban parks · Stormwater management · Community cohesion · Air quality

9.1 Introduction

Green infrastructure (GI) refers to a strategically planned network of natural and semi-natural areas that provide a wide range of ecosystem services, including air and water purification climate regulation, and biodiversity conservation. It encompasses various elements such as parks, green roofs, urban forests, wetlands, and green corridors that integrate nature into urban environments (Thompson et al. 2016). Green infrastructure is essential because of its ability to enhance urban resilience, mitigate the impacts of climate change, and improve the overall quality of life for urban residents. Green infrastructure is increasingly recognized as a vital component of sustainable urban planning. By incorporating natural elements into the urban landscape, cities can address environmental challenges like flooding, heat islands, and air pollution while promoting public health and well-being (Đorđević 2023). The multifunctional benefits of GI make it an essential strategy for creating healthier and more livable urban environments.

Urban areas face numerous health challenges, including air (Izah et al. 2021; Uzoekwe et al. 2021) and water pollution (Jacob et al. 2024a, b; Izah et al. 2024; Iyiola et al. 2024; Ogamba et al. 2021), heat stress, and the spread of vector-borne diseases. Rapid urbanization, population growth, and climate change exacerbated these issues, increasing health risks for urban residents. For instance, poor air quality is linked to respiratory diseases, cardiovascular conditions, and premature mortality (Dushkova and Haase 2020). Additionally, urban heat islands can lead to heat-related illnesses, particularly among vulnerable populations such as the elderly and those with preexisting health conditions (Dadvand et al. 2015). Furthermore, access to green spaces is often limited in densely populated urban areas, contributing to mental health issues, social isolation, and reduced physical activity levels (Gascón et al. 2015). The need for effective interventions to address these urban health challenges has never been more pressing, highlighting the importance of integrating nature-based solutions into urban planning and public health strategies. Nature-based solutions (NbS) are approaches that harness the power of nature to address societal challenges, including climate change adaptation, disaster risk reduction, and public health improvement (Đorđević 2023). In urban settings, NbS can enhance health outcomes by improving air and water quality, reducing heat stress, and promoting physical and mental well-being. Implementing green infrastructure as a nature-based solution can provide multiple co-benefits, including improved biodiversity, enhanced ecosystem services, and increased community resilience (Dushkova and Ignatieva 2020). For example, urban parks and green spaces offer recreational opportunities and contribute to air purification, temperature regulation, and stormwater management. By prioritizing nature-based solutions in urban planning, cities can create healthier environments that promote the well-being of their residents while addressing pressing environmental challenges.

Green infrastructure can significantly improve urban air quality by filtering pollutants and reducing the concentration of harmful airborne particles. Vegetation, such as trees and shrubs, acts as a natural filter, absorbing carbon dioxide and releasing oxygen while trapping particulate matter and other pollutants (Luca et al. 2021). Studies have shown that increased green space in urban areas is associated with lower levels of air pollution and improved respiratory health outcomes (Dorst et al. 2019). Implementing air quality monitoring systems incorporating green infrastructure can enhance public health responses to air pollution. For instance, cities can use real-time data from air quality sensors to inform residents about pollution levels and issue health advisories when necessary. By integrating green infrastructure into urban planning, cities can create buffer zones around high-traffic areas, reducing exposure to pollutants and improving overall air quality (Kabisch and Bosch 2017). Urban heat islands (UHIs) are a significant concern in densely populated areas, where built environments absorb and retain heat, leading to elevated temperatures. Green infrastructure can mitigate the effects of UHIs by providing shade, cooling the air through evapotranspiration, and reducing surface temperatures (Triguero-Mas et al. 2015). Implementing green roofs, urban forests, and shaded public spaces can significantly reduce heat exposure for urban residents, particularly during heat waves. Studies have demonstrated that access to green spaces can lower the risk of heat-related illnesses and improve overall health outcomes (Bočkarjova and Kačalová 2021). For example, cities like Chicago have implemented urban greening initiatives to combat heat stress and promote public health, resulting in measurable improvements in community well-being (Parker and Simpson 2018). Access to green spaces has been linked to improved mental health and well-being. Nature-based solutions can provide opportunities for physical activity, social interaction, and relaxation, all of which contribute to better mental health outcomes (Kabisch et al. 2017).

Research has shown that individuals living near green spaces report lower levels of stress, anxiety, and depression (Gallos et al. 2022). For instance, a study conducted in urban areas of Wisconsin found a positive correlation between access to neighborhood green space and mental health, highlighting the importance of integrating green infrastructure into urban planning to promote mental well-being (Beyer et al. 2014). Green infrastructure can also foster social cohesion and community resilience by providing spaces for social interaction and community engagement. Parks, community gardens, and green corridors are gathering places where residents can connect, share resources, and participate in community activities (MacKinnon et al. 2019). Community gardens, in particular, have gained popularity as a nature-based solution that promotes food security, social interaction, and mental well-being. These gardens provide fresh produce and create opportunities for community members to collaborate, learn, and engage with nature (Richardson et al. 2013). Green infrastructure can enhance social resilience and improve overall health outcomes by fostering a sense of community and belonging.

This chapter examined the role of green infrastructure and nature-based solutions in improving urban health by addressing cities' environmental and public health challenges. It explored how green infrastructure such as urban parks, green roofs, and rain gardens provides essential ecosystem services that benefit physical, mental, and social health. The chapter also highlighted the health benefits of improved air quality, urban cooling, flood control, and enhanced social cohesion made possible through green infrastructure. By presenting examples and analyzing policy frameworks, this chapter gave urban planners, policymakers, and community leaders insights into effectively integrating green infrastructure into urban areas to promote healthier, more resilient cities. It further addressed the challenges and opportunities of implementing these solutions, emphasizing the need for cross-sector collaboration, innovation, and long-term planning to sustain urban health and well-being.

9.2 Understanding Green Infrastructure

Green infrastructure refers to a network of natural and semi-natural spaces that deliver a range of ecosystem services, improve urban environmental quality, and enhance the quality of life in cities. With urbanization posing various health, environmental, and social challenges, green infrastructure has emerged as a critical solution for creating sustainable and livable urban spaces. Some types of green infrastructure (Fig. 9.1) include:

Urban Parks and Green Spaces: Urban parks and green spaces are foundational components of green infrastructure. They serve as public areas for recreation, social interaction, and relaxation while supporting biodiversity by providing habitats for various species. These spaces help improve air quality by filtering pollutants such as particulate matter and nitrogen dioxide, which harm human health. Additionally, they play a significant role in mitigating the urban heat island effect by providing shade and reducing surface temperatures. Moreover, urban parks and green spaces are essential in promoting climate resilience, particularly in extreme weather events exacerbated by climate change. Studies have shown that access to green spaces is linked to increased physical activity, reduced stress, and improved mental well-being among urban residents (Russell et al. 2013). Parks also contribute to community

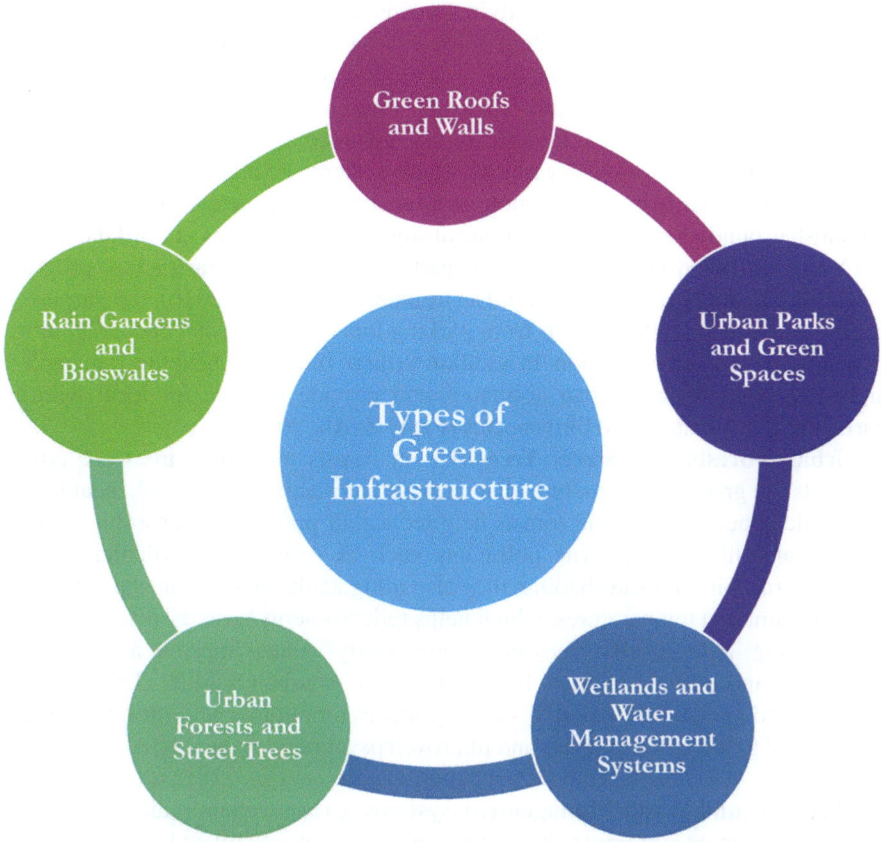

Fig. 9.1 Types of green infrastructure

cohesion by offering places for cultural and recreational activities, enhancing the quality of life in cities, and promoting social interaction.

Green Roofs and Walls: Green roofs, also known as living roofs, are wholly or partially covered with vegetation and a growing medium. These installations provide multiple benefits, including stormwater management, energy savings, and enhanced urban aesthetics. Green roofs act as insulators, reducing the need for heating and cooling, which lowers energy consumption and contributes to climate mitigation. For example, AAK Headquarters and the U.S. Embassy are two buildings in Nairobi that have used green roofs to manage rainfall and offer insulation, demonstrating a dedication to green architecture in an urban environment. Vertical green or living walls are vegetative installations on building exteriors that help improve air quality, provide shade, and contribute to urban cooling. Green roofs and walls help enhance urban biodiversity by creating microhabitats for pollinators such as bees and butterflies, essential for pollinating urban gardens and nearby agricultural areas (Aerts et al. 2018).

Rain Gardens and Bioswales: Rain gardens and bioswales are green infrastructure for managing stormwater. Rain gardens are shallow depressions planted with native vegetation that capture and filter rainwater, allowing it to infiltrate the ground rather than run into storm drains. Bioswales are similar to rain gardens but typically feature gently sloped sides and are often planted along streets or parking lots to capture and filter stormwater runoff. These systems help reduce urban flooding, improve water quality by filtering pollutants such as heavy metals and nutrients, and replenish groundwater supplies. To control stormwater, the cities of Portland and the USA have installed many bioswales as part of their green infrastructure program. According to the Portland Bureau of Environmental Services (2016), the purpose of the bioswales is to collect runoff from parking lots and streets and encourage filtration through soil and vegetation. In addition to these functional benefits, rain gardens and bioswales enhance urban aesthetics and provide habitats for local wildlife, particularly pollinators and bird species (Fathy & Ali, 2023).

Urban Forests and Street Trees: Urban forests and street trees are crucial elements of green infrastructure that provide substantial ecological, health, and economic benefits. Trees and forests help reduce air pollution by absorbing carbon dioxide and filtering airborne pollutants such as ozone and particulate matter. Through transpiration and shading, they also mitigate the urban heat island effect by lowering ambient temperatures, which helps reduce energy costs associated with air conditioning. Additionally, trees increase property values, improve neighborhood aesthetics, and provide habitats for birds and other urban wildlife. Urban forests contribute to mental well-being by offering spaces for outdoor activities, which have been linked to lower stress levels and improved psychological health (Akanbi 2020a, b).

Wetlands and Water Management Systems: Urban wetlands, along with other water management systems such as retention ponds and constructed wetlands, play a critical role in managing stormwater, controlling floods, and purifying water. These systems capture excess rainwater, prevent flooding, and reduce the risk of pollutants reaching rivers and lakes. Wetlands are also highly productive ecosystems that support various plant and animal species, contributing to urban biodiversity. Additionally, they provide recreational and educational opportunities for city residents to engage with local wildlife and learn about wetland ecosystems. As climate change increases the frequency of extreme weather events, urban wetlands and similar water management systems are essential for building urban resilience and reducing the impact of flooding on communities (Roelvink et al. 2021a, b).

Green infrastructure offers substantial environmental benefits by enhancing air quality, managing stormwater, and regulating urban climates. Vegetation in green spaces, urban forests, and green roofs absorbs pollutants such as carbon dioxide and nitrogen dioxide, improving overall air quality and reducing health risks associated with respiratory issues. Water management features such as rain gardens, bioswales, and wetlands capture and filter stormwater, thereby reducing the concentration of contaminants entering urban water bodies and protecting drinking water quality. Furthermore, green infrastructure mitigates the urban heat island effect by providing shade and facilitating evaporative cooling. Trees, plants, and vegetative surfaces lower surface and air temperatures, making urban areas more resilient to rising temperatures and contributing to climate regulation (Erisman et al. 2011). In addition to environmental benefits, green infrastructure enhances social cohesion and generates economic value in urban areas. Urban parks, green spaces, and community gardens offer public places for recreation, physical activity, and social engagement. These spaces have been linked to improved mental health, lower stress levels, and increased physical activity, contributing to overall well-being and public health. Economically, green infrastructure can increase property values, as neighborhoods with more trees and green spaces are often perceived as more desirable. Green roofs, by reducing energy costs, also offer economic savings to building owners. Furthermore, green infrastructure projects create jobs in landscape design, installation, and maintenance, supporting local economies. By fostering community interactions and strengthening social networks, green infrastructure contributes to more cohesive, resilient, and connected urban communities (Russell et al. 2013).

9.3 The Role of Nature-Based Solutions in Urban Health

Nature-based solutions (NbS) are actions that protect, sustainably manage, and restore natural or modified ecosystems to address societal challenges effectively while providing human well-being and biodiversity benefits. These solutions leverage natural processes and ecosystems to mitigate environmental issues, enhance resilience, and improve public health outcomes. NbS encompasses various interventions, including green infrastructure (GI), urban forestry, wetland restoration, and creating green spaces in urban areas. The importance of NbS has gained recognition in recent years as cities face increasing challenges related to climate change, urbanization, and public health (Table 9.1). By integrating nature into urban planning and development, cities can create healthier environments that promote physical and mental well-being, enhance biodiversity, and improve resilience to environmental stressors (Nieuwenhuijsen (2021). The multifaceted benefits of NbS make them a vital component of sustainable urban development. Nature-based solutions contribute to urban health in several ways, addressing physical and mental health outcomes. The following sections outline the fundamental mechanisms through which NbS enhances urban health.

One of the primary benefits of NbS is its ability to improve air quality. Vegetation, such as trees and shrubs, acts as a natural filter, absorbing pollutants and particulate matter while releasing oxygen (Berman et al. 2012). Urban green spaces can significantly reduce the concentration of harmful air pollutants, improving residents' respiratory health outcomes. Research has shown that increased access to green spaces is associated with lower rates of respiratory diseases and improved overall health (O'Regan et al. 2021). For example, cities that have invested in urban greening initiatives have reported reductions in asthma rates and other respiratory conditions, particularly among vulnerable populations such as children and the elderly (Anderson and Gough 2021). Urban areas are particularly susceptible to the urban heat island

Table 9.1 Environmental and human health benefits of nature-based solutions

Nature-based solution	Health benefits	Environmental benefits	Examples
Urban Parks and Green Spaces	Mental and physical health benefits and, community well-being	Provides spaces for recreation, physical activity, and social interaction, while improving mental health through exposure to nature	City parks, community gardens, public green spaces
Green Roofs and Walls	Improved air quality, urban cooling, and energy savings	Vegetated surfaces on buildings that insulate against temperature extremes, absorb pollutants and reduce energy costs	Green roofs on office buildings, vertical gardens on residential complexes
Rain Gardens and Bioswales	Water quality improvement, flood mitigation, and biodiversity support	Capture and filter stormwater runoff, reducing pollution in waterways and recharging groundwater; provide habitats for pollinators and other wildlife	Rain gardens in parking lots, bioswales along urban streets
Urban Forests and Street Trees	Air quality improvement, reduced urban heat, and mental well-being	Trees and forests in urban areas that absorb pollutants, provide shade, lower temperatures, and create environments that promote mental health	Tree-lined streets, urban forests, city arboretums
Constructed Wetlands	Water filtration, habitat creation, and flood control	Engineered wetlands that filter pollutants from runoff, manage stormwater, and offer habitats for wildlife while reducing urban flooding	Wetlands near urban water bodies, retention ponds in residential areas
Greenways and Ecological Corridors	Encourages active transport, connects habitats, and supports mental health	Pathways are integrated with natural elements that link green spaces, promote walking and cycling, and provide wildlife movement corridors	Riverwalks, urban greenways, biking paths integrated with greenery

(continued)

Table 9.1 (continued)

Nature-based solution	Health benefits	Environmental benefits	Examples
Community Gardens	Improved nutrition, social interaction, and physical activity	Spaces where residents can grow their food, fostering a sense of community and providing access to fresh produce	Neighborhood allotments, school gardens
Living Shorelines	Erosion control, flood prevention, and ecosystem support	Natural buffers along coastlines and waterways stabilize soil, protect against erosion, and support marine and terrestrial biodiversity	Mangrove restoration areas planted marshlands along urban coasts

(UHI) effect, where built environments absorb and retain heat, leading to elevated temperatures. NbS, such as green roofs, urban forests, and shaded parks, can mitigate UHI effects by providing shade, cooling the air through evapotranspiration, and reducing surface temperatures (Ji et al. 2023).

By lowering urban temperatures, NbS can help reduce heat-related illnesses and mortality, particularly during extreme heat events. Studies have demonstrated that access to green spaces can lower the risk of heat-related health issues, enhancing the resilience of urban populations to climate change (Hsu and Chao 2022). Access to nature has been linked to improved mental health outcomes, including reduced stress, anxiety, and depression. Nature-based solutions provide opportunities for physical activity, social interaction, and relaxation, all contributing to better mental well-being (Xie et al. 2020). Research indicates that individuals living near green spaces report higher levels of life satisfaction and lower levels of psychological distress (Sturiale and Scuderi 2019). For instance, a study found that exposure to natural environments significantly improved attention and cognitive functioning in individuals with attention deficit hyperactivity disorder (ADHD) (Herman and Drozda 2021). By promoting mental health and well-being, NbS can enhance urban residents' overall quality of life. Nature-based solutions can also foster community resilience by providing spaces for social interaction and community engagement. Green spaces, community gardens, and urban parks serve as gathering places where residents can connect, share resources, and participate in community activities (Anderson et al. 2021). The social cohesion fostered by these spaces enhances community resilience, enabling neighborhoods to withstand environmental and social challenges better. For example, during the COVID-19 pandemic, access to green spaces was essential for maintaining mental health and social connections among residents (Kazak et al. 2018). By integrating NbS into urban planning, cities can create environments that support community well-being and resilience.

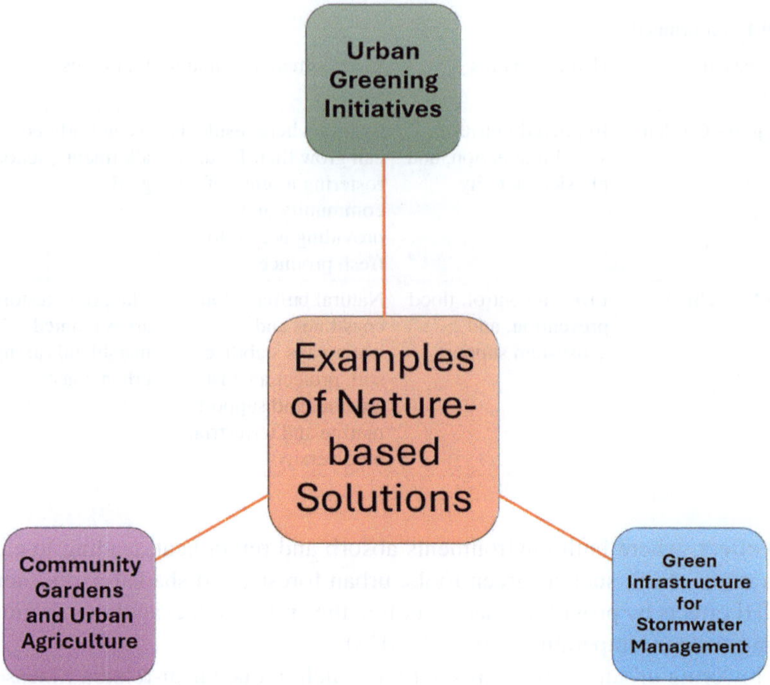

Fig. 9.2 Nature-based solutions for urban environments

Examples of NBS Implementations in Urban Environments

Numerous cities worldwide have successfully implemented Nature-Based Solutions to enhance urban health. The following examples illustrate the diverse applications of NbS in urban settings (Fig. 9.2).

Urban Greening Initiatives: Cities such as Singapore and Melbourne have prioritized urban greening initiatives to improve air quality, reduce heat stress, and enhance biodiversity. Singapore's "Garden City" vision has led to extensive green spaces, vertical gardens, and green roofs throughout the city, improving air quality and enhancing public health (Đorđević, 2023). Similarly, Melbourne's Urban Forest Strategy aims to increase tree canopy cover, providing shade and cooling while improving mental health outcomes for residents (Bratman et al., 2015). These initiatives demonstrate the potential of NbS to create healthier urban environments.

Green Infrastructure for Stormwater Management: Nature-based solutions can also be employed for effective stormwater management, reducing flooding risks and improving water quality. To manage stormwater runoff, cities like Philadelphia have implemented green infrastructure strategies, including rain gardens, permeable pavements, and green roofs (Roy et al. 2012). These solutions mitigate flooding, enhance urban biodiversity, and improve water quality by filtering pollutants. Integrating green infrastructure into stormwater management has resulted in significant public health benefits, including reduced risks of waterborne diseases and improved community resilience to extreme weather events (Rodrigues et al., 2018). Additionally, applying NbS in stormwater management frequently enhances urban recreational options. According to Ranish and Laurent (2021), parks and green spaces created for stormwater management offer communal areas that encourage social integration and physical activity, which is essential for urban public health.

Community Gardens and Urban Agriculture: Community gardens are a prominent example of NbS that promote food security, social interaction, and mental well-being. Cities like Toronto and New York have embraced community gardening initiatives, providing residents access to fresh produce while fostering community engagement (Atanasova et al. 2021). These gardens serve as valuable green spaces that enhance urban biodiversity and provide education and skill development opportunities. By promoting local food production and community involvement, community gardens contribute to healthier urban environments and improved public health outcomes (Vargas et al. 2023). Community gardens can also be highly beneficial in food shortages or economic downturns by offering substitute food sources and fostering urban populations' resilience. Their dual usage as places for food production and social gatherings emphasizes the multiple benefits of NbS in urban health programs (Mejia et al. 2020).

9.4 Human Health Benefits of Green Infrastructure

Green infrastructure often includes a network of natural and semi-natural features, including parks, green roofs, urban forests, and wetlands, that provide essential ecosystem services and enhance urban health. Integrating green infrastructure into urban planning is increasingly recognized as a vital strategy for addressing public health challenges, particularly in densely populated areas. One of green infrastructure's most significant health benefits is its ability to improve air quality. Vegetation, particularly trees and shrubs, acts as a natural filter, absorbing pollutants such as carbon dioxide (CO2), nitrogen oxides (NOx), and particulate matter (PM) (Nieuwenhuijsen 2021). Studies have shown that urban areas with higher greenery levels experience lower air pollutants, improving residents' respiratory health outcomes (Nieuwenhuijsen 2021). For instance, a systematic review found that increased exposure to urban green spaces is associated with reduced rates of respiratory diseases, such as asthma and chronic obstructive pulmonary disease (COPD) (Bratman et al. 2015). By enhancing air quality, green infrastructure contributes to healthier urban environments and mitigates the health risks associated with air pollution.

Urban heat islands (UHIs) are typical in cities, where built environments absorb and retain heat, leading to elevated temperatures. Green infrastructure plays a crucial role in mitigating the effects of UHIs by providing shade, cooling the air through evapotranspiration, and reducing surface temperatures (Dushkova and Ignatieva 2020). Research indicates that urban areas with extensive green cover experience lower temperatures, which can significantly reduce heat-related illnesses and mortality, particularly during heat waves (Thompson et al. 2016). For example, cities like Chicago have implemented urban greening initiatives that have successfully lowered surface temperatures and improved public health outcomes during extreme heat events (Ferreira et al. 2020). Access to green spaces encourages physical activity, essential for maintaining a healthy weight and preventing obesity-related health issues. Parks, trails, and recreational areas provide opportunities for walking, jogging, cycling, and other forms of exercise (Đorđević 2023). Studies have shown that individuals living near green spaces are more likely to engage in regular physical activity, leading to lower obesity rates and improved overall health (Đorđević 2023). For instance, a study conducted in New Zealand found that increased access to green spaces was associated with higher levels of physical activity and lower obesity rates among residents (Anderson et al. 2021). By promoting active lifestyles, green infrastructure improves physical health outcomes in urban populations. Access to nature has been linked to improved mental health outcomes, including reduced stress, anxiety, and depression. Green infrastructure provides relaxation, recreation, and social interaction opportunities, all of which contribute to better mental well-being (Andreucci et al. 2021).

Research has demonstrated that spending time in natural environments can lower cortisol levels (a stress hormone) and improve mood (Marin et al. 2021). For example, a study found that individuals who engaged in nature walks reported lower stress levels and enhanced psychological well-being than those who walked in urban environments (Olszewska et al. 2016). Nature-based interventions, such as gardening and outdoor activities, have been shown to have therapeutic effects on mental health. Community gardens and urban parks provide spaces for individuals to connect with nature, engage in physical activity, and foster social connections (Fisher and Grima 2020). A case study conducted in a community garden in Toronto highlighted the positive impact of gardening on mental health, with participants reporting increased feelings of happiness, reduced anxiety, and improved social interactions (Coutts and Hahn 2015). By providing access to nature, green infrastructure can serve as a valuable resource for improving mental health outcomes in urban populations. Numerous studies have explored the role of parks and green spaces in reducing depression and anxiety. For instance, a systematic review of the literature found that access to urban green spaces is associated with lower levels of depression and anxiety among residents (Felappi 2024). In a notable case study, researchers examined the impact of a newly developed urban park in a low-income neighborhood. The study found that residents who utilized the park reported significant reductions in symptoms of depression and anxiety, highlighting the importance of green spaces in promoting mental health (Wolch et al. 2014). This evidence underscores the potential of green infrastructure as a therapeutic resource for urban communities.

Green infrastructure fosters social interaction and community cohesion by providing spaces for residents to gather, socialize, and engage in community activities. Parks, community gardens, and green corridors are vital social hubs promoting residents' connections (Parker and Simpson 2018). Research has shown that access to green spaces is associated with increased social interactions and community engagement, which is essential for building social capital and resilience (Andreucci et al. 2019). For example, community gardens have been found to enhance social ties among participants, fostering a sense of belonging and community identity (Gallos et al. 2022). Green infrastructure significantly contributes to the overall quality of life in urban areas. Access to green spaces enhances aesthetic values, provides recreational opportunities, and improves mental and physical health outcomes (Roe et al. 2013). Moreover, green infrastructure can help address social inequities by ensuring all community members can access quality green spaces. Research indicates that neighborhoods with higher levels of greenery tend to have better health outcomes, emphasizing the importance of equitable access to green infrastructure for promoting public health (MacKinnon et al. 2019).

9.5 Environmental Services Provided by Green Infrastructure

Green infrastructure plays a crucial role in enhancing urban environments by providing a wide range of ecosystem services that contribute to public health and well-being, including (Fig. 9.3).

Air Quality Improvement

Trees and plants are essential components of green infrastructure that significantly improve air quality. They absorb carbon dioxide (CO2) and other pollutants, such as nitrogen oxides (NOx), sulfur dioxide (SO2), and particulate matter (PM), through their leaves and bark (Sun et al., 2019). Photosynthesis allows trees to convert CO2 into oxygen, while their foliage is a physical barrier that captures airborne particles, effectively filtering the air. Research has shown that urban areas with higher tree canopy cover experience lower levels of air pollution and improved respiratory health outcomes among residents. For instance, a study conducted in Santo Domingo, Dominican Republic, found that urban forests significantly contributed to the ecosystem service of air quality improvement, demonstrating the potential of GI to mitigate pollution (Bonilla-Duarte et al. 2021). By enhancing air quality, green infrastructure benefits the environment and promotes public health by reducing the incidence of respiratory diseases. The health impacts of improved air quality resulting from green infrastructure are profound. Exposure to air pollutants is linked to various health issues, including asthma, chronic obstructive pulmonary disease (COPD), cardiovascular diseases, and premature mortality (Strohbach et al. 2012). By reducing the concentration of harmful pollutants, green infrastructure can bring significant health benefits to urban populations. For example, cities that have implemented extensive urban greening initiatives have reported reductions in asthma rates and other respiratory conditions, particularly among vulnerable populations such as children and the elderly (Wang et al. 2023). The positive correlation between increased

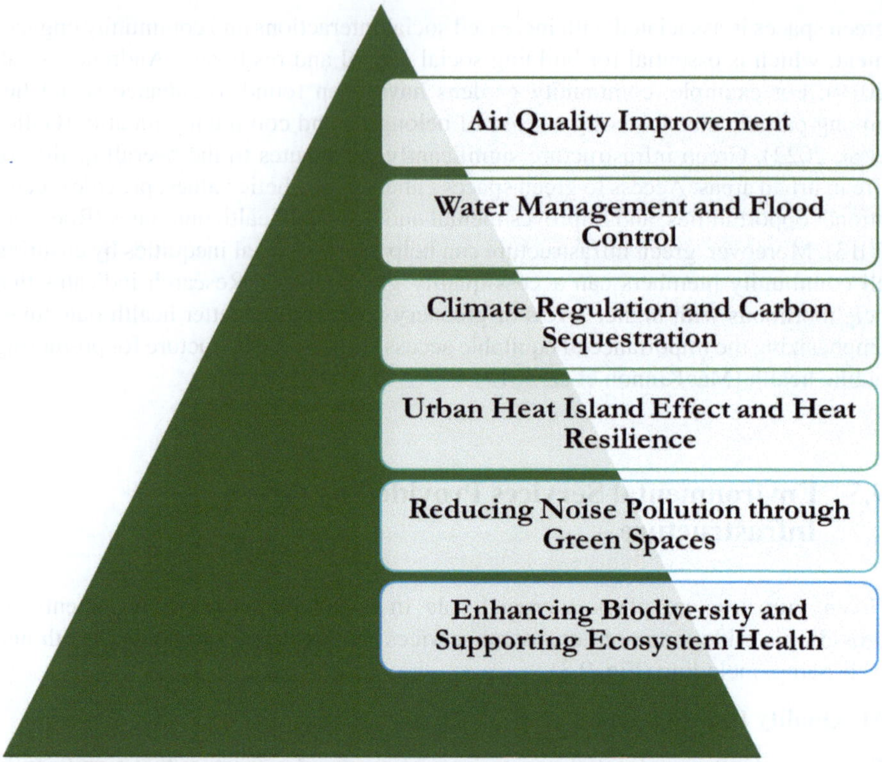

Fig. 9.3 Environmental Health benefits of Green Infrastructures

green space and improved air quality underscores the importance of integrating green infrastructure into urban planning to enhance public health outcomes.

Water Management and Flood Control

Green infrastructure is instrumental in managing stormwater and mitigating flood risks in urban areas. Traditional gray infrastructure, such as storm drains and concrete channels, often increases runoff and flooding during heavy rainfall. In contrast, green infrastructure employs natural processes to absorb, infiltrate, and filter stormwater, reducing the volume and velocity of runoff (Pugh et al. 2012). Techniques such as rain gardens, permeable pavements, and green roofs allow for the absorption of rainwater, promoting groundwater recharge and reducing the risk of urban flooding. For instance, a study in Philadelphia demonstrated that implementing green infrastructure practices significantly reduced stormwater runoff and improved water quality in local waterways. By enhancing stormwater management, green infrastructure contributes to urban resilience and protects public health from the impacts of flooding. Urban wetlands and rain gardens are practical examples of green infrastructure for flood mitigation. These systems are designed to capture and retain stormwater, allowing it to infiltrate the ground and reduce surface runoff. For example, in New Orleans, the restoration of urban wetlands has been shown to enhance flood protection while providing additional benefits such as habitat for wildlife and recreational opportunities for residents (Sharma et al. 2020). Rain gardens, which are shallow depressions planted with native vegetation, are another effective solution for managing stormwater in urban areas. They can be integrated into streetscapes, parks, and residential areas to capture runoff from impervious surfaces, reducing the risk of flooding and improving water quality (Jennings et al. 2016). The successful implementation of these green infrastructure solutions highlights their potential to address urban flooding challenges while enhancing community resilience.

Climate Regulation and Carbon Sequestration

Green infrastructure contributes to climate regulation by mitigating the urban heat island (UHI) effect, where urban areas experience higher temperatures than surrounding rural areas due to human activities and built environments. Vegetation provides shade and cools the air through evapotranspiration, effectively lowering surface and ambient temperatures (Li 2018). Research has demonstrated that urban green spaces can significantly reduce city temperatures, improving residents' thermal comfort. For instance, a study in Los Angeles found that increasing tree canopy cover could lower surface temperatures by several degrees, reducing the need for air conditioning and associated energy consumption (Şenik and Uzun 2022). By mitigating heat stress, green infrastructure enhances public health and well-being, particularly during extreme heat events. Green infrastructure plays a vital role in carbon sequestration, helping to mitigate climate change by absorbing CO2 from the atmosphere. Trees and vegetation store carbon in their biomass, contributing to the overall carbon balance of urban ecosystems (Sun et al. 2019). A study conducted in Beijing estimated that urban green spaces sequester significant amounts of carbon, highlighting their potential to contribute to climate change mitigation efforts (Suhardono 2024). By optimizing the design and management of urban green spaces, cities can enhance their carbon sequestration capacity and reduce greenhouse gas emissions. Moreover, implementing green infrastructure can reduce emissions by promoting sustainable practices such as urban agriculture and reducing reliance on fossil fuels for heating and cooling (Marchi et al. 2015). For example, green roofs can lower energy consumption in buildings by providing insulation and reducing the need for air conditioning, thereby decreasing overall emissions (Song et al. 2020).

Urban Heat Island Effect and Heat Resilience–Enhancing Heat Resilience

The urban heat island (UHI) effect occurs when urban areas experience significantly higher temperatures than their rural surroundings due to human activities and the prevalence of impervious surfaces, such as asphalt and concrete. This phenomenon can lead to increased energy consumption, elevated emissions of air pollutants, and heightened health risks, particularly during heatwaves. Green infrastructure is vital in mitigating the UHI effect by providing shade, cooling the air through evapotranspiration, and reducing surface temperatures. Vegetation, including trees and green roofs, can lower ambient temperatures by several degrees, creating more comfortable urban environments. For example, a study conducted in Los Angeles found that increasing tree canopy cover could significantly reduce surface temperatures, alleviating residents' heat stress. By integrating green infrastructure into urban planning, cities can enhance their resilience to heat-related health risks and improve overall quality of life. Green infrastructure can enhance heat resilience by creating cooler microclimates and providing refuge for residents during extreme heat events. Urban parks, green roofs, and shaded public spaces are critical resources for vulnerable populations, such as the elderly and those with pre-existing health conditions. Cities like Chicago have implemented urban greening initiatives to combat heat stress, resulting in measurable improvements in community well-being during heat waves. By prioritizing green infrastructure, urban areas can effectively manage heat-related health risks and promote resilience in the face of climate change.

Reducing Noise Pollution through Green Spaces

Noise pollution is a significant environmental stressor in urban areas, contributing to various health issues, including sleep disturbances, cardiovascular diseases, and mental health disorders. Urbanization and increased traffic contribute to elevated noise levels, which can negatively impact the quality of life for residents. Green infrastructure can reduce noise pollution by acting as a natural sound barrier. Vegetation, such as trees and shrubs, absorbs, deflects, and refracts sound waves, leading to lower noise levels in urban environments. Research has shown that urban green spaces can significantly reduce noise pollution, particularly in areas adjacent to busy roads or industrial zones. For example, a study conducted in Berlin found that parks and green corridors effectively mitigated noise levels, contributing to improved health outcomes for nearby residents. Green infrastructure can enhance community well-being and promote healthier urban environments by reducing noise pollution. Access to quiet green spaces provides residents with opportunities for relaxation and recreation, contributing to improved mental health and overall quality of life. Cities that prioritize green infrastructure as a means of noise reduction can create more livable environments that support the health and well-being of their residents.

Enhancing Biodiversity and Supporting Ecosystem Health in Cities

Biodiversity is essential for maintaining ecosystem health and resilience. Urban areas often face habitat loss and fragmentation challenges, leading to declines in species diversity and ecosystem services. Green infrastructure can play a crucial role in enhancing biodiversity by providing habitats for various species and promoting ecological connectivity. Green infrastructure, such as urban parks, green roofs, and community gardens, can serve as critical habitats for wildlife, including birds, insects, and other organisms. By incorporating native plant species and creating diverse habitats, cities can support local biodiversity and enhance ecosystem health. For example, a study conducted in Toronto found that urban green spaces with diverse plant species supported higher levels of bird and insect diversity, contributing to the overall health of urban ecosystems. By prioritizing biodiversity in urban planning, cities can create resilient ecosystems that provide essential services and enhance public health. Implementing green infrastructure can improve ecosystem health by enhancing soil quality, water management, and nutrient cycling. For instance, rain gardens and bioswales can filter pollutants from stormwater runoff, improving water quality and promoting healthy aquatic ecosystems. Additionally, urban forests can enhance soil health and reduce erosion, contributing to the overall resilience of urban ecosystems. By integrating green infrastructure into urban planning, cities can promote ecosystem health and ensure the sustainability of urban environments.

9.5.1 Case Studies of Successful Green Infrastructure Projects

Green Roof Initiatives in Major Cities (e.g., Toronto, Chicago): Green roof initiatives have gained popularity in cities like Toronto and Chicago as effective strategies for enhancing urban resilience. In Toronto, the Green Roof Bylaw mandates that new developments include green roofs, contributing to improved air quality, reduced energy consumption, and enhanced biodiversity. Similarly, Chicago's green roof initiatives have transformed the urban landscape and provided numerous environmental benefits while improving the quality of life for residents. Studies have shown that green roofs in Chicago significantly reduce stormwater runoff and contribute to energy savings, showcasing the potential of green infrastructure to address urban challenges.

Urban Parks and Public Health Improvements (e.g., New York City's High Line): The High Line in New York City is a prime example of how urban parks can enhance public health and well-being. This elevated linear park, built on a former railway line, has transformed a neglected space into a vibrant green corridor that provides recreational opportunities and promotes social interaction. Research has demonstrated that the High Line has improved mental health outcomes for visitors, with increased access to green space associated with lower stress and anxiety levels. The success of the High Line highlights the potential of urban parks to promote public health and enhance the livability of cities.

Community-Led Green Infrastructure Projects in Low-Income Areas: Community-led green infrastructure projects have emerged as effective strategies for addressing environmental and social challenges in low-income areas. For example, community gardens in cities like Detroit and Los Angeles have provided residents access to fresh produce while fostering social connections and community engagement. These projects improve food security and enhance mental health and well-being by providing physical activity and social interaction opportunities. Cities can promote resilience and improve public health by actively empowering communities to participate in green infrastructure initiatives.

Rain Gardens and Sustainable Water Management in Cities (e.g., Copenhagen, Portland): Rain gardens have been successfully implemented in cities like Copenhagen and Portland as part of sustainable water management strategies. These green infrastructure features capture and filter stormwater runoff, reducing the risk of flooding and improving water quality. In Portland, the implementation of rain gardens has been linked to significant reductions in stormwater runoff and improved water quality in local waterways. By integrating rain gardens into urban landscapes, cities can enhance their resilience to climate change while promoting public health through improved water management.

9.6 Policy and Planning for Implementing Green Infrastructure in Urban Areas

Integrating GI into urban planning is essential for addressing environmental challenges and enhancing public health. As cities face increasing pressures from climate change, urbanization, and public health crises, the need for effective policies and strategies to implement green infrastructure has never been more critical. Urban planning strategies must prioritize the integration of green infrastructure into the design of urban spaces. This involves identifying opportunities for incorporating green elements, such as parks, green roofs, and urban forests, into existing and new developments. A comprehensive approach to urban design can enhance the functionality of green infrastructure while maximizing its benefits for public health and environmental sustainability Zhu et al. (2019). For example, cities can adopt zoning regulations that encourage the development of green spaces in residential and commercial areas. By requiring a certain percentage of green space in new developments, urban planners can ensure that residents have access to nature, essential for promoting physical and mental health (Li et al. 2017). Additionally, integrating green infrastructure into transportation planning can create green corridors that enhance connectivity and promote active transportation options, such as walking and cycling (Anderson and Gough 2021).

Engaging communities in the planning process is vital for successfully implementing green infrastructure. Public participation ensures that the needs and preferences of residents are considered, leading to more effective and accepted solutions. Community engagement can take various forms, including public consultations, workshops, and participatory design processes (Lee and Kim 2022). For instance, involving community members in designing local parks and green spaces can foster a sense of ownership and stewardship, encouraging residents to maintain and enhance these areas actively. This participatory approach improves the quality of green infrastructure and strengthens social ties and community cohesion (Tayouga and Gagné 2016). Successfully implementing green infrastructure requires supportive policy frameworks at both national and local levels. Governments can establish policies promoting green infrastructure integration into urban planning and development. For example, national policies prioritizing sustainability and climate resilience can provide a foundation for local governments to adopt green infrastructure initiatives (Thorne et al. 2015). In many countries, policies such as the European Union's Green Infrastructure Strategy aim to enhance the integration of nature into urban environments. These policies provide guidelines and funding opportunities for municipalities to develop and implement green infrastructure projects (Zhang et al. 2020). Local governments can also create specific regulations that mandate the inclusion of green infrastructure in new developments, ensuring that urban areas are designed with sustainability in mind.

To encourage the adoption of green infrastructure, policymakers can provide financial incentives, such as grants, tax credits, or subsidies, for developers and property owners who incorporate green elements into their projects. These incentives can help offset the initial costs associated with green infrastructure implementation, making it more accessible for a broader range of stakeholders (Anderson and Gough 2021). For example, cities like Portland, Oregon, have implemented incentive programs that reward property owners for installing green roofs, rain gardens, and permeable pavements. These programs promote sustainable practices and contribute to improved urban resilience and public health (Willems et al. 2020). Integrating green infrastructure into urban development plans requires a holistic approach considering the interconnectedness of social, environmental, and economic factors. Urban planners should assess the potential benefits of green infrastructure concerning public health, climate resilience, and biodiversity conservation (Lennon et al. 2016). For instance, comprehensive urban development plans can incorporate green infrastructure as a critical component of stormwater management strategies, ensuring natural systems are utilized to mitigate flooding and improve water quality. By recognizing the multifunctional benefits of green infrastructure, cities can create more sustainable and resilient urban environments ("The Determination of Green Infrastructure Components of Van City Center and Its Near Surroundings," 2019).

To ensure the effectiveness of green infrastructure initiatives, it is essential to establish monitoring and evaluation frameworks that assess their impacts on public health and environmental quality. These frameworks can provide valuable data on the performance of green infrastructure projects, informing future planning and investment decisions (Nuryanti et al. 2021). For example, cities can implement performance metrics that track changes in air quality, temperature, and biodiversity in areas with newly implemented green infrastructure. By evaluating the outcomes of these projects, urban planners can identify best practices and areas for improvement, ultimately enhancing the effectiveness of green infrastructure in promoting urban health (Afionis et al. 2020). Local governments play a crucial role in implementing green infrastructure by developing policies, regulations, and programs that support its integration into urban planning. They coordinate efforts among various stakeholders, including community organizations, NGOs, and private sector partners, to implement green infrastructure initiatives effectively (Li et al. 2017). Local governments can also facilitate public engagement in planning, ensuring community voices are heard and considered in decision-making. By fostering stakeholder collaboration, local governments can create a supportive environment for green infrastructure development (Fields et al. 2016). Non-governmental organizations (NGOs) and community organizations are essential partners in promoting green infrastructure initiatives. These organizations often have the expertise, resources, and community connections to implement successful projects. They can engage residents in planning and maintaining green infrastructure, fostering a sense of ownership and stewardship

(Dou et al. 2014). For example, NGOs can lead community-led green infrastructure projects, such as community gardens and urban reforestation, that enhance local ecosystems while providing social and health benefits. By collaborating with local governments and other stakeholders, NGOs can help ensure that green infrastructure initiatives are aligned with community needs and priorities.

9.7 Challenges and Barriers to Implementing Green Infrastructure in Urban Areas

Implementing green infrastructure (GI) in urban areas presents numerous challenges and barriers that can hinder its effectiveness and sustainability. These challenges encompass financial and funding issues, space constraints in densely populated areas, maintenance and sustainability concerns, and the need for public awareness and community engagement. Some of these challenges include.

Financial and Funding Challenges

One of the primary barriers to implementing green infrastructure is the high initial costs associated with its development. Installing green roofs, rain gardens, and urban parks often requires significant financial investment, which can deter municipalities and private developers Zaballos et al. (2021). While green infrastructure can provide long-term savings through reduced energy costs and improved public health outcomes, the upfront costs can be prohibitive, particularly in low-income areas with limited funding. Access to the financing for green infrastructure projects can also be a challenge. Many cities rely on government grants, public–private partnerships, and philanthropic contributions to finance green infrastructure initiatives. However, competition for these funding sources can be intense, and not all projects receive the necessary financial support (Janiszek and Krzysztofik 2023).

Additionally, the lack of dedicated funding streams for green infrastructure can lead to uncertainty and delays in project implementation. In developing regions, financial constraints can be even more pronounced. Rapid urbanization often outpaces the availability of resources for infrastructure development, making it challenging to prioritize green infrastructure initiatives (Tayouga and Gagné 2016). As cities in developing countries face pressing challenges related to poverty, inadequate housing, and essential services, green infrastructure may be viewed as a lower priority than immediate needs.

Space Constraints in Densely Populated Areas

Urban areas are often characterized by high population density and limited available land, posing significant challenges to implementing green infrastructure. In many cities, space is at a premium, and competing land uses—such as residential, commercial, and industrial development—can limit opportunities for creating green spaces (Ge et al. 2023). For example, in cities like Tokyo and Hong Kong, where land is scarce, the integration of green infrastructure into existing urban fabric requires innovative design solutions that maximize the use of available space (Wang et al. 2022). This challenge necessitates careful planning and collaboration among stakeholders to identify suitable locations for green infrastructure projects. In densely populated areas, balancing the interests of various stakeholders can complicate the implementation of green infrastructure. Developers, property owners, and local communities may have differing land-use priorities and concerns, leading to conflicts and delays in project approval (Tudorie et al. 2019). For instance, while community members may advocate for creating parks and green spaces, developers may prioritize commercial or residential projects that generate immediate economic returns. Effective communication and negotiation among stakeholders are essential for finding common ground and ensuring the successful implementation of green infrastructure initiatives.

Maintenance and Sustainability of Green Infrastructure

The long-term success of green infrastructure relies on proper maintenance and management. Unlike traditional gray infrastructure, which may require less frequent upkeep, green infrastructure elements such as green roofs, rain gardens, and urban forests require regular maintenance to ensure their effectiveness (Brzoska and Spāģe 2020). This includes irrigation, weeding, pruning, and monitoring for pests and diseases. The need for ongoing maintenance can pose challenges for municipalities and property owners, particularly in areas with limited resources. Without adequate maintenance, green infrastructure can become less effective, reducing biodiversity, poor water management, and diminished aesthetic value (Ou 2023). Ensuring the sustainability and resilience of green infrastructure is critical for its long-term effectiveness. Climate change, urbanization, and environmental degradation can impact the health and functionality of green infrastructure elements (Artmann et al. 2019). For example, extreme weather events, such as heavy rainfall or prolonged drought, can affect the performance of green roofs and rain gardens, necessitating adaptive management strategies. To enhance the resilience of green infrastructure, it is essential to incorporate climate considerations into the design and planning processes. This may involve selecting plant species well-suited to local climate conditions, implementing water-efficient irrigation systems, and designing green infrastructure to withstand extreme weather events (Privitera and Rosa 2018).

Public Awareness and Community Engagement

Public awareness of the benefits of green infrastructure is crucial for garnering support for its implementation. However, many residents may not fully understand the value of green infrastructure in enhancing urban health, improving air and water quality, and providing recreational opportunities (Black et al. 2016). To address this challenge, effective communication strategies are needed to educate the public about the benefits of green infrastructure. Public outreach campaigns, workshops, and community events can help raise awareness and foster a sense of ownership and stewardship among residents (Matthews et al. 2015). Community engagement is essential for the successful implementation of green infrastructure. Involving residents in the planning and design processes can ensure that green infrastructure projects meet the needs and preferences of the community. For example, participatory design workshops allow community members to share their ideas and concerns regarding green infrastructure initiatives. By incorporating community input, urban planners can create green spaces that are not only functional but also culturally relevant and accessible (Kim and Tran 2018).

9.8 Future Directions and Innovations in Green Infrastructure for Urban Health

As urban areas continue to grow and face increasing environmental challenges, implementing green infrastructure (GI) has emerged as a critical strategy for enhancing urban health and resilience. Technological advances, innovative applications, and collaborative efforts among various stakeholders will shape future directions in green infrastructure. The evolution of sensor technology is paving the way for more accurate and affordable monitoring solutions in green infrastructure. Advances in materials science and nanotechnology enable the creation of susceptible sensors capable of detecting a wide range of environmental parameters, including air and water quality indicators, soil health metrics, and even biological contaminants (Aldegheishem et al. 2022). These innovations are particularly significant for urban areas, where the demand for real-time data on environmental conditions is high. The miniaturization of sensors and the reduction in manufacturing costs are making it feasible to deploy large networks of sensors for comprehensive ecological monitoring (Wang et al. 2021). For example, low-cost air quality sensors can be distributed throughout urban areas to provide real-time data on pollution levels, empowering communities to take action against air quality issues.

Integrating artificial intelligence (AI) with environmental monitoring technologies enhances our ability to predict ecological health risks. Machine learning algorithms can analyze vast datasets from various sources, including sensor data, satellite imagery, and historical health records, to identify patterns and predict future environmental conditions (El-Kassar and Singh 2019). For instance, AI models can forecast pollution levels based on meteorological data, traffic patterns, and industrial activities, allowing public health officials to issue timely alerts and implement mitigation strategies (Monteiro et al. 2021). Additionally, AI can be used to model the potential impacts of climate change on biodiversity and public health, providing valuable insights for policymakers and conservationists. Innovations in bio-sensing technologies and environmental DNA (eDNA) monitoring are opening new avenues for environmental health assessment. Biosensors can detect specific biological indicators, such as pathogens or toxins, in environmental samples, providing real-time data on ecosystem health (Chan and Marafa 2018). eDNA monitoring involves analyzing genetic material shed by organisms into their environment and is a powerful tool for assessing biodiversity and ecosystem health. This method detects species presence and abundance without direct observation, making it particularly useful for monitoring elusive or endangered species (Hussein et al. 2020). By integrating eDNA data with traditional monitoring methods, researchers can better understand ecosystem dynamics and their implications for public health.

Blockchain technology is a promising solution for enhancing data security and transparency in environmental monitoring systems. By providing a decentralized and immutable ledger for data storage, blockchain can ensure the integrity and authenticity of ecological data collected from various sources (Aldegheishem et al. 2022). This technology is particularly relevant for environmental monitoring, where data accuracy is critical for informed decision-making. For instance, blockchain can be used to track the provenance of ecological data, ensuring that it has not been tampered with or altered. Additionally, blockchain can facilitate secure data sharing among stakeholders, promoting collaboration and transparency in environmental health initiatives (Wang et al. 2021). Public–private partnerships (PPPs) are increasingly recognized as vital for advancing green infrastructure technologies. These collaborations can leverage the strengths of both sectors, combining public-sector resources and regulatory frameworks with private-sector innovation and expertise (Luketich et al. 2019). For example, partnerships between government agencies and technology companies can facilitate developing and deploying advanced monitoring systems that address specific environmental health challenges. By pooling resources and expertise, PPPs can accelerate the adoption of smart technologies and enhance the effectiveness of environmental monitoring efforts (Newton and Rogers 2020).

As environmental health issues transcend national borders, international collaboration is essential for effective monitoring and management. Collaborative efforts among countries, international organizations, and research institutions can enhance data sharing, standardize monitoring protocols, and promote best practices in environmental health monitoring ("Cloud-Based Data Analytics for Smart Cities: Enhancing Urban Infrastructure and Services", 2023). For instance, initiatives such as the Global Environmental Monitoring System (GEMS) aim to provide a framework for international cooperation in monitoring water quality and other environmental parameters. By fostering collaboration and knowledge exchange, these initiatives can strengthen global efforts to address environmental health challenges and promote sustainable practices (Campos et al. 2019). To ensure the long-term effectiveness of green infrastructure, it is essential to establish monitoring frameworks that assess their impacts on public health and environmental quality over time. These frameworks can provide valuable data on the performance of green infrastructure projects, informing future planning and investment decisions (Guido et al. 2022). For example, cities can implement performance metrics that track changes in air quality, temperature, and biodiversity in areas with newly implemented green infrastructure. By evaluating the outcomes of these projects, urban planners can identify best practices and areas for improvement, ultimately enhancing the effectiveness of green infrastructure in promoting urban health (Livesley et al. 2021). Ongoing research is crucial for understanding the long-term health impacts of green infrastructure. Studies that assess the relationships between green space exposure and health outcomes can provide insights into the mechanisms through which green infrastructure benefits public health (Hubl et al., 2018). For instance, longitudinal studies that track changes in health indicators among residents before and after the implementation of green infrastructure can help establish causal relationships and quantify health benefits. This evidence is essential for advocating for integrating green infrastructure into urban planning and public health policies (Du et al. 2019).

9.9 Conclusion

The diverse types of green infrastructure and their associated benefits demonstrate the importance of incorporating nature-based solutions into urban planning. Green infrastructure provides a multitude of environmental services that significantly enhance urban health. Green infrastructure plays a vital role in promoting public health and well-being by improving air quality, managing stormwater, mitigating heat stress, and contributing to climate regulation and carbon sequestration. As cities face environmental challenges, prioritizing integrating green infrastructure into urban planning will be essential for creating healthier, more sustainable urban environments that benefit all residents. Green infrastructure addresses ecological challenges and supports public health, social well-being, and economic development, making cities more sustainable and livable. Integrating green infrastructure and nature-based solutions into urban planning is essential for addressing the health challenges urban populations face. Nature-based solutions are vital in enhancing urban health by improving air quality, mitigating heat stress, promoting mental well-being, and fostering community resilience. Integrating NbS into urban planning and development is essential for addressing the health challenges urban populations face. The health benefits of green infrastructure are multifaceted, encompassing physical, mental, and social health outcomes. Green infrastructure plays a vital role in improving urban health by improving air quality, reducing heat stress, promoting physical activity, and enhancing community cohesion. As cities continue to face environmental and public health challenges, prioritizing integrating nature-based solutions into urban planning will be essential for creating healthier, more sustainable urban environments that promote the well-being of all residents. As cities continue to grow and confront environmental challenges, prioritizing nature-based solutions will be crucial for creating healthier, more sustainable urban environments that promote the well-being of all residents. By improving air quality, managing heat-related risks, promoting mental well-being, and enhancing community resilience, green infrastructure can significantly contribute to urban health. As cities grow and face environmental challenges, prioritizing nature-based solutions will be crucial for creating healthier, more sustainable urban environments. Implementing green infrastructure in urban areas faces several challenges, including financial and funding constraints, space limitations, maintenance requirements, and the need for public awareness and community engagement. Addressing these barriers is essential for maximizing the benefits of green infrastructure and promoting sustainable urban development. By fostering collaboration among stakeholders, prioritizing education and outreach, and developing innovative funding solutions, cities can overcome these challenges and create healthier, more resilient urban environments. The future of green infrastructure is characterized by technological advances, innovative applications, and collaborative efforts that enhance urban health and resilience. Cities can effectively address environmental challenges and promote public health by prioritizing integrating intelligent technologies, such as AI and blockchain, and fostering stakeholder partnerships. As urban areas grow and face increasing climate change and urbanization pressures, implementing nature-based solutions will be essential for creating healthier, more sustainable urban environments that benefit all residents.

References

Aerts R, Honnay O, Nieuwenhuyse A (2018) Biodiversity and human health: mechanisms and evidence of the positive health effects of diversity in nature and green spaces. British Med Bull

Afionis S, Mkwambisi D, Dallimer M (2020) Lack of cross-sector and cross-level policy coherence and consistency limits urban green infrastructure implementation in malawi. Front Environ Sci 8. https://doi.org/10.3389/fenvs.2020.558619

Akanbi A (2020) Estemd: a distributed processing framework for environmental monitoring based on apache kafka streaming engine. https://doi.org/10.1145/3445945.3445949

Akanbi A (2020b) Science–policy challenges for biodiversity, public health, and urbanization: examples from Belgium. Environ Res Lett 8(2):025015. https://doi.org/10.1088/1748-9326/8/2/025015

Aldegheishem A, Alrajeh N, García L, Lloret J (2022) Swap: smart water protocol for the irrigation of urban gardens in smart cities. IEEE Access 10:39239–39247. https://doi.org/10.1109/access.2022.3165579

Anderson V, Gough W (2021) Harnessing the four horsemen of climate change: a framework for deep resilience, decarbonization, and planetary health in Ontario, Canada. Sustainability 13(1):379. https://doi.org/10.3390/su13010379

Anderson V, Gough W, Agic B (2021) Nature-based equity: an assessment of the public health impacts of green infrastructure in Ontario Canada. Int J Environ Res Public Health 18(11):5763. https://doi.org/10.3390/ijerph18115763

Andreucci M, Loder A, Brown M, Brajković J (2021) Exploring challenges and opportunities of biophilic urban design: evidence from research and experimentation. Sustainability 13(8):4323. https://doi.org/10.3390/su13084323

Andreucci M, Russo A, Olszewska-Guizzo A (2019) Designing urban green blue infrastructure for mental health and elderly wellbeing. Sustainability 11(22):6425. https://doi.org/10.3390/su11226425

Artmann M, Kohler M, Meinel G, Gan J, Iojă I (2019) How smart growth and green infrastructure can mutually support each other—conceptual framework for compact and green cities. Ecol Ind 96:10–22. https://doi.org/10.1016/j.ecolind.2017.07.001

Atanasova N, Castellar J, Pineda-Martos R, Nika C, Katsou E, Istenič D, Pucher B, Andreucci MB, Langergraber G (2021) Nature-based solutions and circularity in cities. Circular Econ Sustain 1(1):319–332. https://doi.org/10.1007/s43615-021-00024-1

Berman M, Kross E, Krpan K, Askren M, Burson A, Deldin P, Kaplan S, Sherdell L, Gotlib IH, Jonides J (2012) Interacting with nature improves cognition and affect for individuals with depression. J Affect Disord 140(3):300–305. https://doi.org/10.1016/j.jad.2012.03.012

Beyer K, Kaltenbach A, Szabo A, Bogar S, Nieto F, Malecki K (2014) Exposure to neighborhood green space and mental health: evidence from the survey of the health of Wisconsin. In: ISEE conference abstracts, 2014(1). https://doi.org/10.1289/isee.2014.o-280

Black J, Tara K, Pakzad P (2016) Mainstreaming green infrastructure elements into the design of public road reserves: challenges for road authorities. Int J Environ Prot 1–8. https://doi.org/10.5963/ijep0601001

Bočkarjova M, Kačalová A (2021) Greening of European cities: social benefits of urban nature for urban air quality. Eur Stud 8(1):177–204. https://doi.org/10.2478/eustu-2022-0072

Bonilla-Duarte S, González C, Rodríguez L, Jäuregui-Haza U, García-García A (2021) Contribution of urban forests to the ecosystem service of air quality in the city of santo domingo, dominican republic. Forests 12(9):1249. https://doi.org/10.3390/f12091249

Bratman G, Hamilton J, Hahn K, Daily G, Gross J (2015) Nature experience reduces rumination and subgenual prefrontal cortex activation. Proc Natl Acad Sci 112(28):8567–8572. https://doi.org/10.1073/pnas.1510459112

Brzoska P, Spāģe A (2020) From city- to site-dimension: assessing the urban ecosystem services of different types of green infrastructure. Land 9(5):150. https://doi.org/10.3390/land9050150

Campos N, Rocha A, Gondim R, Silva T, Gomes D (2019) Smart & green: an internet-of-things framework for smart irrigation. Sensors 20(1):190. https://doi.org/10.3390/s20010190

Chan C, Marafa L (2018) Knowledge-perception bridge of green-smart integration of cities: an empirical study of Hong Kong. Sustainability 10(1):107. https://doi.org/10.3390/su10010107

Coutts C, Hahn M (2015) Green infrastructure, ecosystem services, and human health. Int J Environ Res Public Health 12(8):9768–9798. https://doi.org/10.3390/ijerph120809768

Dadvand P, Nieuwenhuijsen M, Esnaola M, Forns J, Basagaña X, Álvarez-Pedrerol M, Rivas I, López-Vicente M, De Castro Pascual M, Su J, Jerrett M, Sunyer J (2015) Green spaces and cognitive development in primary schoolchildren. Proc Natl Acad Sci 112(26):7937–7942. https://doi.org/10.1073/pnas.1503402112

Đorđević T (2023) Urban planning based on nature—a nature-based solution. J Build Constr Plann Res 11(01):1–25. https://doi.org/10.4236/jbcpr.2023.111001

Dorst H, Jagt A, Raven R, Runhaar H (2019) Urban greening through nature-based solutions–key characteristics of an emerging concept. Sustain Cities Soc 49:101620. https://doi.org/10.1016/j.scs.2019.101620

Dou X, Wang Z, Yuan H, Jing P (2014) Function, barriers and policies of green infrastructure construction in urban environment. Adv Mater Res 1065–1069:2814–2818. https://doi.org/10.4028/www.scientific.net/amr.1065-1069.2814

Du H, Liu D, Lu Z, Crittenden J, Mao G, Wang S, Zou H (2019) Research development on sustainable urban infrastructure from 1991 to 2017: a bibliometric analysis to inform future innovations. Earth S Future 7(7):718–733. https://doi.org/10.1029/2018ef001117

Dushkova D, Haase D (2020) Not simply green: nature-based solutions as a concept and practical approach for sustainability studies and planning agendas in cities. Land 9(1):19. https://doi.org/10.3390/land9010019

Dushkova D, Ignatieva M (2020) New trends in urban environmental health research: from geography of diseases to therapeutic landscapes and healing gardens. Geograph Environ Sustain 13(1):159–171. https://doi.org/10.24057/2071-9388-2019-99

El-Kassar A, Singh S (2019) Green innovation and organizational performance: the influence of big data and the moderating role of management commitment and hr practices. Technol Forecast Soc Chang 144:483–498. https://doi.org/10.1016/j.techfore.2017.12.016

Erisman J, Galloway J, Seitzinger S, Bleeker A, Butterbach-Bahl K (2011) Reactive nitrogen in the environment and its effect on climate change. Current Opinion in Environmental Sustainability 3(5):281–290. https://doi.org/10.1016/j.cosust.2011.08.012

Fathy C, Ali H (2023) A secure iot-based irrigation system for precision agriculture using the expeditious cipher. Sensors 23(4):2091. https://doi.org/10.3390/s23042091

Felappi J (2024) Urban park qualities driving visitors mental well-being and wildlife conservation in a neotropical megacity. Sci Rep 14(1). https://doi.org/10.1038/s41598-024-55357-2

Ferreira V, Barreira A, Loures L, Antunes M, Panagopoulos T (2020) Stakeholders' engagement on nature-based solutions: a systematic literature review. Sustainability 12(2):640. https://doi.org/10.3390/su12020640

Fields B, Thomas J, Wagner J (2016) Living with water in the era of climate change: lessons from the Lafitte greenway in post-katrina New Orleans. J Plan Educ Res 37(3):309–321. https://doi.org/10.1177/0739456x16655600

Fisher B, Grima N (2020). The importance of urban natural areas and urban ecosystem services during the covid-19 pandemic. https://doi.org/10.31235/osf.io/sd3h6

Gallos P, Menychtas A, Panagopoulos C, Bimpas M (2022). Quantifying citizens' well-being in areas with natural based solutions using mobile computing. https://doi.org/10.3233/shti210958

Gascón M, Triguero-Mas M, Martínez D, Dadvand P, Forns J, Plasència A, Nieuwenhuijsen M (2015) Mental health benefits of long-term exposure to residential green and blue spaces: a systematic review. Int J Environ Res Public Health 12(4):4354–4379. https://doi.org/10.3390/ijerph120404354

Ge M, Yang H, Zhu Y, Kim M, Cui X (2023) Examining the microclimate pattern and related spatial perception of the urban stormwater management landscape: the case of rain gardens. Atmosphere 14(7):1138. https://doi.org/10.3390/atmos14071138

Guido G, Haghshenas S, Haghshenas S, Vitale A, Gallelli V, Astarita V (2022) Prioritizing the potential smartification measures by using an integrated decision support system with sustainable development goals (a case study in southern italy). Safety 8(2):35. https://doi.org/10.3390/safety8020035

Herman K, Drozda Ł (2021) Green infrastructure in the time of social distancing: urban policy and the tactical pandemic urbanism. Sustainability 13(4):1632. https://doi.org/10.3390/su13041632

Hsu K, Chao J (2022) The impact of urban green-infrastructure development on the price of surrounding real estate: a case study of Taichung city's central district. Iop Conf Ser Earth Environ Sci 1006(1):012012. https://doi.org/10.1088/1755-1315/1006/1/012012

Hubl M, Skowron P, Aleithe M (2018) Towards a supportive city with smart urban objects in the internet of things: the case of adaptive park bench and adaptive light, vol 16, pp 51–58. https://doi.org/10.15439/2018f118

Hussein J, Salama M, Kumble P, Hanson H (2020). The impact of the relation between political borders and ecosystems in creating green infrastructure opportunities-the city of prague. https://doi.org/10.20944/preprints202012.0228.v1

Iyiola AO, Afolabi OA, Alimi SK, Akingba OO, Izah SC, Ogwu MC (2024) Climate change and water crisis in the Global South. In: Izah SC, Ogwu MC, Loukas A, Hamidifar H (eds) Water crises and sustainable management in the global south. Springer, Singapore, pp 111–140. https://doi.org/10.1007/978-981-97-4966-9_4

Izah SC, Uzoekwe SA, Aigberua AO (2021). Source, geochemical spreading and risks of trace metals in particulate matter 2.5 within a gas flaring area in Bayelsa State, Nigeria. Adv Environ Technol 7(2):101–118

Izah SC, Jacob DE, Nelson IU, Avez S (2024) Urban water crisis in the global South. In: Izah SC, Ogwu MC, Loukas A, Hamidifar H (eds) Water crises and sustainable management in the global south. Springer, Singapore, pp 45–83. https://doi.org/10.1007/978-981-97-4966-9_2

Jacob DE, Nelson IU, Izah SC (2024b) Indigenous water management strategies in the global south. In: Izah SC, Ogwu MC, Loukas A, Hamidifar H (eds) Water crises and sustainable management in the global south. Springer, pp 487–525. https://link.springer.com/chapter/, https://doi.org/10.1007/978-981-97-4966-9_16

Jacob DE, Nelson IU, Izah SC, Daniel KS (2024a) Rural water crises in the global south: understanding the scope and impact. In: Izah SC, Ogwu MC, Loukas A, Hamidifar H (eds) Water crises and sustainable management in the global south. Springer, Singapore, pp 3–44. https://doi.org/10.1007/978-981-97-4966-9_1

Janiszek M, Krzysztofik R (2023) Green infrastructure as an effective tool for urban adaptation—solutions from a big city in a postindustrial region. Sustainability 15(11):8928. https://doi.org/10.3390/su15118928

Jennings V, Larson L, Yun J (2016) Advancing sustainability through urban green space: cultural ecosystem services, equity, and social determinants of health. Int J Environ Res Public Health 13(2):196. https://doi.org/10.3390/ijerph13020196

Ji C, Sha L, Dou Y (2023) Mapping the research trends of city biodiversity from 1995–2021: a bibliometric analysis. https://doi.org/10.20944/preprints202304.0978.v1

Kabisch N, Bosch M (2017) Urban green spaces and the potential for health improvement and environmental justice in a changing climate, pp 207–220. https://doi.org/10.1007/978-3-319-56091-5_12

Kabisch N, Korn H, Stadler J, Bonn A (2017) Nature-based solutions to climate change adaptation in urban areas. https://doi.org/10.1007/978-3-319-56091-5

Kazak J, Chruściński J, Szewrański S (2018) The development of a novel decision support system for the location of green infrastructure for stormwater management. Sustainability 10(12):4388. https://doi.org/10.3390/su10124388

Kim H, Tran T (2018) An evaluation of local comprehensive plans toward sustainable green infrastructure in US. Sustainability 10(11):4143. https://doi.org/10.3390/su10114143

Lee E, Kim G (2022) Analysis of domestic and international green infrastructure research trends from the esg perspective in South Korea. Int J Environ Res Public Health 19(12):7099. https://doi.org/10.3390/ijerph19127099

Lennon M, Scott M, Collier M, Foley K (2016) The emergence of green infrastructure as promoting the centralisation of a landscape perspective in spatial planning—the case of Ireland. Landsc Res 42(2):146–163. https://doi.org/10.1080/01426397.2016.1229460

Li F (2018) Carbon sequestration and spatial differentiation characteristics of urban forest in china. Appl Ecol Environ Res 16(2):1563–1580. https://doi.org/10.15666/aeer/1602_15631580

Li F, Liu X, Zhang X, Zhao D, Liu H, Zhou C, Wang R (2017) Urban ecological infrastructure: an integrated network for ecosystem services and sustainable urban systems. J Clean Prod 163:S12–S18. https://doi.org/10.1016/j.jclepro.2016.02.079

Livesley S, Marchionni V, Cheung P, Daly E, Pataki D (2021) Water smart cities increase irrigation to provide cool refuge in a climate crisis. Earth S Future 9(1). https://doi.org/10.1029/2020ef001806

Luca C, Naumann S, Davis M, Tondelli S (2021) Nature-based solutions and sustainable urban planning in the European environmental policy framework: analysis of the state of the art and recommendations for future development. Sustainability 13(9):5021. https://doi.org/10.3390/su13095021

Luketich A, Papuga S, Crimmins M (2019) Ecohydrology of urban trees under passive and active irrigation in a semiarid city. PLoS ONE 14(11):e0224804. https://doi.org/10.1371/journal.pone.0224804

MacKinnon K, Ham C, Reilly K, Hopkins J (2019) Nature-based solutions and protected areas to improve urban biodiversity and health, pp 363–380. https://doi.org/10.1007/978-3-030-02318-8_16

Marchi M, Pulselli R, Marchettini N, Pulselli F, Bastianoni S (2015) Carbon dioxide sequestration model of a vertical greenery system. Ecol Model 306:46–56. https://doi.org/10.1016/j.ecolmodel.2014.08.013

Marin A, Kičić M, Vuletić D, Ostoić S (2021) Perception and attitudes of residents towards green spaces in Croatia–an exploratory study. South-East Eur Forestry, 12(2):123–134. https://doi.org/10.15177/seefor.21-12

Matthews T, Lo A, Byrne J (2015) Reconceptualizing green infrastructure for climate change adaptation: barriers to adoption and drivers for uptake by spatial planners. Landsc Urban Plan 138:155–163. https://doi.org/10.1016/j.landurbplan.2015.02.010

Mejia A, Bhattacharya M, Nigon-Crowley A, Kirkpatrick K, Katoch C (2020) Community gardening during times of crisis: Recommendations for community-engaged dialogue, research, and praxis. J Agric Food Syst Commun Dev 10(1):13–19. https://doi.org/10.5304/jafscd.2020.101.006

Monteiro L, Cristina R, Covas D (2021) Water and energy efficiency assessment in urban green spaces. Energies 14(17):5490. https://doi.org/10.3390/en14175490

Newton P, Rogers B (2020) Transforming built environments: towards carbon neutral and blue-green cities. Sustainability 12(11):4745. https://doi.org/10.3390/su12114745

Nieuwenhuijsen M (2021) Green infrastructure and health. Annu Rev Public Health 42(1):317–328. https://doi.org/10.1146/annurev-publhealth-090419-102511

Nuryanti D, Maryono M, Muhammad F (2021) Preliminary study to strengthening strategic environmental assessment to implement green infrastructure concept in waste management. In: E3s web of conferences, vol 317, pp 01077. https://doi.org/10.1051/e3sconf/202131701077

Ogamba EN, Charles EE, Izah SC (2021) Distributions, pollution evaluation and health risk of selected heavy metal in surface water of Taylor creek, Bayelsa State, Nigeria. Toxicol Environ Health Sci 13(2):109–121. https://doi.org/10.1007/s13530-020-00076-0

Olszewska A, Marques P, Ryan R, Barbosa F (2016) What makes a landscape contemplative? Environ Plann B Urban Anal City Sci 45(1):7–25. https://doi.org/10.1177/0265813516660716

O'Regan A, Hunter R, Nyhan M (2021) "Biophilic cities": quantifying the impact of google street view-derived greenspace exposures on socioeconomic factors and self-reported health. Environ Sci Technol 55(13):9063–9073. https://doi.org/10.1021/acs.est.1c01326

Ou J (2023) Planning and design strategies for green stormwater infrastructure from an urban design perspective. Water 16(1):29. https://doi.org/10.3390/w16010029

Parker J, Simpson G (2018) Public green infrastructure contributes to city livability: a systematic quantitative review. Land 7(4):161. https://doi.org/10.3390/land7040161

Privitera R, Rosa D (2018) Reducing seismic vulnerability and energy demand of cities through green infrastructure. Sustainability 10(8):2591. https://doi.org/10.3390/su10082591

Pugh T, MacKenzie A, Whyatt J, Hewitt C (2012) Effectiveness of green infrastructure for improvement of air quality in urban street canyons. Environ Sci Technol 46(14):7692–7699. https://doi.org/10.1021/es300826w

Ranish S, Laurent A (2021) A synthesis of social and economic benefits linked to green infrastructure. MDPI Water 13(24):3651. https://doi.org/10.3390/w13243651

Richardson E, Pearce J, Mitchell R, Kingham S (2013) Role of physical activity in the relationship between urban green space and health. Public Health 127(4):318–324. https://doi.org/10.1016/j.puhe.2013.01.004

Rodrigues V, Rafael S, Sorte S, Coelho S, Vicente B, Leitão J, Lopes M, Miranda AI, Borrego C (2018) Adaptation to climate change at local scale: a cfd study in porto urban area. https://doi.org/10.5772/intechopen.72972

Roe J, Thompson C, Aspinall P, Brewer M, Duff E, Miller D, Mitchell R, Clow A (2013) Green space and stress: evidence from cortisol measures in deprived urban communities. Int J Environ Res Public Health 10(9):4086–4103. https://doi.org/10.3390/ijerph10094086

Roelvink F, Storlazzi C, Dongeren A, Pearson S (2021) Coral reef restorations can be optimized to reduce coastal flooding hazards. Front Marine Sci 8. https://doi.org/10.3389/fmars.2021.653945

Roelvink F, Storlazzi C, Dongeren A, Pearson S (2021). Coral reef restorations can be Romanelli. In: Corvalán C, Cooper H, Manga L, Maiero M, Campbell-Lendrum D (2014). From Manaus to Maputo: toward a public health and biodiversity framework. Ecohealth 11(3):292–299. https://doi.org/10.1007/s10393-014-0959-2

Roy S, Byrne J, Pickering C (2012) A systematic quantitative review of urban tree benefits, costs, and assessment methods across cities in different climatic zones. Urban For Urban Green 11(4):351–363. https://doi.org/10.1016/j.ufug.2012.06.006

Russell R, Guerry A, Balvanera P, Gould R, Basurto X, Chan K, Klain S, Levine J, Tam J (2013) Humans and nature: how knowing and experiencing nature affect well-being. Annu Rev Environ Resour 38(1):473–502. https://doi.org/10.1146/annurev-environ-012312-110838

Şenik B, Uzun O (2022) A process approach to the open green space system planning. Landscape Ecol Eng 18(2):203–219. https://doi.org/10.1007/s11355-021-00492-5

Sharma R, Pradhan L, Kumari M, Bhattacharya P (2020). Assessment of carbon sequestration potential of tree species in amity university campus Noida, vol 365, pp 52. https://doi.org/10.3390/iecf2020-08075

Song P, Kim G, Mayer A, He R, Tian G (2020) Assessing the ecosystem services of various types of urban green spaces based on i-tree eco. Sustainability 12(4):1630. https://doi.org/10.3390/su12041630

Strohbach M, Arnold E, Haase D (2012) The carbon footprint of urban green space—a life cycle approach. Landsc Urban Plan 104(2):220–229. https://doi.org/10.1016/j.landurbplan.2011.10.013

Sturiale L, Scuderi A (2019) The role of green infrastructures in urban planning for climate change adaptation. Climate 7(10):119. https://doi.org/10.3390/cli7100119

Suhardono S (2024) Assessing the ecological function effectiveness of urban parks in Surakarta city, Indonesia. Ecol Eng Environ Technol 25(4):227–240. https://doi.org/10.12912/27197050/183790

Sun Y, Xie S, Zhao S (2019) Valuing urban green spaces in mitigating climate change: a city-wide estimate of aboveground carbon stored in urban green spaces of china's capital. Glob Change Biol 25(5):1717–1732. https://doi.org/10.1111/gcb.14566

Tayouga S, Gagné S (2016) The socio-ecological factors that influence the adoption of green infrastructure. Sustainability 8(12):1277. https://doi.org/10.3390/su8121277

Thompson C, Aspinall P, Roe J, Robertson L, Miller D (2016) Mitigating stress and supporting health in deprived urban communities: the importance of green space and the social environment. Int J Environ Res Public Health 13(4):440. https://doi.org/10.3390/ijerph13040440

Thorne C, O'Donnell E, Ozawa C, Hamlin S, Smith L (2015) Overcoming uncertainty and barriers to adoption of blue-green infrastructure for urban flood risk management. J Flood Risk Manag 11(S2). https://doi.org/10.1111/jfr3.12218

Triguero-Mas M, Dadvand P, Cirach M, Martínez D, Medina A, Mompart A, Basagaña X, Gražulevičienė R, Nieuwenhuijsen M (2015) Natural outdoor environments and mental and physical health: relationships and mechanisms. Environ Int 77:35–41. https://doi.org/10.1016/j.envint.2015.01.012

Tudorie C, Gielen E, Vallés-Planells M, Galiana F (2019) Urban green indicators: a tool to estimate the sustainability of our cities. Int J Des Nat Ecodyn 14(1):19–29. https://doi.org/10.2495/dne-v14-n1-19-29

Uzoekwe SA, Izah SC, Aigberua AO (2021) Environmental and human health risk of heavy metals in atmospheric particulate matter (PM_{10}) around gas flaring vicinity in Bayelsa State, Nigeria. Toxicol Environ Health Sci. 13(4):323–335. https://doi.org/10.1007/s13530-021-00085-7

Vargas R, Torres M, Reinoso E (2023) Analysis of the effects of green infrastructure on flooding in Mexico city. https://doi.org/10.21203/rs.3.rs-2949532/v1

Wang A, Lin W, Liu B, Wang H, Xu H (2021) Does smart city construction improve the green utilization efficiency of urban land? Land 10(6):657. https://doi.org/10.3390/land10060657

Wang H, Feng Y, Ai L (2023) Progress of carbon sequestration in urban green space based on bibliometric analysis. Front Environ Sci 11. https://doi.org/10.3389/fenvs.2023.1196803

Willems J, Kenyon A, Sharp L, Molenveld A (2020) How actors are (dis)integrating policy agendas for multi-functional blue and green infrastructure projects on the ground. J Environ Planning Policy Manag 23(1):84–96. https://doi.org/10.1080/1523908x.2020.1798750

Wolch J, Byrne J, Newell J (2014) Urban green space, public health, and environmental justice: the challenge of making cities 'just green enough.' Landsc Urban Plan 125:234–244. https://doi.org/10.1016/j.landurbplan.2014.01.017

Xie J, Luo S, Furuya K, Sun D (2020) Urban parks as green buffers during the covid-19 pandemic. Sustainability 12(17):6751. https://doi.org/10.3390/su12176751

Zaballos I, Laurenz A, Sádaba J (2021) Urban green infrastructure as a strategy to address urban energy efficiency and sustainability. A case study of milagrosa (pamplona). Sustainability 14(1):28. https://doi.org/10.3390/su14010028

Zhang Y, Zhao W, Chen X, Jun C, Hao J, Tang X, Zhai J (2020) Assessment on the effectiveness of urban stormwater management. Water 13(1):4. https://doi.org/10.3390/w13010004

Zhao J, Wang D, Zhang F, Pan J, Claesson P, Larsson R, Shi Y (2022) Self-powered, long-durable, and highly selective oil–solid triboelectric nanogenerator for energy harvesting and intelligent monitoring. Nano-Micro Lett 14(1). https://doi.org/10.1007/s40820-022-00903-8

Zhu Z, Lang W, Tao X, Feng J, Liu K (2019) Exploring the quality of urban green spaces based on urban neighborhood green index—a case study of guangzhou city. Sustainability 11(19):5507. https://doi.org/10.3390/su11195507

Chapter 10
Sustainable Agricultural Practices: Technological Innovations for Healthier Food Systems

Wisdom Ebiye Sawyer and Saoban Sunkanmi Salimon

Abstract Sustainable agricultural practices are critical to building healthier food systems that can withstand the pressures of global food insecurity, environmental degradation, and climate change. This chapter highlights the role of technological innovations in enhancing sustainability across agricultural practices, focusing on precision agriculture, soil health technologies, and water-efficient systems. Through the integration of renewable energy sources such as solar and wind power, alongside advancements in biotechnology, farmers can optimize resource use, reduce greenhouse gas emissions, and improve crop resilience to the burgeoning myriads of climate challenges. The chapter also addresses the multifaceted barriers to technology adoption, including economic and social factors, emphasizing the need for supportive policies and collaborative efforts among stakeholders. Adopting innovative practices and technologies is essential for fostering a resilient and equitable food future, ensuring food security while protecting vital natural resources. Through continuous research and development, the agricultural sector can leverage these advancements to promote environmental stewardship and economic viability, paving the way for a more sustainable agricultural landscape.

Keywords Sustainable agriculture · Technological innovations · Precision agriculture · Renewable energy · Biotechnology · Food security · Environmental stewardship

W. E. Sawyer
Department of Community Medicine, Faculty of Clinical Sciences, Bayelsa Medical University, Yenagoa, Nigeria

S. S. Salimon (✉)
School of Biological and Medical Physics, Moscow Institute of Physics and Technology, Dolgoprudny, Russian Federation
e-mail: salimonsaoban@yahoo.com

S. Chibueze Izah and M. Chidozie Ogwu (eds.), *Innovative Approaches in Environmental Health Management*, Environmental Science and Engineering,
https://doi.org/10.1007/978-3-031-81966-7_10

10.1 Introduction

Sustainable agriculture has emerged as a critical approach to addressing the multi-faceted challenges of global food systems, particularly in the context of increasing population pressures, environmental degradation, and climate change. This concept encompasses a range of practices aimed at producing environmentally sound, economically viable, and socially responsible food. Sustainable agriculture is important because of its potential to ensure food security while preserving the ecological integrity of agricultural systems. As the global population is projected to reach approximately 9.7 billion by 2050, the demand for food will increase significantly, necessitating innovative solutions to enhance agricultural productivity without compromising environmental health (Tilman et al. 2011). The significance of sustainable agriculture extends beyond mere food production; it encompasses the preservation of ecosystem services, biodiversity, and the planet's overall health. Sustainable practices can mitigate the adverse effects of conventional agriculture, such as soil erosion, water depletion, and biodiversity loss (Adedibu 2023). By integrating ecological principles into farming systems, sustainable agriculture aims to create resilient food systems that can withstand the pressures of climate change while meeting the nutritional needs of a growing population (Qin et al. 2022). Furthermore, sustainable agriculture promotes social equity by improving the livelihoods of farmers and rural communities, thereby contributing to broader economic stability (Rahman 2023).

Sustainable agriculture is characterized by its focus on long-term ecological balance and the efficient use of resources. This approach seeks to optimize agricultural productivity while minimizing negative environmental impacts. Fundamental principles of sustainable agriculture include crop diversification, soil health management, water conservation, and reducing chemical inputs (Qin et al. 2022). These practices enhance the resilience of agricultural systems and contribute to preserving natural resources, which is essential for future food security (Adedibu 2023). The importance of sustainable agriculture in global food systems cannot be overstated. It is pivotal in addressing food insecurity, affecting approximately 820 million people worldwide (Korneeva et al. 2023). Farmers can increase their productivity and profitability by adopting sustainable practices while reducing their reliance on external inputs, such as synthetic fertilizers and pesticides (Liu et al. 2020). This shift benefits individual farmers and contributes to the overall stability of food supply chains, thereby enhancing food security at the national and global levels (Kassie et al. 2015).

Moreover, sustainable agriculture is integral to combating climate change. Agricultural practices are a significant source of greenhouse gas emissions, accounting for approximately 10–12% of global emissions (Greenland et al. 2019). By implementing sustainable practices, such as agroforestry, conservation tillage, and organic farming, the agricultural sector can reduce its carbon footprint and contribute to climate change mitigation efforts (Xie et al. 2019). Additionally, sustainable agriculture promotes biodiversity, which is crucial for maintaining ecosystem resilience and ensuring the availability of diverse food sources (Jena et al. 2022). Technological innovations play a vital role in enhancing the sustainability of agricultural practices. Advances in digital technologies, such as precision agriculture, remote sensing, and data analytics, enable farmers to make informed decisions that optimize resource use and minimize environmental impacts (Singh et al. 2023). For instance, precision agriculture allows for the targeted application of inputs, such as water and fertilizers, based on real-time data, thereby reducing waste and improving crop yields (Scherer et al. 2018).

Furthermore, integrating information and communication technologies (ICT) in agriculture has facilitated knowledge sharing and capacity building among farmers. Mobile applications and web-based platforms provide farmers with up-to-date information on best practices, market trends, and weather forecasts, empowering them to make informed decisions. This increased access to information enhances productivity and promotes equitable access to resources, thereby contributing to sustainable agricultural development (Jayathilake et al. 2018).

Blockchain technology is another innovative solution that has the potential to enhance sustainability in agriculture. Blockchain can improve traceability and accountability in agricultural practices by providing a transparent and secure method for tracking the supply chain. This technology can help reduce food waste, ensure fair trade practices, and enhance consumer trust in food products, ultimately contributing to more sustainable food systems (Akella et al. 2023). Moreover, biotechnology offers promising solutions for sustainable agriculture by developing more resilient crops to environmental stresses, such as drought and pests. Genetic engineering and other biotechnological advancements can lead to the development of crop varieties that require fewer inputs while maintaining high yields, thereby reducing the environmental impact of agricultural production (Das et al. 2023). In addition to these technological advancements, policy frameworks, and collaborative approaches are essential for promoting sustainable practices. Governments and organizations must work together to create supportive policies that incentivize sustainable agriculture and facilitate access to innovative technologies. By fostering an environment conducive to innovation, stakeholders can drive the transition towards more sustainable agricultural systems that meet the needs of current and future generations.

The chapter focuses on critical technological innovations that drive sustainable agriculture, including precision agriculture, soil health technologies, and water-efficient systems, which optimize resource use and enhance productivity. It explores the integration of renewable energy in farming through solar and wind systems, along with advancements in bioenergy, contributing to energy efficiency. Additionally, the chapter discusses biotechnology's role in crop improvement, sustainable livestock and aquaculture practices, the challenges of technology adoption, and future directions, emphasizing the need for collaboration among stakeholders to foster innovation and address barriers to sustainability.

10.2 Key Technological Innovations in Sustainable Agriculture

Sustainable agriculture increasingly relies on critical technological innovations that enhance resource efficiency and environmental stewardship (Table 10.1). Among these innovations, precision agriculture is a transformative approach that utilizes drones, sensors, and artificial intelligence (AI) to optimize resource use. Precision agriculture allows for the collection and analysis of data at a granular level, enabling farmers to make informed decisions about planting, irrigation, and fertilization. For instance, integrating proximal soil mapping services has improved nutrient management planning and soil health monitoring, thereby enhancing overall agricultural productivity and sustainability (Rhymes et al. 2023). Furthermore, optical and electrochemical sensing technologies facilitate real-time tracking of soil nutrients, allowing for precise application of fertilizers and reducing waste (Camargo et al. 2024). This data-driven approach maximizes crop yields and minimizes the environmental impact of over-fertilization and water usage.

Soil health technologies represent another critical area of innovation in sustainable agriculture. The application of biotechnology, including genetically modified organisms (GMOs) and microbial inoculants, has been shown to enhance soil fertility and structure, thereby promoting healthier crop growth (Wang et al. 2015). Additionally, deploying soil sensors enables continuous monitoring of soil moisture, temperature, and nutrient levels, essential for effective nutrient management (Camargo et al. 2024). These technologies contribute to the maintenance of soil organic matter (SOM) and biodiversity, which are vital for sustaining agricultural productivity in the long term (Rhymes et al. 2023). By fostering a better understanding of soil processes and dynamics, these innovations empower farmers to adopt practices that enhance soil health and resilience against climate change.

Water-efficient technologies are paramount in addressing the pressing challenges of water scarcity and climate variability in agriculture. Advanced irrigation systems, such as drip and sprinkler irrigation, have been identified as practical means of conserving water while maintaining crop productivity (Pérez-Blanco et al. 2020). These systems allow for precise water application based on crop needs, significantly reducing water waste compared to traditional irrigation methods (Arouna et al. 2023). Moreover, integrating smart irrigation controllers, which utilize weather data and soil moisture levels to optimize irrigation schedules, has enhanced water use efficiency (Marazky 2015). Research indicates that such technologies conserve water and improve crop yields, making them essential for sustainable agricultural practices (Arouna et al. 2023). Recycling water for farming practices is gaining traction as a viable strategy for water conservation. Studies have demonstrated that using recycled water for irrigation can effectively support crop growth while mitigating the impacts of water scarcity (Nackley et al. 2015). However, farmers' acceptance of recycled water has historically been hindered by concerns over salinity and potential plant damage (DeVincentis et al. 2015). Overcoming these perceptions requires robust education and demonstration of the benefits of recycled water, coupled with

Table 10.1 Overview of the technology-supported innovations in sustainable agriculture

Technology	Focus	Applications	Benefits	Challenges	Refs.
Precision agriculture	Technology for managing field variability in crops	Drones, soil sensors, AI, and GPS mapping	Increases yield, reduces costs and lowers environmental impact	High initial costs, expertise needed, and data privacy	Sharma et al. (2023)
Soil health technologies	Innovations to monitor and improve soil fertility	Soil sensors, biotechnology, and nutrient software	Enhances soil quality, boosts biodiversity, and reduces erosion	Soil variability, complex biology, and limited farmer knowledge	Lehmann et al. (2020)
Water-efficient technologies	Systems to optimize agricultural water use	Drip irrigation, rainwater harvesting, and moisture sensors	Reduces water use, improves drought resilience, and minimizes runoff	High installation costs, maintenance needs, and climate variability	Ward and Pulido-Velazquez (2008)
Biotechnology	Genetic and microbial tech to improve crops	GMOs, biofertilizers, and biopesticides	Increases yields, reduce chemical inputs, and enhances pest resistance	Public acceptance, regulatory issues, biodiversity risks	Mishra et al. (2020)
Automated farming systems	Robotics and automation for farm tasks	Autonomous tractors, and robotic harvesters	Reduces labor costs, increases efficiency, and improves precision	High capital costs, technical training required	Benos et al. (2023)
Data analytics and AI	AI and big data for agricultural decision-making	Predictive analytics, AI and pest management	Optimizes resources, enhances risk management, and improves decisions	Data privacy, and technical infrastructure required	Ahmad et al. (2024)
Agroecology	Sustainable practices that align with ecosystems	Crop rotation, polyculture, and organic farming	Boosts biodiversity, reduces chemicals, and enhances ecosystem services	Requires knowledge, potentially lower short-term yields	Jeanneret et al. (2021), Zeng et al. (2023)

the implementation of efficient irrigation methods to minimize risks (Nackley et al. 2015). As farmers become more aware of the advantages of using recycled water, its adoption will likely increase, contributing to more sustainable water management practices. In addition to technological innovations, socio-economic factors play a significant role in adopting sustainable agricultural practices. Research indicates that farmers' age, education level, and access to information are critical determinants of their willingness to embrace advanced irrigation technologies (Khan et al. 2021). Policymakers must consider these socio-economic variables when designing interventions promoting sustainable practices. For instance, providing training and resources to farmers can enhance their understanding of new technologies and encourage their adoption (Khan et al. 2021). Furthermore, financial incentives and subsidies can alleviate the economic barriers often preventing farmers from investing in water-efficient technologies (Hrozencik et al. 2023). By addressing these socio-economic challenges, stakeholders can facilitate a more widespread transition to sustainable agricultural practices.

The role of policy in promoting sustainable agriculture cannot be overstated. Effective policies that encourage the adoption of water-efficient technologies and sustainable practices are essential for addressing the challenges of climate change and resource scarcity (Gautam et al. 2020). For example, implementing water pricing mechanisms can incentivize farmers to conserve water and adopt more efficient irrigation practices (Zhou et al. 2015). Policies supporting research and development in agricultural technologies can also lead to innovations that enhance productivity while minimizing environmental impacts (Pérez-Blanco et al. 2020). Governments can play a pivotal role in advancing sustainable agriculture by fostering a supportive policy environment.

Integrating conservation agriculture practices is also vital for enhancing sustainability in farming systems. Techniques such as minimum tillage, cover cropping, and crop rotation improve soil health and reduce erosion (Mwangi 2024). These practices enhance soil structure and fertility and promote biodiversity, which is essential for resilient agricultural ecosystems (Rhymes et al. 2023). Moreover, adopting integrated farming systems that combine diverse agricultural activities can lead to synergistic benefits, optimizing resource use and minimizing waste (Mwangi 2024). By promoting these conservation practices, farmers can enhance their resilience to climate variability and improve the sustainability of their operations. Technological advancements in remote sensing and data analytics are revolutionizing how farmers manage their resources. The use of remote sensing technologies allows for monitoring soil nutrient variability and crop health over large areas, enabling farmers to make data-driven decisions (Ji and Liu 2022). This approach not only enhances resource use efficiency but also reduces agricultural practices' environmental footprint (Zu et al. 2015). Furthermore, integrating data analytics into farming operations can lead to more precise management of inputs, ultimately resulting in higher yields and lower costs (Zu et al. 2015). As these technologies evolve, their potential to transform sustainable agriculture will only increase.

10.3 Integrating Renewable Energy in Farming

Integrating renewable energy sources into agriculture is becoming increasingly vital as the sector faces challenges related to energy consumption, environmental sustainability, and food security. Table 10.2 provides an overview of integrating renewable energy in farming. Solar and wind energy systems are at the forefront of this transition, providing farmers with alternative energy sources to reduce reliance on fossil fuels. Adopting solar-powered and wind-powered systems can significantly enhance energy efficiency in agricultural operations, lowering operational costs and reducing greenhouse gas emissions. Bathaei and Štreimikienė (2023) emphasized the importance of shifting towards clean renewable energy sources, noting that sustainable energy in agriculture is essential for meeting operational energy needs while minimizing environmental impacts. Furthermore, implementing solar technologies in milk production has improved economic viability and sustainability in dairy farming (Snegirjovs et al. 2018).

Table 10.2 Overview of integrating renewable energy in farming

Renewable energy source	Focus	Applications in agriculture	Innovations	Benefits	Challenges	Refs.
Solar energy	Harnesses sunlight through photovoltaic and thermal systems	Solar irrigation, solar greenhouses, solar pumps	Solar tracking, integrated panels in structures	Reduces fossil fuel use, lowers costs, improves water management	High installation costs, weather dependency	Ukoba et al. (2024)
Wind energy	Generates electricity with wind turbines	Powering equipment, wind irrigation, on-site energy	Vertical-axis turbines, wind-solar hybrids	Sustainable energy, cuts energy bills, combines with other sources	Site-specific, visual and noise concerns	Solarin and Bello (2022)
Bioenergy	Produces energy from organic materials and byproducts	Biogas from manure, biofuels, crop residue energy	Advanced digestion, biofuel crop innovations	Reduces waste, provides renewable fuel, improves soil health	Land competition, and emissions during production	Aidonojie et al. (2023), Xu et al. (2024)
Energy-efficient equipment	Minimizes energy use with optimized tools and machinery	Precision farming, electric tractors, efficient irrigation	Smart controllers, electric/hybrid machinery	Cuts operational costs, reduces carbon footprint, boosts productivity	High upfront costs, and need for training	Feng et al. (2024)

Innovations in bioenergy also play a crucial role in enhancing energy efficiency within agricultural systems. The utilization of biomass, derived from agricultural waste, can serve as a renewable energy source that not only provides energy but also contributes to waste management. This dual benefit is particularly relevant in mixed farming systems, where energy consumption can be optimized through by-products for energy generation. Uzal (2013) discussed how integrating green energy sources like biomass and wind can substantially improve energy use efficiency (EUE), thereby reducing greenhouse gas emissions associated with traditional farming practices. Additionally, developing energy-efficient agricultural systems is essential for achieving sustainability goals, as it allows for producing the same output with reduced energy input (Uzal 2013). The energy efficiency of farming operations can vary significantly based on several factors, including farm size, crop type, and management practices. Zea et al. (2020) found that smallholder commercial vegetable farms in Ecuador exhibited varying levels of energy efficiency influenced by local conditions and crop types. This finding underscores the importance of tailored energy management strategies that consider the unique characteristics of each farming operation.

Moreover, integrating renewable energy sources can help smaller farms overcome energy efficiency challenges by providing a stable and cost-effective energy supply. In addition to solar and wind energy, advancements in energy-efficient farming equipment are crucial for enhancing overall agricultural productivity. The adoption of precision agriculture technologies, which utilize data-driven approaches to optimize resource use, can significantly reduce energy consumption. For example, Ilyas et al. (2020) applied data envelopment analysis to evaluate energy efficiency in New Zealand dairy farming systems, demonstrating that precision farming techniques can enhance energy productivity. Similarly, automatic milking systems (AMS) have gained popularity in modern dairy farms, as these systems can optimize energy consumption while improving milk production efficiency (Höhendinger et al. 2021). The integration of such technologies supports energy efficiency and contributes to the economic sustainability of farming operations.

The circular economy concept is increasingly relevant in sustainable agriculture, as it promotes the efficient use of resources and minimizes waste. Zăpucioiu et al. (2023) emphasized that agriculture can play a pivotal role in implementing circular economy principles by adopting sustainable practices that reduce energy and water consumption while enhancing soil quality. This approach aligns with the broader goals of sustainable development, as it encourages the recycling of materials and using renewable energy sources. By integrating circular economy principles, farmers can create more resilient systems that are better equipped to adapt to the challenges posed by climate change and resource scarcity. Furthermore, the financial services sector is critical in facilitating the transition to sustainable agriculture. Financial institutions can help farmers invest in energy-efficient technologies and practices by providing funding and support for renewable energy projects. Micu et al. (2022) highlighted the importance of financial resources in enabling farmers to adopt sustainable practices that mitigate the impacts of climate change. This financial backing is essential for promoting innovations in renewable energy and energy-efficient farming equipment, ultimately leading to more sustainable agricultural systems. The impacts of climate change on agriculture necessitate a proactive approach to energy management. Farmers must adapt to extreme weather events while ensuring food security and sustainable resource use (Izah et al. 2024, 2022; Yarkwan et al. 2024; Ogwu et al. 2024; Odubo et al. 2024). Pelletier et al. (2011) emphasized that the finite nature of energy resources and the environmental implications of energy use in agriculture warrant careful consideration. As the agricultural sector seeks to meet the demands of a growing population, integrating renewable energy sources and energy-efficient practices will be crucial in achieving long-term sustainability.

10.4 Role of Biotechnology in Crop Improvement

Biotechnology has emerged as a transformative force in agriculture, particularly crop improvement (Ogidi and Izah 2024). The global demand for food is projected to double by 2050, necessitating significant advancements in agricultural productivity to meet this challenge (Tilman et al. 2011). Table 10.3 provides an overview of the role of biotechnology in crop improvement. While effective to some extent, traditional crop production methods are insufficient to achieve the required yield increases, estimated at approximately 2.4% per year for staple crops such as maize, rice, wheat, and soybean (Ray et al. 2013). In this context, biotechnology offers innovative solutions through genetically modified organisms (GMOs) and genome editing techniques, which can enhance crop yields and resilience against biotic and abiotic stresses.

Genetically modified organisms (GMOs) represent a significant advancement in agricultural biotechnology, allowing for the precise alteration of plant genomes to enhance desirable traits. Techniques such as CRISPR genome editing enable scientists to modify specific genes, improving yield, disease resistance, and stress tolerance (Marone et al. 2023). This precision in genetic engineering accelerates the breeding process and minimizes the unintended consequences often associated with traditional breeding methods. Integrating these advanced biotechnological tools can revolutionize crop improvement, making it possible to develop varieties better suited to the changing climate and increasing pest pressures (Varshney et al. 2011). Enhancing pest and disease resistance through biotechnological advancements is critical for sustainable agriculture. Insect pests and diseases account for significant losses in crop yields, with estimates suggesting that 20–40% of global crop production is lost due to these factors (Douglas 2018; Abdelkhalek et al. 2023). Biotechnological approaches, such as RNA interference (RNAi) and the development of transgenic crops, have shown promise in providing effective pest control solutions (Mamta and Rajam 2017). By silencing specific genes in pests or pathogens, these technologies can reduce their ability to damage crops, thereby improving overall yield and reducing reliance on chemical pesticides (Marques et al. 2021). Also, developing crops with inherent disease resistance can lead to lower production costs and reduced environmental impact (Das et al. 2023).

Table 10.3 Overview of the role of biotechnology in crop improvement

Biotechnological approach	Description	Applications	Benefits	Challenges	Examples	Refs.
Genetically modified organisms (GMOs)	Altering genetic material to express desired traits	Herbicide-resistant, insect-resistant crops	Higher yields, reduced pesticide use, improved nutrition	Public acceptance, environmental concerns	Bt cotton, Roundup Ready soybeans	Raman (2017)
Genome editing (CRISPR/Cas9)	Precise DNA modification for trait enhancement	Drought tolerance, improved nutrition	Fast, precise, allows multiple trait edits	Ethical concerns, unintended genome effects	CRISPR-edited drought-resistant wheat	Rai et al. (2023)
Biotechnology for pest resistance	Developing crops that naturally resist pests	Bt crops, disease-resistant varieties	Reduces chemical use, lowers costs	Pest resistance development, regulatory hurdles	Bt corn, virus-resistant papaya	Kumari et al. (2022)
Marker-assisted selection (MAS)	Molecular markers used in selective breeding	Disease resistance, quality traits	Speeds breeding, enhances precision	High costs, requires training and infrastructure	MAS for rice blast resistance	Collard and Mackill (2008), Hasan et al. (2021)
Transgenic approaches	Introducing genes from different species	Nutrient-rich crops, stress resistance	Addresses deficiencies, improves resilience	Gene flow risks, regulatory challenges	Golden rice, virus-resistant sweet potato	Abdul Aziz et al. (2022)
Synthetic biology	Engineering organisms to create new traits	Enhanced crop traits, biofuel development	Customized solutions, new crop potential	Technical complexity, ethical issues	Engineered yeast for biofuels	Anderson et al. (2012)
Microbial biotechnology	Using beneficial microbes for crop health	Biofertilizers, biopesticides	Boosts soil fertility, reduces chemical use	Microbial variability, need for field testing	Rhizobium for legumes, Trichoderma for disease control	Kumar et al. (2022)

Integrated pest management (IPM) strategies, which combine biotechnological innovations with traditional agricultural practices, are essential for sustainable crop production. These strategies emphasize preventive measures and ecological solutions, such as crop rotation and resistant varieties, to manage pest populations effectively (Deguine et al. 2021). The integration of precision agriculture and geospatial techniques further enhances IPM by providing data-driven insights into pest dynamics and crop health, enabling farmers to make informed decisions about pest control interventions (Roberts et al. 2021). This holistic approach improves crop resilience and promotes environmental sustainability by reducing the need for chemical inputs. The role of biotechnology in enhancing crop resilience extends beyond pest and disease resistance to include improvements in abiotic stress tolerance. Climate change poses significant challenges to agriculture, with increased temperatures, altered precipitation patterns, and extreme weather events threatening crop production (Mittler and Blumwald 2010). Biotechnological advancements, such as developing drought-resistant and heat-tolerant crop varieties, are crucial for ensuring food security in these challenges (Varshney et al. 2011). By leveraging molecular breeding techniques and genetic engineering, researchers are creating crops that can thrive under adverse conditions, safeguarding yields and supporting farmers' livelihoods (Mamata and Lamichhane 2021).

Moreover, biotechnology plays a pivotal role in improving the nutritional quality of crops and addressing global malnutrition issues. Biofortification, the process of enhancing the nutritional content of crops through genetic modification, has led to the development of varieties such as Golden Rice, fortified with beta-carotene to combat vitamin A deficiency (Das et al. 2023). This approach can help improve food security. The potential for biotechnology to address both agricultural productivity and nutritional quality underscores its importance in contemporary food systems. As biotechnology continues to evolve, it faces challenges related to public perception, regulatory frameworks, and market acceptance.

The commercialization of GM crops has been met with resistance in some regions, driven by concerns over safety, environmental impact, and ethical considerations (Huesing et al. 2016). However, as scientific understanding of biotechnology improves and regulatory frameworks adapt to accommodate new technologies, there is potential for broader acceptance of GMOs and genome-edited crops. Public engagement and education will be critical in addressing misconceptions and fostering a more informed dialogue about the benefits and risks of biotechnological agriculture advancements (Khan et al. 2017).

10.5 Sustainable Livestock and Aquaculture Practices

Sustainable livestock and aquaculture practices are increasingly critical in addressing global food security and environmental challenges (Table 10.4). Integrating technological innovations in feed efficiency and waste management is essential for enhancing the sustainability of these sectors. The demand for animal protein is projected to rise significantly, necessitating improvements in production efficiency while minimizing environmental impacts. Sustainable practices in livestock and aquaculture involve optimizing feed utilization and managing waste effectively to reduce pollution and resource depletion (Alkemade et al. 2013; Pelletier and Tyedmers 2010).

Technological advancements have led to significant improvements in feed efficiency, which is a critical factor in sustainable livestock and aquaculture production. For instance, the development of precision feeding techniques, which utilize real-time data to adjust feed formulations based on the specific needs of animals, has shown promise in enhancing feed conversion ratios (Chahid et al. 2021; Glencross et al. 2023). Additionally, the use of alternative protein sources, such as fermented plant proteins, has gained traction as a sustainable substitute for fishmeal in aquaculture diets, addressing both economic and ecological concerns associated with traditional feed ingredients (Mugwanya et al. 2023; Shine et al. 2023). These innovations improve growth rates and reduce the overall environmental footprint of animal production systems. Waste management is another critical aspect of sustainable livestock and aquaculture practices. Discharging nutrients from aquaculture operations can lead to eutrophication and degradation of aquatic ecosystems (Amirkolaie 2011; Zhang et al. 2015). Implementing integrated multi-trophic aquaculture (IMTA) strategies can enhance nutrient recycling and minimize waste discharge. IMTA systems utilize the natural interactions between species to absorb excess nutrients, improving overall system efficiency and sustainability (Zhang et al. 2015). Furthermore, applying biodegradable polymers in aquaculture systems has enhanced water quality and shifted microbial communities, contributing to better waste management practices (Kokkuar et al. 2021).

Monitoring systems are vital in improving animal health and reducing environmental impacts on livestock and aquaculture. Implementing digital technologies, such as machine vision systems and sensor networks, allows for real-time monitoring of animal health, growth trajectories, and environmental conditions (Saberioon et al. 2017). These systems provide valuable insights into feeding behavior, water quality, and overall system performance, enabling producers to make informed decisions that enhance sustainability. For example, optical sensors can help detect changes in water quality parameters, allowing for timely interventions that prevent detrimental effects on fish health and growth (Saberioon et al. 2017). In addition to technological innovations, genetic advancements have been pivotal in improving feed efficiency and overall production in livestock and aquaculture. Selective breeding programs focused on traits such as growth rate and feed conversion efficiency have demonstrated significant potential to enhance productivity (Dvergedal et al. 2019; Osmond and Colombo 2019). The integration of genetic engineering techniques may further optimize these traits, allowing for the development of species that require less feed and produce fewer waste products, thereby reducing the environmental impact of animal production (Osmond and Colombo 2019). This genetic approach and improved feeding strategies can lead to more sustainable livestock and aquaculture systems.

The importance of animal welfare in sustainable practices cannot be overstated. Research indicates improved animal welfare is linked to better health outcomes

Table 10.4 Sustainable livestock and aquaculture practices

Practice	Focus	Technological innovations	Monitoring systems	Benefits	Challenges	Examples	Refs.
Feed efficiency improvements	Enhancing feed formulation and utilization for growth	Precision feeding, feed additives, genetic selection	Smart feeders, nutritional monitoring	Cuts feed costs, lowers emissions, boosts growth	High initial costs, training needed	Smart feed systems, dairy supplements	Kause et al. (2022)
Waste management innovations	Efficient livestock waste management solutions	Anaerobic digestion, recycling, composting, biofilters	Waste tracking, real-time monitoring	Produces biogas, improves soil health, reduces odors	High setup costs, technical maintenance	Farm digesters, and composting setups	Almansa et al. (2023)
Animal health monitoring	Systems to track livestock and aquaculture health	Wearable devices, automated diagnostics, EID	IoT devices, health monitoring apps	Early disease detection, better welfare, lower vet costs	Data privacy, tech dependence	RFID tags, and health tracking apps	Herlin et al. (2021)
Aquaculture technologies	Innovations for sustainable fish farming	RAS, biofiltration, automated feeding	Water quality and environmental monitoring	Efficient water use, waste reduction, improved fish health	High operational costs, technical expertise needed	RAS in tilapia, multi-trophic systems	Clough et al. (2020)

(continued)

Table 10.4 (continued)

Practice	Focus	Technological innovations	Monitoring systems	Benefits	Challenges	Examples	Refs.
Integrated systems	Combining crop and livestock production	Agroecology, silvopasture, intercropping	Farm management, yield monitoring	Maximizes resources, boosts biodiversity, reduces pests	Management complexity, knowledge integration	Agroforestry, crop-livestock systems	Kremen (2020)
Water-efficient technologies	Optimizing water use in livestock and aquaculture	Drip irrigation, water-efficient feeding, rain harvesting	Soil moisture, water quality sensors	Cuts water use, improves welfare, increases crop yield	High installation, maintenance needs	Rainwater systems, smart pasture irrigation	Ward and Pulido-Velazquez (2008), Marimuthu et al. (2024)

and increased productivity (Oliva-Teles 2012; Hixson 2014). Producers can enhance growth performance and reduce stress-related health issues by focusing on the behavioral needs of livestock and aquaculture species. For instance, understanding fish's social dynamics and environmental preferences can lead to better management practices that optimize feeding and reduce competition for resources (Bardera et al. 2019; Hancz 2020). This holistic approach to animal welfare contributes to the sustainability of production systems by ensuring that animals are healthy and productive.

Furthermore, the role of policy and regulatory frameworks in promoting sustainable livestock and aquaculture practices is crucial. Governments and organizations must establish guidelines encouraging innovative technologies and sustainable practices while incentivizing producers to reduce their environmental impact (Pelletier and Tyedmers 2010; Herrero et al. 2013). Collaborative efforts among stakeholders, including researchers, producers, and policymakers, can facilitate the dissemination of best practices and foster a culture of sustainability within the agricultural sector. Consumer awareness and demand for sustainably produced animal products also drive changes in livestock and aquaculture practices. As consumers become more informed about the environmental and ethical implications of their food choices, there is a growing market for certified products certified as sustainable (Hixson 2014). This shift in consumer behavior encourages producers to adopt more sustainable practices as they seek to meet the expectations of environmentally conscious consumers. By aligning production methods with consumer preferences, the livestock and aquaculture industries can enhance their sustainability while maintaining profitability.

10.6 Challenges and Barriers to Adoption of Technology

Technology adoption across various sectors is often impeded by challenges and barriers that can be categorized into economic, social, and infrastructural dimensions (Fig. 10.1). Economic barriers, such as high initial costs and ongoing maintenance expenses, significantly hinder the adoption of new technologies, particularly in developing countries with limited financial resources. For instance, in the agricultural sector of Bangladesh, technological barriers, infrastructural constraints, and a lack of awareness have been recurrent impediments to adopting modern farming technologies, which resonate with challenges faced in similar contexts globally (Mia and Monir 2023). The high cost of technology, coupled with the fear of unemployment due to automation, further exacerbates resistance to adopting modern technologies, particularly in sectors like manufacturing and agriculture (Javaid et al. 2022).

Social barriers also play a critical role in technology adoption, encompassing individuals' and organizations' attitudes, beliefs, and behaviors toward new technologies. Research indicates that a lack of digital literacy and awareness among potential users can significantly impede the integration of Information and Communication Technology (ICT) in various sectors, including education and healthcare (Vrasidas

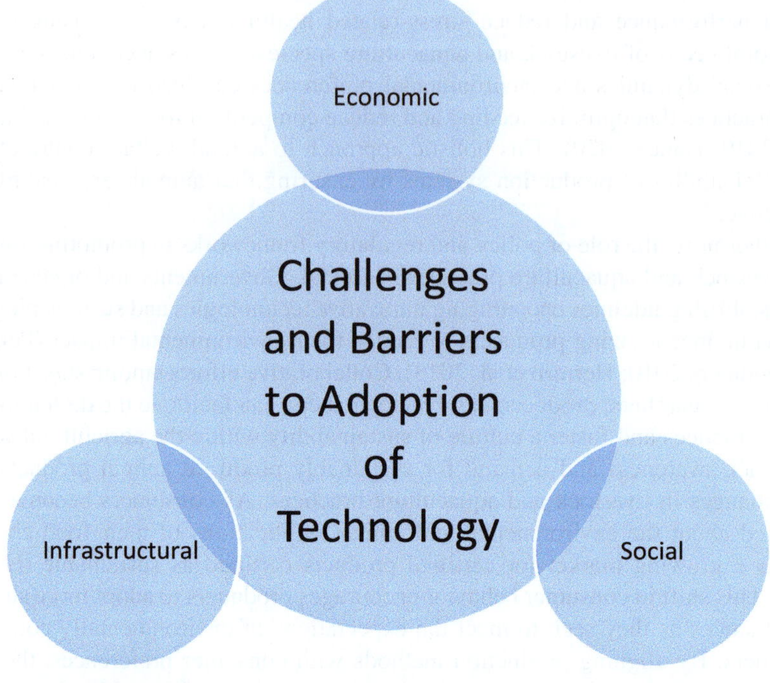

Fig. 10.1 Dimensions of the challenges and barriers to adoption of technology

2014; Tegegne and Wubante 2022). For instance, in the healthcare sector of Ethiopia, barriers such as inadequate training opportunities and poor computer literacy were identified as significant obstacles to ICT adoption (Tegegne and Wubante 2022). Furthermore, the cultural context and social norms surrounding technology use can influence the willingness of individuals and organizations to embrace new technologies, as seen in the reluctance of some farmers to adopt modern agricultural practices due to traditional beliefs (Secinaro et al. 2022). Infrastructural barriers are another significant challenge to technology adoption, particularly in developing regions lacking the necessary physical and digital infrastructure. The absence of reliable internet connectivity, inadequate transportation networks, and insufficient energy supply can severely limit the ability of organizations to implement and utilize new technologies effectively (Megavitry and Harahap 2023; Bae et al. 2024). For example, the electrification of public transport systems, such as buses, faces significant hurdles due to the lack of charging infrastructure, which is essential for successfully adopting electric vehicles (Blynn and Attanucci 2019). Similarly, in the context of renewable energy technologies, infrastructural challenges, including the availability of suitable sites for installation and maintenance, have been identified as critical barriers (Karakaya and Sriwannawit 2015).

Ethical considerations also emerge as crucial in adopting new technologies, particularly concerning data privacy and security. As organizations increasingly rely on digital technologies, concerns about the moral implications of data collection and usage have come to the forefront. For instance, adopting technologies such as blockchain in agriculture raises questions about data ownership and the potential for misuse of sensitive information (Kamilaris et al. 2019). Moreover, the ethical implications of automation and artificial intelligence, particularly regarding job displacement and the potential for exacerbating social inequalities, must be carefully considered when implementing new technologies (Raj et al. 2020). Environmental considerations are equally important in the discourse surrounding technology adoption. Implementing new technologies often involves environmental costs, such as increased energy consumption and electronic waste generation. For example, adopting photovoltaic systems for renewable energy generation faces challenges related to the environmental impact of manufacturing solar panels and the disposal of outdated technologies (Karakaya and Sriwannawit 2015). Additionally, the push for sustainable practices in sectors such as agriculture and construction necessitates carefully evaluating the environmental implications of adopting new technologies, ensuring that they contribute positively to sustainability goals rather than detracting from them (Bevan and Nolan 2019).

The interplay between these barriers highlights the complexity of technology adoption and the need for a multifaceted approach to address the challenges faced by various sectors. Policymakers and stakeholders must adopt a holistic strategy encompassing economic incentives, social awareness, and infrastructural development to facilitate the widespread adoption of new technologies (Benz et al. 2022). For instance, initiatives to improve digital literacy and provide financial support for technology adoption can help bridge the digital divide and promote equitable access to technological advancements (Megavitry and Harahap 2023). Furthermore, fostering collaboration between the public and private sectors can enhance the development of the necessary infrastructure to support technology adoption (Cassou and Oliveira 2011).

10.7 Future Directions and Innovations

The future of food systems sustainability is increasingly intertwined with emerging technologies such as vertical farming, artificial intelligence (AI), and robotics. These innovations are reshaping agricultural practices and addressing the challenges of urbanization, climate change, and food security. Vertical farming, characterized by the cultivation of crops in stacked layers, is gaining traction as a viable solution to the limitations of traditional agriculture. It allows food production in urban settings, reducing food miles and enhancing access to fresh produce (Jürkenbeck et al. 2019; Al-Kodmany 2018). As urban populations grow, the demand for locally sourced food is becoming more pronounced, making vertical farming an attractive option for city dwellers (Jürkenbeck et al. 2019; Kabir et al. 2023). Furthermore, advancements in controlled environment agriculture (CEA) have positioned vertical farming as a sustainable alternative that minimizes the environmental footprint associated with conventional farming methods (Gan et al. 2022; Vatistas et al. 2022).

Integrating AI and robotics into agricultural practices is revolutionizing food production systems. AI technologies facilitate precision agriculture, enabling farmers to optimize resource use and improve crop yields (Giri et al. 2020; Smith 2018). Robotics enhances operational efficiency by automating labor-intensive tasks such as planting, harvesting, and monitoring crop health (Duong et al. 2020). Combining these technologies increases productivity and addresses labor shortages in the agricultural sector, which is becoming increasingly critical as the global population continues to rise (Saad et al. 2021). Moreover, AI-driven analytics can provide valuable insights into market trends and consumer preferences, allowing producers to adapt their strategies accordingly (Kumar et al. 2021). This technological synergy is essential for creating resilient food systems that can withstand the pressures of climate change and fluctuating market conditions.

Collaboration among governments, industry stakeholders, and research institutions is vital for scaling these innovative solutions. Public–private partnerships can facilitate the development and implementation of policies that support the adoption of vertical farming and AI technologies in agriculture (Bhaskar et al. 2020). Governments can play a crucial role by providing funding, regulatory support, and incentives for research and development in sustainable agricultural practices (Saad et al. 2021). Industry stakeholders, including agribusinesses and technology firms, can contribute their expertise and resources to drive innovation and improve the efficiency of food production systems (Smith 2018; Duong et al. 2020). Research institutions are essential for advancing knowledge and developing new technologies that can be integrated into existing agricultural practices (Bhaskar et al. 2020). By fostering collaboration across these sectors, stakeholders can create a more robust ecosystem that promotes sustainable food systems.

Consumer acceptance of these emerging technologies is also critical to their successful implementation. Research indicates that consumers are increasingly aware of the benefits of vertical farming, particularly in terms of sustainability and food quality (Jürkenbeck et al. 2019; Specht et al. 2019). However, concerns regarding the perceived unnaturalness of some farming technologies may hinder widespread adoption (Son and Hwang 2023). Addressing these concerns through education and transparent communication about the benefits and safety of vertical farming and AI-driven agriculture is essential for fostering consumer trust (Son and Hwang 2023; Specht et al. 2019). Additionally, engaging consumers in the development process can help ensure that their preferences and values are considered, ultimately leading to greater acceptance and support for these innovations (Jürkenbeck et al. 2019; Specht et al. 2019). The ethical implications of deploying AI and robotics in agriculture must also be carefully considered. As these technologies become more integrated into food production systems, questions arise regarding their impact on labor, environmental sustainability, and animal welfare (van der Burg et al. 2024; Ryan et al. 2022). Establishing ethical guidelines that govern AI and robotics in agriculture is essential to ensure that these technologies contribute positively to society and the environment (van der Burg et al. 2024).

Furthermore, ongoing research into the ethical dimensions of agricultural technology can help identify potential risks and develop strategies to mitigate them (van der Burg et al. 2024; Ryan et al. 2022). By prioritizing ethical considerations, stakeholders can build a more sustainable and equitable food system that benefits all members of society. In addition to addressing ethical concerns, developing innovative farming systems that leverage AI and robotics can enhance the resilience of food production systems. These systems can adapt to changing environmental conditions and optimize resource use, reducing waste and improving sustainability (Chen and Hengjinda 2019; Saad et al. 2021). For instance, AI algorithms can analyze data from sensors deployed in vertical farms to optimize irrigation and nutrient delivery, ensuring that crops receive the resources needed for optimal growth (Chen and Hengjinda 2019; Smith 2018). This level of precision improves crop yields and minimizes the environmental impact of farming practices (Vatistas et al. 2022; Saad et al. 2021). Moreover, integrating IoT technologies can facilitate real-time monitoring and management of agricultural operations, further enhancing efficiency and sustainability (Singh 2023; Duong et al. 2020).

The role of education and training in promoting the adoption of these technologies cannot be overstated. As the agricultural landscape evolves, it is crucial to equip farmers and agricultural workers with the skills and knowledge necessary to effectively utilize AI and robotics (Shyr et al. 2019; Giri et al. 2020). Educational programs focusing on integrating technology in agriculture can help bridge the gap between traditional farming practices and modern innovations (Shyr et al. 2019; Giri et al. 2020). Additionally, fostering a culture of innovation within the agricultural sector can encourage the exploration of new technologies and practices that enhance sustainability (Smith 2018; Saad et al. 2021). By investing in education and training, stakeholders can ensure that the workforce is prepared to embrace the future of food production. The potential for vertical farming, AI, and robotics to contribute to global food security is significant. As the world faces the dual challenges of a growing population and climate change, innovative agricultural practices will be essential for meeting food demand sustainably (Saad et al. 2021; Giri et al. 2020). Vertical farming can provide a reliable source of fresh produce in urban areas, while AI and robotics can enhance productivity and efficiency across the agricultural supply chain (Smith 2018; Duong et al. 2020). By prioritizing research, collaboration, and ethical considerations, stakeholders can harness the power of these technologies to create a more sustainable and resilient food system that meets the needs of future generations.

10.8 Conclusion

Sustainable agricultural practices are increasingly recognized as vital for developing healthier food systems that address pressing global challenges such as food security, climate change, and environmental degradation. Technological innovations are at the forefront of this transformation, offering diverse tools and methodologies that enhance resource efficiency and environmental stewardship. Critical advancements in precision agriculture, soil health technologies, and water-efficient systems empower farmers to optimize operations while minimizing their ecological footprints. Moreover, integrating renewable energy sources, such as solar and wind power, alongside innovations in biotechnology and energy-efficient farming equipment supports sustainable agricultural practices by reducing greenhouse gas emissions and improving economic viability. These innovations collectively foster resilience in farming systems and enhance food security by ensuring that agricultural practices are productive and sustainable.

Looking ahead, the future of sustainable agriculture will hinge on the successful adoption and integration of these technological innovations, driven by collaboration among farmers, policymakers, and research institutions. Addressing barriers to technology adoption, including socio-economic factors and ethical considerations, will be essential in creating an enabling environment for change. As the agricultural sector evolves, emerging technologies like vertical farming, artificial intelligence, and robotics hold significant potential to revolutionize food production. By emphasizing education, consumer awareness, and support for sustainable practices, stakeholders can foster a robust ecosystem that promotes healthier food systems. Ultimately, the synergistic combination of innovative technologies and sustainable practices offers a pathway toward achieving global food security while protecting vital natural resources for future generations.

References

Abdelkhalek ST, Moussa MA, Gomaa SI, Qiu CL, Wang MQ (2023) Agrochemicals: safety evaluation and characterization for humans and biodiversity. In: One Health implications of agrochemicals and their sustainable alternatives. Springer Nature Singapore, Singapore, pp 3–51. https://doi.org/10.1007/978-981-99-3439-3_1

Abdul Aziz M, Brini F, Rouached H, Masmoudi K (2022) Genetically engineered crops for sustainably enhanced food production systems. Front Plant Sci 13:1027828. https://doi.org/10.3389/fpls.2022.1027828

Adedibu P (2023) Ecological problems of agriculture: impacts and sustainable solutions. ScienceOpen Preprints. https://doi.org/10.14293/pr2199.000145.v1

Ahmad A, Liew AXW, Venturini F, Kalogeras A, Candiani A, Di Benedetto G, Ajibola S, Cartujo P, Romero P, Lykoudi A et al (2024) AI can empower agriculture for global food security: challenges and prospects in developing nations. Front Artif Intell 7:1328530. https://doi.org/10.3389/frai.2024.1328530

Aidonojie PA, Ukhurebor KE, Oaihimire IE, Ngonso BF, Egielewa PE, Akinsehinde BO, Kusuma HS, Darmokoesoemo H (2023) Bioenergy revamping and complimenting the global environmental legal framework on the reduction of waste materials: a facile review. Heliyon 9(1):e12860. https://doi.org/10.1016/j.heliyon.2023.e12860

Akella G, Wibowo S, Grandhi S, Mubarak S (2023) A systematic review of blockchain technology adoption barriers and enablers for smart and sustainable agriculture. Big Data Cogn Comput 7(2):86. https://doi.org/10.3390/bdcc7020086

Alkemade R, Reid RS, van den Berg M, de Leeuw J, Jeuken M (2013) Assessing the impacts of livestock production on biodiversity in rangeland ecosystems. Proc Natl Acad Sci 110(52):20900–20905. https://doi.org/10.1073/pnas.1011013108

Al-Kodmany K (2018) The vertical farm: a review of developments and implications for the vertical city. Buildings 8(2):24. https://doi.org/10.3390/buildings8020024

Almansa XF, Starostka R, Raskin L, Zeeman G, De Los Reyes III F, Waechter J, Yeh D, Radu T (2023) Anaerobic digestion as a core technology in addressing the global sanitation crisis: challenges and opportunities. Environ Sci Technol 57(48):19078–19087. https://doi.org/10.1021/acs.est.3c05291

Amirkolaie AK (2011) Reduction in the environmental impact of waste discharged by fish farms through feed and feeding. Rev Aquac 3(1):19–26. https://doi.org/10.1111/j.1753-5131.2010.01040.x

Anderson J, Strelkowa N, Stan GB, Douglas T, Savulescu J, Barahona M, Papachristodoulou A (2012) Engineering and ethical perspectives in synthetic biology. Rigorous, robust and predictable designs, public engagement and a modern ethical framework are vital to the continued success of synthetic biology. EMBO Rep 13(7):584–590. https://doi.org/10.1038/embor.2012.81

Arouna A, Dzomeku IK, Shaibu AG, Nurudeen AR (2023) Water management for sustainable irrigation in rice (Oryza sativa L.) production: a review. Agronomy 13(6):1522. https://doi.org/10.3390/agronomy13061522

Bae Y, Rindt CR, Mitra SK, Ritchie SG (2024) Fleet operator perspectives on heavy-duty vehicle alternative fueling infrastructure. Transp Res Rec 2678(1):490–506. https://doi.org/10.1177/03611981231171150

Bardera G, Usman N, Owen M, Pountney D, Sloman KA, Alexander ME (2019) The importance of behaviour in improving the production of shrimp in aquaculture. Rev Aquac 11(4):1104–1132. https://doi.org/10.1111/raq.12282

Bathaei A, Štreimikienė D (2023) Renewable energy and sustainable agriculture: review of indicators. Sustainability 15(19):14307. https://doi.org/10.3390/su151914307

Benos L, Moysiadis V, Kateris D, Tagarakis AC, Busato P, Pearson S, Bochtis D (2023) Human-robot interaction in agriculture: a systematic review. Sensors 23(15):6776. https://doi.org/10.3390/s23156776

Benz S, Jaax A, Yotov YV (2022) Shedding light on the drivers of services tradability over two decades. In: OECD trade policy papers (No. 264). OECD Publishing, Paris. https://doi.org/10.1787/d5f3c149-en

Bevan WJ, Nolan L (2019). Barriers for renewable technology in commercial buildings. In: Proceedings of the world conference on sustainability, pp 116–124. https://doi.org/10.31705/wcs.2019.12

Bhaskar S, Bradley S, Sakhamuri S, Moguilner S, Chattu VK, Pandya S, Schroeder S, Ray D, Banach M (2020) Designing futuristic telemedicine using artificial intelligence and robotics in the COVID-19 era. Front Public Health 8:556789. https://doi.org/10.3389/fpubh.2020.556789

Blynn K, Attanucci J (2019) Accelerating bus electrification: a mixed methods analysis of barriers and drivers to scaling transit fleet electrification. Transp Res Rec 2673(8):577–587. https://doi.org/10.1177/0361198119842117

Camargo JR, Orzari LO, de Souza Rodrigues J, de Lima LF, Paixão TR, Fraceto LF, Janegitz BC (2024) Advancements in disposable electrochemical systems for sustainable agriculture monitoring: trends, gaps, and applied examples. TrAC-Trends Anal Chem 180:117968. https://doi.org/10.1016/j.trac.2024.117968

Cassou SP, Xavier de Oliveira E (2011) Barriers to technological adoption in Spain and Portugal. Port Econ J 10:189–209. https://doi.org/10.1007/s10258-010-0069-1

Chahid A, N'Doye I, Majoris JE, Berumen ML, Laleg-Kirati TM (2021) Fish growth trajectory tracking via reinforcement learning in precision aquaculture. arXiv preprint arXiv:2103.07251. https://doi.org/10.48550/arxiv.2103.07251

Chen JIZ, Hengjinda P (2019) Applying AI technology to the operation of smart farm robots. Sens Mater 31(5):1777–1788. https://doi.org/10.18494/sam.2019.2389

Clough S, Mamo J, Hoevenaars K, Bardocz T, Petersen P, Rosendorf P, Atiye T, Gukelberger E, Guya E, Hoinkis J (2020) Innovative technologies to promote sustainable recirculating aquaculture in Eastern Africa—a case study of a Nile Tilapia (*Oreochromis niloticus*) hatchery in Kisumu, Kenya. Integr Environ Assess Manag 16(6):934–941. https://doi.org/10.1002/ieam.4295

Collard BC, Mackill DJ (2008) Marker-assisted selection: an approach for precision plant breeding in the twenty-first century. Philos Trans R Soc London Ser B, Biol Sci 363(1491):557–572. https://doi.org/10.1098/rstb.2007.2170

Das S, Ray M, Panday D, Mishra P (2023) Role of biotechnology in creating sustainable agriculture. PLOS Sustain Transforma 2(7):e0000069. https://doi.org/10.1371/journal.pstr.0000069

Deguine J-P, Aubertot J-N, Flor R, Lescourret F, Wyckhuys KA, Ratnadass A (2021) Integrated pest management: good intentions, hard realities. A review. Agron Sustain Dev 41(3). https://doi.org/10.1007/s13593-021-00689-w

DeVincentis A, Brumfield R, Gottlieb P, Johnson J (2015) Cost analysis of using recycled water in container production: a case study of southern New Jersey. HortScience 50(8):1196–1201. https://doi.org/10.21273/hortsci.50.8.1196

Douglas A (2018) Strategies for enhanced crop resistance to insect pests. Annu Rev Plant Biol 69(1):637–660. https://doi.org/10.1146/annurev-arplant-042817-040248

Duong LT, Al-Fadhli M, Jagtap S, Bader F, Martindale W, Swainson M, Paoli A (2020) A review of robotics and autonomous systems in the food industry: from the supply chains perspective. Trends Food Sci Technol 106:355–364. https://doi.org/10.1016/j.tifs.2020.10.028

Dvergedal H, Ødegård J, Øverland M, Mydland L, Klemetsdal G (2019) Selection for feed efficiency in Atlantic salmon using individual indicator traits based on stable isotope profiling. Genet Sel Evol 51(1). https://doi.org/10.1186/s12711-019-0455-9

Feng G, Zhang J, Yan X, Dong C, Liu M, Xu L (2024) Research on energy-saving control of agricultural hybrid tractors integrating working condition prediction. PLoS ONE 19(3):e0299658. https://doi.org/10.1371/journal.pone.0299658

Gan C, Soukoutou R, Conroy D (2022) Sustainability framing of controlled environment agriculture and consumer perceptions: a review. Sustainability 15(1):304. https://doi.org/10.3390/su15010304

Gautam T, Paudel K, Guidry K (2020) An evaluation of irrigation water use efficiency in crop production using a data envelopment analysis approach: a case of Louisiana, USA. Water 12(11):3193. https://doi.org/10.3390/w12113193

Giri A, Saxena R, Saini P, Rawte S (2020) Role of artificial intelligence in advancement of agriculture. Int J Chem Stud 8(2):375–380. https://doi.org/10.22271/chemi.2020.v8.i2f.8796

Glencross BD, Fracalossi DM, Hua K, Izquierdo MS, Ma K, Øverland M, Yakupityage A (2023) Harvesting the benefits of nutritional research to address global challenges in the 21st century. J World Aquaculture Soc 54(2):343–363. https://doi.org/10.1111/jwas.12948

Greenland S, Levin E, Dalrymple J, O'Mahony B (2019) Sustainable innovation adoption barriers: water sustainability, food production and drip irrigation in Australia. Soc Responsib J 15(6):727–741. https://doi.org/10.1108/srj-07-2018-0181

Hancz C (2020) Feed efficiency, nutrient sensing and feeding stimulation in aquaculture: a review. Acta Agraria Kaposváriensis 24(1):35–54. https://doi.org/10.31914/aak.2375

Hasan N, Choudhary S, Naaz N, Sharma N, Laskar RA (2021) Recent advancements in molecular marker-assisted selection and applications in plant breeding programmes. J Genet Eng Biotechnol 19(1):128. https://doi.org/10.1186/s43141-021-00231-1

Herlin A, Brunberg E, Hultgren J, Högberg N, Rydberg A, Skarin A (2021) Animal welfare implications of digital tools for monitoring and management of cattle and sheep on pasture. Animals 11(3):829. https://doi.org/10.3390/ani11030829

Herrero M, Grace D, Njuki J, Johnson N, Enahoro D, Silvestri S, Rufino MC (2013) The roles of livestock in developing countries. Animal 7:3–18. https://doi.org/10.1017/s1751731112001954

Hixson SM (2014) Fish nutrition and current issues in aquaculture: the balance in providing safe and nutritious seafood, in an environmentally sustainable manner. J Aquac Res Dev 3(3). https://doi.org/10.4172/2155-9546.1000234
Höhendinger M, Krieg HJ, Dietrich R, Rauscher S, Stumpenhausen J, Bernhardt H (2021) Impacts of divergent moving drives on energy efficiency and performance of various AMS in operative conditions. Agriculture 11(9):806. https://doi.org/10.3390/agriculture11090806
Hrozencik RA, Suter JF, Ferraro PJ, Hendricks N (2023) Social comparisons and groundwater use: evidence from Colorado and Kansas. Am J Agr Econ 106(2):946–966. https://doi.org/10.1111/ajae.12415
Huesing JE, Andrés D, Braverman MP, Burns A, Felsot A, Harrigan G, Hellmich RL, Reynolds A, Shelton AM, Jansen van Rijssen W et al (2016) Global adoption of genetically modified (GM) crops: challenges for the public sector. J Agric Food Chem 64(2):394–402. https://doi.org/10.1021/acs.jafc.5b05116
Ilyas HMA, Safa M, Bailey A, Rauf S, Khan A (2020) Energy efficiency outlook of New Zealand dairy farming systems: an application of data envelopment analysis (DEA) approach. Energies 13(1):251. https://doi.org/10.3390/en13010251
Izah SC, Aigberua AO, Ogwu MO (2022) Trace element composition of Gallus gallus domesticus eggs and health risks associated with their consumption in Port Harcourt, Nigeria. J Food Saf Hyg 8(3):202–222
Izah SC, Odubo TC, Ogwu MC, Hait M (2024) Aspects of Microorganisms in the food industry. In: Ogwu MC, Izah SC, Ntuli NR (eds) Food safety and quality in the global south. Springer, Singapore. pp 399–425. https://doi.org/10.1007/978-981-97-2428-4_13
Javaid M, Khan S, Haleem A, Rab S (2022) Adoption of modern technologies for implementing Industry 4.0: an integrated MCDM approach. Benchmarking: Int J 30(10):3753–3790. https://doi.org/10.1108/bij-01-2021-0017
Jayathilake HACK, Jayasinghe-Mudalige UK, Perera LDRD, Gow GA, Waidyanatha N (2018) Use of free open source software technologies to enhance knowledge mobilization in smallholder agricultural communities in Sri Lanka. Trop Agric Res 29(2):147–157. https://doi.org/10.4038/tar.v29i2.8285
Jeanneret P, Aviron S, Alignier A, Lavigne C, Helfenstein J, Herzog F, Kay S, Petit S (2021) Agroecology landscapes. Landsc Ecol 36(8):2235–2257. https://doi.org/10.1007/s10980-021-01248-0
Jena PR, Lippe RS, Stellmacher T (2022) Editorial: sustainable certification standards: environmental and social impacts. Front Sustain Food Syst 6:922672. https://doi.org/10.3389/fsufs.2022.922672
Ji W, Liu Y (2022) Research on quantitative evaluation of remote sensing and statistics based on wireless sensors and farmland soil nutrient variability. Comput Intell Neurosci 2022(1):3646264. https://doi.org/10.1155/2022/3646264
Jürkenbeck K, Heumann A, Spiller A (2019) Sustainability matters: consumer acceptance of different vertical farming systems. Sustainability 11(15):4052. https://doi.org/10.3390/su11154052
Kabir MSN, Reza MN, Chowdhury M, Ali M, Samsuzzaman, Ali MR, Lee KY, Chung SO (2023) Technological trends and engineering issues on vertical farms: a review. Horticulturae 9(11):1229. https://doi.org/10.3390/horticulturae9111229
Kamilaris A, Fonts A, Prenafeta-Boldú FX (2019) The rise of blockchain technology in agriculture and food supply chains. Trends Food Sci Technol 91:640–652. https://doi.org/10.1016/j.tifs.2019.07.034
Karakaya E, Sriwannawit P (2015) Barriers to the adoption of photovoltaic systems: the state of the art. Renew Sustain Energy Rev 49:60–66. https://doi.org/10.1016/j.rser.2015.04.058
Kassie M, Teklewold H, Jaleta M, Marenya P, Erenstein O (2015) Understanding the adoption of a portfolio of sustainable intensification practices in Eastern and Southern Africa. Land Use Policy 42:400–411. https://doi.org/10.1016/j.landusepol.2014.08.016

Kause A, Nousiainen A, Koskinen H (2022) Improvement in feed efficiency and reduction in nutrient loading from rainbow trout farms: the role of selective breeding. J Anim Sci 100(8):skac214. https://doi.org/10.1093/jas/skac214

Khan F, Ahmad K, Ahmed A, Haider S (2017) Applications of biotechnology in agriculture—review article. World J Biol Biotechnol 2(1):139–142. https://doi.org/10.33865/wjb.002.01.0013

Khan MAA, Mahmood Ch K, Ashraf I, Siddiqui MT, Knox JW (2021) Evaluating socio-economic and environmental factors influencing farm-level water scarcity in Punjab, Pakistan. Irrig Drain 70(4):797–808. https://doi.org/10.1002/ird.2552

Kokkuar N, Li L, Srisapoome P, Dong S, Tian X (2021) Application of biodegradable polymers as carbon sources in ex situ biofloc systems: water quality and shift of microbial community. Aquac Res 52(8):3570–3579. https://doi.org/10.1111/are.15200

Korneeva E, Alamanova C, Orozonova A, Parmanasova A, Krayneva R (2023) Sustainable development of the agricultural sector of the economy. E3S Web Conf 431:01030. https://doi.org/10.1051/e3sconf/202343101030

Kremen C (2020) Ecological intensification and diversification approaches to maintain biodiversity, ecosystem services and food production in a changing world. Emerg Top Life Sci 4(2):229–240. https://doi.org/10.1042/ETLS20190205

Kumar I, Rawat J, Mohd N, Husain S (2021) Opportunities of artificial intelligence and machine learning in the food industry. J Food Qual 2021:1–10. https://doi.org/10.1155/2021/4535567

Kumar S, Sindhu SS, Kumar R (2022) Biofertilizers: an ecofriendly technology for nutrient recycling and environmental sustainability. Curr Res Microb Sci 3:100094. https://doi.org/10.1016/j.crmicr.2021.100094

Kumari P, Jasrotia P, Kumar D, Kashyap PL, Kumar S, Mishra CN, Kumar S, Singh GP (2022) Biotechnological approaches for host plant resistance to insect pests. Front Genet 13:914029. https://doi.org/10.3389/fgene.2022.914029

Lehmann J, Bossio DA, Kögel-Knabner I, Rillig MC (2020) The concept and future prospects of soil health. Nat Rev Earth Environ 1(10):544–553. https://doi.org/10.1038/s43017-020-0080-8

Liu J, Wang M, Yang L, Rahman S, Sriboonchitta S (2020) Agricultural productivity growth and its determinants in South and Southeast Asian countries. Sustainability 12(12):4981. https://doi.org/10.3390/su12124981

Mamata K, Lamichhane A (2021) Advances in agricultural biotechnology. Nepal J Biotechnol 9(1):85–92. https://doi.org/10.3126/njb.v9i1.38643

Mamta B, Rajam MV (2017) RNAi technology: a new platform for crop pest control. Physiol Mol Biol Plants 23(3):487–501. https://doi.org/10.1007/s12298-017-0443-x

Marazky M (2015) Effect of smart irrigation controllers units on the performance and productivity of subsurface and surface drip irrigation systems for tomato crop in arid regions. J Soil Sci Agric Eng 6(1):27–46. https://doi.org/10.21608/jssae.2015.41173

Marimuthu S, Kannan SV, Pazhanivelan S, Geethalakshmi V, Raju M, Sivamurugan AP, Karthikeyan M, Byrareddy VM, Mushtaq S, Surendran U (2024) Harnessing rain hose technology for water-saving sustainable irrigation and enhancing blackgram productivity in garden land. Sci Rep 14(1):18692. https://doi.org/10.1038/s41598-024-69655-2

Marone D, Mastrangelo AM, Borrelli GM (2023) From transgenesis to genome editing in crop improvement: applications, marketing, and legal issues. Int J Mol Sci 24(8):7122. https://doi.org/10.3390/ijms24087122

Marques V, Angelotti-Mendonça J, Roberto SR (2021) Advances and challenges in RNA interference technology for citrus huanglongbing vector control. Horticulturae 7(9):277. https://doi.org/10.3390/horticulturae7090277

Megavitry R, Harahap RH (2023) Revolutionizing the future: the importance of utilizing food technology for sustainable nutrition and global prosperity. West Sci Interdiscip Stud 1(09):764–772. https://doi.org/10.58812/wsis.v1i09.210

Mia MR, Monir JI (2023) Technological advancement and its impact on international trade dynamics: a comprehensive study of Bangladesh. Malays J Bus Econ Manag 2(2):75–83. https://doi.org/10.56532/mjbem.v2i2.22

Micu MM, Dinu TA, Fîntîneru G, Tudor VC, Stoian E, Dumitru EA, Stoicea P, Iorga A (2022) Climate change—between "myth and truth" in Romanian farmers' perception. Sustainability 14(14):8689. https://doi.org/10.3390/su14148689

Mishra J, Dutta V, Arora NK (2020) Biopesticides in India: technology and sustainability linkages. 3 Biotech 10(5):210. https://doi.org/10.1007/s13205-020-02192-7

Mittler R, Blumwald E (2010) Genetic engineering for modern agriculture: challenges and perspectives. Annu Rev Plant Biol 61(1):443–462. https://doi.org/10.1146/annurev-arplant-042809-112116

Mugwanya M, Dawood MAO, Kimera F, Sewilam H (2023) Replacement of fish meal with fermented plant proteins in the aquafeed industry: a systematic review and meta-analysis. Rev Aquac 15(1):62–88. https://doi.org/10.1111/raq.12701

Mwangi J (2024) Role of crop rotation practices on soil health. Am J Agric 6(1):32–45. https://doi.org/10.47672/aja.1758

Nackley LL, Barnes CM, Oki LR (2015) Investigating the impacts of recycled water on long-lived conifers. AoB Plants 7:plv035. https://doi.org/10.1093/aobpla/plv035

Odubo TC, Izah SC, Ogwu MC (2024) Mycological safety in Tiger nut (Cyperus esculentus L.) drinks: cultivation, contaminants, and compliance strategies. In: Ogwu MC, Izah SC, Ntuli NR (eds) Food safety and quality in the global south. Springer, Singapore, pp 463–487. https://doi.org/10.1007/978-981-97-2428-4_15

Ogidi OI, Izah SC (2024) Emerging technologies in the global south food industry: prospects and challenges. In: Ogwu MC, Izah SC, Ntuli NR (eds) Food safety and quality in the global south. Springer, Singapore, pp 197–231. https://doi.org/10.1007/978-981-97-2428-4_7

Ogwu MC, Izah SC, Ntuli NR, Odubo TC (2024) Food security complexities in the global south. In: Ogwu MC, Izah SC, Ntuli NR (eds) Food safety and quality in the global south. Springer, Singapore, pp 3–33. https://doi.org/10.1007/978-981-97-2428-4_1

Oliva-Teles A (2012) Nutrition and health of aquaculture fish. J Fish Dis 35(2):83–108. https://doi.org/10.1111/j.1365-2761.2011.01333.x

Osmond AT, Colombo SM (2019) The future of genetic engineering to provide essential dietary nutrients and improve growth performance in aquaculture: advantages and challenges. J World Aquac Soc 50(3):490–509. https://doi.org/10.1111/jwas.12595

Pelletier N, Tyedmers P (2010) Forecasting potential global environmental costs of livestock production 2000–2050. Proc Natl Acad Sci 107(43):18371–18374. https://doi.org/10.1073/pnas.1004659107

Pelletier N, Audsley E, Brodt S, Garnett T, Henriksson P, Kendall A, Kramer KJ, Murphy D, Nemecek T, Troell M (2011) Energy intensity of agriculture and food systems. Annu Rev Environ Resour 36(1):223–246. https://doi.org/10.1146/annurev-environ-081710-161014

Pérez-Blanco CD, Hrast-Essenfelder A, Perry C (2020) Irrigation technology and water conservation: a review of the theory and evidence. Rev Environ Econ Policy 14(2):216–239. https://doi.org/10.1093/reep/reaa004

Qin Y, He J, Wei M, Du X (2022) Challenges threatening agricultural sustainability in central Asia: status and prospect. Int J Environ Res Public Health 19(10):6200. https://doi.org/10.3390/ijerph19106200

Rahman A (2023) Implementing agricultural policies: driving sustainable growth in the farming sector in Indonesia. Jurnal Ilmiah Ilmu Administrasi Publik 13(1):441. https://doi.org/10.26858/jiap.v13i1.50352

Rai GK, Khanday DM, Kumar P, Magotra I, Choudhary SM, Kosser R, Kalunke R, Giordano M, Corrado G, Rouphael Y, Pandey S (2023) Enhancing crop resilience to drought stress through CRISPR-Cas9 genome editing. Plants (Basel) 12(12):2306. https://doi.org/10.3390/plants12122306

Raj A, Dwivedi G, Sharma A, de Sousa Jabbour ABL, Rajak S (2020) Barriers to the adoption of industry 4.0 technologies in the manufacturing sector: an inter-country comparative perspective. Int J Prod Econ 224:107546. https://doi.org/10.1016/j.ijpe.2019.107546

Raman R (2017) The impact of genetically modified (GM) crops in modern agriculture: a review. GM Crop Food 8(4):195–208. https://doi.org/10.1080/21645698.2017.1413522

Ray DK, Mueller ND, West PC, Foley JA (2013) Yield trends are insufficient to double global crop production by 2050. PLoS ONE 8(6):e66428. https://doi.org/10.1371/journal.pone.0066428

Rhymes J, Chadwick DR, Williams AP, Harris IM, Lark RM, Jones DL (2023) Evaluating the accuracy and usefulness of commercially-available proximal soil mapping services for grassland nutrient management planning and soil health monitoring. Precision Agric 24(3):898–920. https://doi.org/10.1007/s11119-022-09979-z

Roberts DP, Short NM, Sill J, Lakshman DK, Hu X, Buser M (2021) Precision agriculture and geospatial techniques for sustainable disease control. Indian Phytopathol 74(2):287–305. https://doi.org/10.1007/s42360-021-00334-2

Ryan M, van der Burg S, Bogaardt MJ (2022) Identifying key ethical debates for autonomous robots in agri-food: a research agenda. AI Ethics 2(3):493–507. https://doi.org/10.1007/s43681-021-00104-w

Saad MHM, Hamdan NM, Sarker MR (2021) State of the art of urban smart vertical farming automation system: advanced topologies, issues and recommendations. Electronics 10(12):1422. https://doi.org/10.3390/electronics10121422

Saberioon M, Gholizadeh A, Císař P, Pautsina A, Urban J (2017) Application of machine vision systems in aquaculture with emphasis on fish: state-of-the-art and key issues. Rev Aquac 9(4):369–387. https://doi.org/10.1111/raq.12143

Scherer LA, Verburg PH, Schulp CJ (2018) Opportunities for sustainable intensification in European agriculture. Glob Environ Chang 48:43–55. https://doi.org/10.1016/j.gloenvcha.2017.11.009

Secinaro S, Dal Mas F, Massaro M, Calandra D (2022) Exploring agricultural entrepreneurship and new technologies: academic and practitioners' views. Br Food J 124(7):2096–2113. https://doi.org/10.1108/bfj-08-2021-0905

Sharma A, Sharma A, Tselykh A, Bozhenyuk A, Choudhury T, Alomar MA, Sánchez-Chero M (2023) Artificial intelligence and internet of things oriented sustainable precision farming: towards modern agriculture. Open Life Sci 18(1):20220713. https://doi.org/10.1515/biol-2022-0713

Shine F, Joseph SS, Thomas AA, Raj D (2023) Growth promotion efficacy of gold nanoparticle supplementation diet in *Oreochromis mossambicus*. Ecol Environ Conserv 29(Special):07–14. https://doi.org/10.53550/eec.2023.v29isp2.002

Shyr WJ, Yang FC, Liu PW, Hsieh YM, You CS, Chen DC (2019) Development of assessment indicators for measuring the student learning effects of artificial intelligence-based robot design. Comput Appl Eng Educ 27(4):863–868. https://doi.org/10.1002/cae.22118

Singh NK, Sunitha NH, Tripathi G, Saikanth DRK, Sharma A, Jose AE, Mary MV (2023) Impact of digital technologies in agricultural extension. Asian J Agric Ext, Econ Sociol 41(9):963–970. https://doi.org/10.9734/ajaees/2023/v41i92127

Singh P (2023) Security and privacy in fog/cloud-based IoT systems for AI and robotics. EAI Endorsed Trans AI Robot 2:e5. https://doi.org/10.4108/airo.3616

Smith MJ (2018) Getting value from artificial intelligence in agriculture. Anim Prod Sci 60(1):46–54. https://doi.org/10.1071/an18522

Snegirjovs A, Ļebedeva K, Kashkarova G, Shipkovs P (2018) Solar technologies application in milk production. In: Proceedings of the 17th international scientific conference "Engineering for Rural Development", Jelgava, Latvia, pp 23–25. https://doi.org/10.22616/erdev2018.17.n097

Solarin SA, Bello MO (2022) Wind energy and sustainable electricity generation: evidence from Germany. Environ Dev Sustain 24(7):9185–9198. https://doi.org/10.1007/s10668-021-01818-x

Son J, Hwang K (2023) How to make vertical farming more attractive: effects of vegetable growing conditions on consumer assessment. Psychol Mark 40(8):1466–1483. https://doi.org/10.1002/mar.21823

Specht K, Zoll F, Schümann H, Bela J, Kachel J, Robischon M (2019) How will we eat and produce in the cities of the future? From edible insects to vertical farming—a study on the perception and acceptability of new approaches. Sustainability 11(16):4315. https://doi.org/10.3390/su11164315

Tegegne MD, Wubante SM (2022) Identifying barriers to the adoption of information communication technology in Ethiopian healthcare systems: a systematic review. Adv Med Educ Pract 13:821–828. https://doi.org/10.2147/amep.s374207

Tilman D, Balzer C, Hill J, Befort BL (2011) Proc Natl Acad Sci 108(50):20260–20264. https://doi.org/10.1073/pnas.1116437108

Ukoba K, Yoro KO, Eterigho-Ikelegbe O, Ibegbulam C, Jen TC (2024) Adaptation of solar energy in the Global South: prospects, challenges and opportunities. Heliyon 10(7):e28009. https://doi.org/10.1016/j.heliyon.2024.e28009

Uzal S (2013) Comparison of the energy efficiency of dairy production farms using different housing systems. Environ Prog Sustain Energy 32(4):1202–1208. https://doi.org/10.1002/ep.11727

van der Burg S, Giesbers E, Bogaardt MJ, Ouweltjes W, Lokhorst K (2024) Ethical aspects of AI robots for agri-food: a relational approach based on four case studies. AI Soc 39:541–555. https://doi.org/10.1007/s00146-022-01429-8

Varshney RK, Bansal KC, Aggarwal PK, Datta SK, Craufurd PQ (2011) Agricultural biotechnology for crop improvement in a variable climate: hope or hype? Trends Plant Sci 16(7):363–371. https://doi.org/10.1016/j.tplants.2011.03.004

Vatistas C, Avgoustaki DD, Bartzanas T (2022) A systematic literature review on controlled-environment agriculture: how vertical farms and greenhouses can influence the sustainability and footprint of urban microclimate with local food production. Atmosphere 13(8):1258. https://doi.org/10.3390/atmos13081258

Vrasidas C (2014) The rhetoric of reform and teachers' use of ICT. Br J Edu Technol 46(2):370–380. https://doi.org/10.1111/bjet.12149

Wang Y, Huang T, Liu J, Lin Z, Li S, Wang R, Ge Y (2015) Soil pH value, organic matter and macronutrients contents prediction using optical diffuse reflectance spectroscopy. Comput Electron Agric 111:69–77. https://doi.org/10.1016/j.compag.2014.11.019

Ward FA, Pulido-Velazquez M (2008) Water conservation in irrigation can increase water use. Proc Natl Acad Sci 105(47):18215–18220. https://doi.org/10.1073/pnas.0805554105

Xie H, Huang Y, Chen Q, Zhang Y, Wu Q (2019) Prospects for agricultural sustainable intensification: a review of research. Land 8(11):157. https://doi.org/10.3390/land8110157

Xu Y, Smith P, Qin Z (2024) Sustainable bioenergy contributes to cost-effective climate change mitigation in China. iScience 27(7):110232. https://doi.org/10.1016/j.isci.2024.110232

Yarkwan B, Isaac TO, Okopi A, Izah SC (2024) Evidence of the toxic potentials of agrochemicals on human health and biodiversity: carcinogens and mutagens. In: Ogwu MC, Izah SC, Ntuli NR (eds) Food safety and quality in the Global South. Springer, Singapore, pp 331–359. https://doi.org/10.1007/978-981-97-2428-4_11

Zăpucioiu LF, Sterie MC, Dimitriu TA (2023) The relevance of the circular economy in the context of sustainable development. Proc Int Conf Bus Excel 17(1):1534–1543. https://doi.org/10.2478/picbe-2023-0138

Zea P, Chilpe J, Sánchez D, Chica EJ (2020) Energy efficiency of smallholder commercial vegetable farms in Cuenca (Ecuador). Trop Subtrop Agroecosyst 23(2). https://doi.org/10.56369/tsaes.2995

Zeng S, Li J, Wanger TC (2023) Agroecology, technology, and stakeholder awareness: implementing the UN food systems summit call for action. iScience 26(9):107510. https://doi.org/10.1016/j.isci.2023.107510

Zhang Y, Bleeker A, Liu J (2015) Nutrient discharge from China's aquaculture industry and associated environmental impacts. Environ Res Lett 10(4):045002. https://doi.org/10.1088/1748-9326/10/4/045002

Zhou Q, Wu F, Zhang Q (2015) Is irrigation water price an effective leverage for water management? An empirical study in the middle reaches of the Heihe River basin. Phys Chem Earth, Parts ABC 89:25–32. https://doi.org/10.1016/j.pce.2015.09.002

Zu D, Su Z, Yang X, Gu X, Wang Y (2015). Design and implementation of soil nutrient monitoring system based on "3S" technology. Int J Smart Home 9(5):153–164. https://doi.org/10.14257/ijsh.2015.9.5.15.

Part III
Strategic Management Approaches for Sustainable Environmental Health

Chapter 11
Participatory Approaches to Environmental Health Management

Tsholofelo Lori, Mildred Osei-Kwarteng, Fredrick Ojija, Aliu Olugbemiga Ojo, and Matthew Chidozie Ogwu

Abstract Participatory approaches in environmental health have emerged as essential strategies for effectively addressing the complexities of ecological challenges through direct community involvement. This chapter examines the theoretical underpinnings and practical applications of participatory methods, emphasizing their ability to engage communities in meaningful ways. By focusing on community-based participatory research, environmental justice, and social-ecological systems theory, this chapter highlights how these approaches provide frameworks for equitable, community-driven health solutions. It also examines the benefits of participatory methods, including improved health outcomes, increased community empowerment,

T. Lori (✉)
Department of Safety Management, Faculty of the Built Environment, Arts and Science, BA ISAGO University, Gaborone, Botswana
e-mail: tsholofelo.lori@baisago.ac.bw

M. Osei-Kwarteng
Department of Horticulture, Faculty of Agriculture, Food and Consumer Sciences, University for Development Studies, Tamale, Ghana
e-mail: osei.kwarteng@uds.edu.gh

F. Ojija
Department of Earth Sciences, College of Science and Technical Education, Mbeya University of Science and Technology, Mbeya, Tanzania

Department of Research and Publications, Mbeya University of Science and Technology, Mbeya, Tanzania

F. Ojija
e-mail: fredrick.ojija@must.ac.tz

A. O. Ojo
Department of Plant Biology and Biotechnology, Faculty of Life Sciences, University of Benin, Benin City, Nigeria

Department of Environmental Biology, Auchi Polytechnic, Auchi, Nigeria

M. C. Ogwu
Goodnight Family Department of Sustainable Development, Appalachian State University, Boone, USA
e-mail: ogwumc@appstate.edu

S. Chibueze Izah and M. Chidozie Ogwu (eds.), *Innovative Approaches in Environmental Health Management*, Environmental Science and Engineering,
https://doi.org/10.1007/978-3-031-81966-7_11

enhanced environmental literacy, and strengthened social cohesion. Various engagement techniques, such as workshops, focus groups, community mapping, and citizen science initiatives, are discussed to illustrate how these methods can be tailored to specific environmental health contexts. The discussion includes case studies that showcase the successes and challenges of community engagement in environmental health, emphasizing the need for inclusive and adaptive approaches. The chapter concludes by presenting strategies for creating effective participatory ecological health programs that promote sustainability, build local capacity, and advocate for policy changes. By integrating participatory methods with broader environmental health initiatives, communities can enhance resilience, foster collaboration, and improve ecological health outcomes.

Keywords Environmental health · Community engagement · Environmental justice · Social-ecological systems · Citizen science · Community empowerment · Sustainable solutions · Health outcomes

11.1 Introduction

Participatory approaches in environmental health refer to strategies involving community members in the decision-making processes related to environmental health issues. These approaches emphasize collaboration between researchers, policymakers, and community stakeholders to ensure that the voices and perspectives of those affected by environmental health challenges are heard and considered Gray (2018). Participatory methods can take various forms, including community-based participatory research (CBPR), focus groups, workshops, and public consultations, all aimed at fostering engagement and empowerment among community members (Ramírez-Andreotta et al. 2016). Community engagement is crucial for addressing environmental health challenges effectively. Engaging communities in planning and implementing ecological health initiatives fosters a sense of ownership and responsibility, leading to more sustainable and culturally relevant solutions (Sørensen et al. 2012).

Moreover, community involvement can enhance the effectiveness of public health interventions by ensuring that they are tailored to the specific needs and priorities of the population (Hoover 2016). Research has shown that participatory approaches can improve health outcomes by increasing awareness of environmental health risks and empowering individuals to act (Souza et al. 2013). For example, involving community members in monitoring environmental exposures can enhance their understanding of the risks and motivate them to advocate for policy changes that protect their health (Ravenscroft et al. 2014). Community engagement is essential for building trust, promoting social equity, and ensuring that environmental health initiatives are effective and sustainable.

Participatory strategies for sustainable solutions encompass a range of methods that facilitate collaboration between communities and stakeholders. These strategies can include CBPR, Public Engagement Campaigns, Collaborative Policy Development, and Capacity Building and Training (Martínez et al. 2011; Madrigal et al. 2016; Lichtveld et al. 2019; Sprague 2023). These participatory strategies can foster stakeholder collaboration, enhance community resilience, and promote sustainable solutions to environmental health challenges. Nature-based solutions (NbS) can significantly improve urban air quality by providing green spaces that filter pollutants and reduce the concentration of harmful airborne particles. Trees and vegetation absorb carbon dioxide and other pollutants, contributing to cleaner air and better respiratory health for urban residents (Allen et al. 2019). Research has shown that urban areas with higher levels of greenery experience lower rates of respiratory diseases, highlighting the importance of integrating NbS into urban planning (Cláudio et al. 2018). For instance, cities that have invested in urban greening initiatives have reported reductions in asthma rates and other respiratory conditions, particularly among vulnerable populations such as children and the elderly (Marsili 2023). Urban heat islands (UHIs) are a significant concern in densely populated areas, where built environments absorb and retain heat, leading to elevated temperatures. NbS, such as green roofs, urban forests, and shaded parks, can mitigate the effects of UHIs by providing shade, cooling the air through evapotranspiration, and reducing surface temperatures (Bernhard et al. 2013). Research also indicates that urban areas with extensive green cover experience lower temperatures, which can significantly reduce heat-related illnesses and mortality, particularly during heatwaves (Schure et al. 2013). For example, cities like Chicago have implemented urban greening initiatives that have successfully lowered surface temperatures and improved public health outcomes during extreme heat events (Ramírez et al. 2019). Access to nature has been linked to improved mental health outcomes, including reduced stress, anxiety, and depression. NbS provides opportunities for physical activity, social interaction, and relaxation, all of which contribute to better mental well-being (English et al. 2017). Studies have shown that individuals living near green spaces report higher levels of life satisfaction and lower levels of psychological distress (Sloot et al. 2018). For instance, a study conducted in a community garden in Toronto highlighted the positive impact of gardening on mental health, with participants reporting increased feelings of happiness, reduced anxiety, and improved social interactions (Franz et al. 2022). NbS can also foster community resilience by providing social interaction and community engagement spaces. Green spaces, community gardens, and urban parks are vital social hubs that promote connections among residents (Pasetto et al. 2019). The social cohesion fostered by these spaces enhances community resilience, enabling neighborhoods to withstand environmental and social challenges better. For example, during the COVID-19 pandemic, access to green spaces was essential for maintaining mental health and social connections among residents (Galway 2016). By integrating NbS into urban planning, cities can create environments that support community well-being and resilience.

This chapter examines the significance of participatory approaches in addressing environmental health challenges, emphasizing the importance of community engagement for creating sustainable, practical solutions. This chapter defines vital participatory methods—CBPR, citizen science, and participatory action research—and explores how they empower communities to take ownership of environmental health initiatives. By outlining the theoretical foundations of participatory approaches, the chapter aims to provide readers with a framework for understanding how engaging stakeholders, especially marginalized groups, can improve health outcomes, enhance environmental stewardship, and build social cohesion.

11.2 Theoretical Foundations of Participatory Approaches

Participatory approaches in environmental health are grounded in theoretical frameworks emphasizing collaboration, empowerment, and equity. These frameworks underscore the necessity of community involvement in identifying, addressing, and sustaining solutions for environmental health challenges. Critical theories include CBPR, environmental justice and equity principles, and Social-Ecological Systems Theory. These frameworks guide participatory methods that foster inclusive, community-driven actions and improved environmental health outcomes.

Community-Based Participatory Research

CBPR is a collaborative research approach that actively involves community members, researchers, and other stakeholders throughout all phases of the research process. CBPR is unique in fostering co-learning, shared power, and mutual respect, making it particularly valuable for environmental health projects where local knowledge is crucial (Israel et al. 2010; Minkler and Wallerstein 2011). Unlike traditional research, CBPR encourages communities to identify research priorities, co-design interventions, and collaboratively analyze and disseminate results, ensuring that the findings benefit the communities involved. This participatory approach has been shown to enhance the relevance and cultural sensitivity of environmental health interventions, addressing local communities' unique needs and priorities (Horowitz et al. 2009). CBPR is particularly effective in contexts where ecological exposures and health risks are heightened, such as communities near industrial sites or those affected by air and water pollution. For instance, communities dealing with pollution from industrial facilities can collaborate with researchers to explore links between exposure and health outcomes, using the results to advocate for policy changes (Minkler and Wallerstein 2011). By combining scientific inquiry with local expertise, CBPR transforms research into a tool for social change, enabling communities to leverage data and scientific evidence to advocate for healthier environments (Israel et al. 2010). Furthermore, CBPR has effectively promoted sustainable health interventions, as community members are more likely to support initiatives they have helped design (Schulz et al. 2018).

Principles of Environmental Justice and Equity

Environmental justice and equity principles are central to participatory approaches in environmental health, emphasizing the fair treatment and meaningful involvement of all individuals, regardless of race, color, nationality, or income. Ecological justice is rooted in recognizing that marginalized communities often experience disproportionate exposure to environmental hazards, resulting in increased health risks (Bullard 2000). Participatory approaches informed by environmental justice principles focus on engaging these communities to address disparities and to advocate for equitable access to healthy environments (Brulle and Pellow 2006). Equity, a fundamental aspect of environmental justice, requires that all individuals participate in decision-making processes that impact their health and well-being. Participatory approaches facilitate this by actively involving community members in all aspects of environmental health initiatives, from risk assessment to solution implementation (Agyeman 2005). For example, participatory approaches have engaged Indigenous communities in discussions about land use and pollution, ensuring their voices are heard and their perspectives integrated into environmental policies (Taylor 2014). By prioritizing community input, participatory methods help to counteract historical injustices and empower marginalized communities to shape the outcomes that affect their health (Pellow 2018). This approach helps build trust, strengthens community resilience, and supports sustained advocacy for environmental justice (Brulle and Pellow 2006).

Social-Ecological Systems Theory in Community Engagement

Social-Ecological Systems Theory offers a holistic perspective on the interconnectedness of social and ecological systems, emphasizing the complex and dynamic relationships between humans and their environments (Berkes et al. 2000). This theory underpins participatory approaches by recognizing that environmental health challenges are influenced by both natural and human factors, which means that solutions must consider ecological, cultural, and social dimensions (Ostrom 2009). By fostering community engagement, participatory approaches rooted in Social-Ecological Systems Theory encourage local populations to manage and sustain their environments actively. Social-Ecological Systems Theory also highlights the importance of adaptive capacity, or the ability of communities to respond to environmental changes. By involving community members in the planning and decision-making processes, participatory approaches help build adaptive capacity and enable communities to address complex issues such as climate change and ecosystem degradation (Folke et al. 2005). In coastal regions, for example, participatory approaches are used to engage residents in planning and implementing flood mitigation strategies, drawing on their knowledge of local conditions to develop more effective and resilient solutions (Adger 2000). These participatory methods promote shared responsibility, enhance social cohesion, and contribute to sustainable environmental health outcomes by fostering an integrated approach to community and ecosystem health (Berkes et al. 2000).

11.3 Benefits of Participatory Approaches to Environmental Health Management

Participatory methods offer numerous benefits for environmental health by promoting community ownership, enhancing the relevance and accuracy of data, and fostering sustainable solutions (Table 11.1). Community ownership is strengthened through collaborative processes that align with local values and priorities, leading to higher levels of engagement and more successful project outcomes (Horowitz et al. 2009). Additionally, the involvement of community members in data collection and analysis increases the quality and contextual relevance of environmental health data, which is critical for effective intervention design (Israel et al. 2010). Moreover, participatory methods support the long-term sustainability of ecological health initiatives by building local capacity and empowering communities to address challenges independently. For example, training community members in environmental monitoring equips them with skills to continue these activities beyond the initial project, fostering resilience and self-sufficiency (Frumkin 2010). When residents feel invested in the process, they are more likely to sustain environmental health improvements and advocate for further change. The social cohesion that emerges from these collaborative efforts also strengthens community networks, enabling communities to mobilize resources and support for future environmental health projects (Schulz et al. 2018).

Participatory approaches in environmental health have proven to be highly effective in fostering community engagement, improving health outcomes, and promoting sustainability. By involving community members directly in environmental health projects, these approaches empower individuals, enhance local knowledge, and build a sense of ownership. Participatory approaches in environmental health directly contribute to better health metrics and outcomes. By involving community members in identifying environmental hazards, assessing health risks, and implementing solutions, these approaches ensure that interventions are more targeted, culturally relevant, and responsive to local needs (Horowitz et al. 2009). Community engagement helps highlight specific health concerns, such as poor air quality or contaminated water sources, that may otherwise go unnoticed by external researchers. By addressing these issues collaboratively, communities can reduce health risks and improve environmental quality, improving overall health outcomes. For instance, in areas with significant air pollution, community-led air quality monitoring programs can identify pollution sources, provide data for advocacy, and lead to policy changes that reduce emissions. When residents participate in such programs, they better understand how environmental factors impact their health, leading to increased support for mitigation measures and healthier communities.

Additionally, participatory approaches have improved compliance with health recommendations, as individuals are likelier to adopt practices they helped develop and implement (Israel et al. 2010). In sum, participatory approaches facilitate interventions better tailored to community-specific issues, resulting in meaningful improvements in environmental health metrics. Empowerment is a central goal of

Table 11.1 Benefits of participatory methods for environmental health outcomes

Benefit	Example of participatory-based method	Examples of environmental health
Improved health outcomes	Community health workshops where residents learn about local environmental hazards and collaboratively identify solutions	Community-led air quality monitoring programs that lead to local policy changes, reducing asthma rates in affected areas
Empowerment and ownership	Involving community members in the design and implementation of waste management systems	Community members co-designing waste management systems, increasing compliance and sustainable practices
Increased awareness and education	Conducting educational campaigns led by local leaders on the health impacts of pollution and preventive behaviors	Workshops on the health impacts of pollution empower communities to advocate for cleaner air and water
Social cohesion and community building	Organizing community forums to discuss and plan green spaces, allowing residents to collaborate on shared goals	Community forums on green space development that bring residents together, fostering a sense of community and belonging
Culturally relevant solutions	Engaging indigenous groups in co-creating conservation projects that integrate traditional knowledge	Involving indigenous groups in conservation projects that respect traditional knowledge and practices
Sustainable and long-term impact	Training residents in environmental monitoring, enabling them to continue these activities independently	Training residents to monitor water quality, ensuring ongoing vigilance and adaptive management
Enhanced data relevance and accuracy	Involving residents in data collection, such as mapping pollution sources and tracking environmental changes	Residents identify pollution sources in their neighborhoods, leading to precise monitoring and cleanup efforts
Increased community advocacy and policy influence	Facilitating community-led advocacy groups that address environmental policy concerns with local governments	Community-led campaigns to regulate industrial emissions, leading to stricter local environmental regulations

(continued)

Table 11.1 (continued)

Benefit	Example of participatory-based method	Examples of environmental health
Adaptability and resilience	Collaborating with communities on climate adaptation strategies, such as developing flood prevention plans	Community involvement in flood prevention planning, leading to adaptive strategies for climate resilience
Economic benefits	Supporting community-driven eco-tourism projects that promote environmental conservation while generating local income	Community-driven eco-tourism initiatives that generate income while promoting conservation

participatory approaches in environmental health. These methods enhance community autonomy and ownership by involving community members in decision-making and giving them a voice in shaping environmental policies and interventions. Participatory approaches allow communities to manage their resources actively, fostering a sense of responsibility for local environmental health (Minkler and Wallerstein 2011). This empowerment is essential in marginalized or underserved communities with historically limited control over decisions affecting their environments. For example, communities involved in water quality monitoring initiatives are empowered to assess their water sources and act when contamination is detected. By having the skills and knowledge to manage and protect their resources, community members become stewards of their environment, strengthening resilience to future challenges. This autonomy fosters a long-term commitment to environmental health, as communities with a sense of ownership over projects are more likely to sustain and expand them over time (Schulz et al. 2018). Empowerment also builds capacity, enabling communities to independently address new environmental health issues.

Participatory projects serve as powerful tools for building environmental literacy within communities. Community members gain a deeper understanding of the connections between environmental conditions and health outcomes through involvement in environmental health initiatives. This hands-on experience increases their awareness of local environmental issues and equips them with the knowledge to make informed decisions (Agyeman 2005). Education and capacity-building efforts embedded in participatory approaches help residents develop skills in environmental monitoring, data collection, and health risk assessment. For instance, workshops and training sessions on pollution, waste management, and biodiversity conservation raise awareness about environmental challenges and encourage behavior changes that promote health. As community members learn to identify and monitor local hazards, they become more proactive in protecting their environment and advocating for healthier living conditions. Participatory education efforts improve individual knowledge and inspire collective action, as informed communities are more likely to mobilize around shared environmental goals. This increased environmental literacy supports sustainable practices, as residents are empowered to make decisions that benefit their health and the environment. Participatory approaches strengthen social cohesion by bringing community members together to address common environmental health challenges. Collaborative projects allow individuals to build relationships, share knowledge, and work toward shared goals. This collective action fosters a sense of unity and purpose, reinforcing the community's social fabric (Frumkin 2010). By working together on projects that enhance the environment, such as creating green spaces or cleaning up local waterways, community members develop a sense of pride and connection to their surroundings. For example, when neighborhoods come together to design and implement urban gardens or green spaces, they improve their local environment and create gathering places for social interaction and recreation. These shared spaces enhance community well-being by providing physical activity, relaxation, and social engagement opportunities. Furthermore, the relationships built through participatory environmental health projects create networks of mutual support, increasing community resilience to environmental and health-related challenges (Horowitz et al. 2009). Stronger community ties make it easier for residents to mobilize for future initiatives, advocate for resources, and maintain ongoing environmental health efforts.

11.4 Key Components of Community Engagement in Environmental Health

Community engagement in environmental health is a multifaceted process that involves several key components, including stakeholder identification and involvement, participatory planning and decision-making, education and capacity building, and community-led monitoring and evaluation (Table 11.2). Each element ensures that environmental health initiatives are effective, equitable, and sustainable.

Stakeholder Identification and Involvement: Identifying and involving key community stakeholders is essential for effective community engagement. Stakeholders can include residents, community organizations, government agencies, and vulnerable populations. Engaging marginalized groups is particularly important, as they often bear the brunt of environmental health issues. For instance, Cuerrier et al. (2015) emphasize the importance of incorporating indigenous knowledge into environmental decision-making, highlighting how local insights can inform strategies that address community-specific challenges. Similarly, Qasim et al. (2023) discuss the principles of social democracy and equity in stakeholder engagement, which are vital for ensuring that the voices of marginalized populations are heard and considered in the decision-making process.

Participatory Planning and Decision-Making: Participatory planning and decision-making involve co-designing solutions with the community, ensuring that interventions are tailored to local needs. The World Health Organization recognizes human-centered design as a critical strategy for addressing health challenges, as it

Table 11.2 Essential components of community engagement in environmental health management

Component	Description	Examples
Stakeholder identification and involvement	Identifying and involving a diverse range of stakeholders, including vulnerable and marginalized groups, and to ensure inclusive participation	Engaging residents, health organizations, environmental groups, and policymakers in environmental health projects
Participatory planning and decision-making	Involving the community in setting goals, planning actions, and making decisions about environmental health interventions	Community meetings to co-design a park for flood mitigation, health improvement, and recreation
Education and capacity building	Providing information and training to improve community knowledge on environmental issues and build skills for active participation	Workshops on air quality monitoring or waste management practices, with training on the use of simple sensors
Community-led monitoring and evaluation	Enabling communities to actively monitor environmental factors and evaluate the outcomes of health initiatives	Resident-led water testing and quality assessments in local rivers to track pollution levels over time
Resource sharing and accessibility	Ensuring that community members have access to necessary resources, tools, and information to participate effectively	Providing access to equipment, data, and information for air quality monitoring in local neighborhoods
Communication and feedback mechanisms	Creating channels for ongoing communication, feedback, and sharing of progress with the community	Regular newsletters, community bulletins, or digital platforms to update residents on project milestones
Building trust and transparency	Establishing trust by being transparent about project goals, processes, and outcomes to foster stronger community ties	Hosting open forums to discuss project results and community concerns, with opportunities for residents to ask questions
Collaboration with local institutions	Partnering with local schools, universities, and NGOs to leverage expertise and support for environmental health projects	Collaborating with local universities on research projects that involve students and community members
Empowerment through decision-making	Empowering community members to take an active role in decision-making processes, giving them a stake in outcomes	Allowing residents to vote on project priorities, such as where to place green spaces or which pollutants to monitor
Sustainable practices and knowledge transfer	Promoting long-term sustainability by transferring knowledge and skills that enable the community to maintain initiatives independently	Training residents on waste recycling methods and sustainable gardening practices to reduce local pollution

promotes the involvement of community members in the planning process (Watkins 2023). Setiawan et al. (2022) further illustrate this by demonstrating how community members' social and cultural contexts must be considered when designing frameworks for disaster mitigation, ensuring that solutions are relevant and practical. This approach fosters ownership and enhances the likelihood of successful implementation.

Education and Capacity Building: Education and capacity building are critical for raising awareness about environmental health issues and equipping communities with the skills needed for effective environmental management. Anwar et al. highlight the importance of health literacy in low-resource settings, advocating for interventions informed by local wisdom to improve health equity (Anwar 2023). Additionally, Yadav et al. note that co-design methods can enhance stakeholder awareness and shared ownership of health interventions, leading to improved health outcomes (Yadav et al. 2021). By empowering communities through education and skill-building, stakeholders can become active participants in managing their environmental health.

Community-Led Monitoring and Evaluation: Establishing community-led monitoring and evaluation programs is essential for assessing the effectiveness of environmental health initiatives. Carr et al. (2021) describe how community members can lead the design of programs that fit their lifestyles, thereby increasing participation and ownership. Furthermore, integrating participatory evaluation techniques allows communities to assess the sustainability of solutions and adapt them as necessary. This approach is supported by the findings of Oetzel et al. (2015), who emphasize the importance of community governance in enhancing stewardship of research and ensuring that community needs are met. As discussed by Ellingsen et al., long-term environmental monitoring is also crucial for understanding changes and impacts over time, ensuring that interventions remain relevant and practical (Ellingsen et al. 2017).

11.5 Approaches and Techniques for Community Engagement

Community engagement in environmental health is essential for fostering collaboration between residents, researchers, and policymakers. Various approaches and techniques can be employed to facilitate this engagement, ensuring that community voices are heard and integrated into environmental health initiatives (Table 11.3). Critical methodologies for community engagement include.

Workshops and Focus Groups

Workshops and focus groups effectively gather community input, allowing in-depth discussions on environmental health issues. These settings enable community

Table 11.3 Approaches to community engagement for environmental health management

Approach	Focal activities	Purpose/benefits	Examples
Workshops and focus groups	Structured group discussions that gather input from community members on specific issues	Facilitates dialogue, gathers diverse perspectives, and allows for in-depth discussion on community concerns	Town hall meetings, environmental impact workshops
Surveys and community mapping	Data collection tools are used to gather quantitative and spatial data on community health and environmental issues	Provides measurable data, identifies key areas of concern, and helps visualize environmental health impacts	Health surveys, mapping of pollution sources or green spaces
Participatory action research (PAR)	Collaborative research method where community members are involved in the entire research process	Empowers communities, enhances the relevance of research, and supports actionable solutions	Community-led studies on air quality, water safety assessments
Citizen science	Involves residents in collecting and sharing environmental health data, often using simple tools or apps	Builds local knowledge, raises environmental awareness, and gathers extensive data on environmental changes	Air quality monitoring, water testing, biodiversity tracking
Community forums and public meetings	Open meetings that allow community members to voice concerns and discuss solutions with stakeholders	Enhances transparency, builds trust, and fosters collective problem-solving	Public hearings on environmental projects, health forums
Storytelling and digital media	Uses narratives and digital tools to share community experiences related to environmental health	Amplifies community voices raises awareness, and builds empathy among broader audiences	Video documentaries, community podcasts, social media campaigns
Focus groups and role-playing exercises	Interactive sessions where participants explore issues from different perspectives, often through role play	Encourages empathy, deepens understanding of stakeholder roles, and helps identify community priorities	Role-playing in planning exercises, scenario-based workshops
Environmental health education programs	Provides information on environmental issues and health impacts, often with interactive activities	Builds community capacity, enhances health literacy, and promotes preventive health behaviors	School programs on pollution, workshops on safe waste disposal

(continued)

Table 11.3 (continued)

Approach	Focal activities	Purpose/benefits	Examples
Collaborative design and co-creation workshops	Engages community members in designing interventions and solutions for local environmental health challenges	Ensures solutions are culturally relevant, increases community ownership, and supports sustainable outcomes	Co-designing public parks, planning sustainable waste systems
Community-based environmental monitoring	Involves residents in ongoing data collection to track changes in the environment over time	Enables real-time tracking of environmental health, empowers residents with skills, and builds long-term vigilance	Monitoring air or water quality, tracking wildlife habitats

members to express their concerns, share experiences, and propose solutions. Facilitators are crucial in leading productive talks; they must create an inclusive environment that encourages participation from all attendees. Zulu et al. (2021) emphasize the importance of identifying community leaders and champions who can facilitate these discussions and effectively represent the community's needs. Additionally, Ishom et al. highlight that facilitators should be trained to empower community members, enabling them to articulate their challenges and collaboratively develop solutions (Ishom et al. 2023). This participatory approach enhances the quality of the discussions and fosters a sense of ownership among participants, which is vital for the success of environmental health initiatives.

Surveys and Community Mapping

Surveys and community mapping are valuable tools for gathering quantitative and spatial data from residents. Surveys can capture a wide range of information, including residents' perceptions of environmental hazards and their health impacts. This data can be instrumental in identifying priority areas for intervention. On the other hand, community mapping allows residents to represent environmental hazards and resources within their neighborhoods visually. This technique has been shown to enhance community engagement by providing a tangible representation of local issues. For instance, Wang et al. describe how community mapping was utilized to understand environmental health concerns in both urban and rural settings, effectively engaging residents in the identification of local priorities (Wang et al. 2021a, b). Moreover, Adhikari et al. note that systematic reviews of community engagement highlight the need for detailed survey methodologies to ensure comprehensive community participation (Adhikari et al. 2016).

Participatory Action Research

Participatory Action Research (PAR) involves communities in the research process, empowering them to identify problems and develop solutions actively. This approach is efficient in environmental health, fostering collaboration between researchers and community members. Case studies have demonstrated the effectiveness of PAR in addressing environmental health issues. For example, Sahan et al. (2017) discuss how PAR was employed in Myanmar to engage communities in malaria treatment initiatives, illustrating the importance of understanding local social contexts and building trust. Similarly, Cantu et al. (2015) highlight the role of community-based research in Texas borderland colonias, emphasizing how community participation can lead to more relevant and effective health interventions. By involving communities in the research process, PAR enhances the relevance of findings and builds local capacity for addressing environmental health challenges.

Citizen Science and Crowdsourcing

Citizen science and crowdsourcing empower residents to collect and share data related to environmental health. These approaches leverage the knowledge and skills of community members, enabling them to contribute to scientific research and decision-making processes. Platforms for citizen science, such as mobile applications and online databases, facilitate data collection on environmental hazards, pollution levels, and health outcomes. Bernhard et al. (2013) illustrate how citizen science initiatives can identify ecological health priorities in underserved populations, demonstrating the potential for community-driven data collection to inform public health strategies. Furthermore, Nathan et al. (2013) emphasize the role of community representatives in health services, showcasing how citizen science can enhance community engagement and improve health outcomes. By integrating citizen science into environmental health initiatives, communities can play a pivotal role in monitoring their environments and advocating for necessary changes.

11.6 Case Studies of Community Engagement in Environmental Health Initiatives

Community engagement in environmental health initiatives is critical for addressing local concerns and fostering sustainable practices. This response presents four case studies that exemplify effective community engagement strategies in environmental health: Community Air Quality Monitoring, Water Quality Monitoring by Local Communities, Urban Green Space Development through Participatory Planning, and Waste Management and Recycling Programs.

Community Air Quality Monitoring

Engaging residents in tracking and reducing air pollution has become increasingly important as urban areas face rising pollution levels. Community-led air quality monitoring initiatives empower residents to actively participate in data collection and analysis actively, thereby enhancing their awareness of local air quality issues. For instance, Haklay and Eleta (2018) discuss how community-led air quality monitoring projects can complement government efforts by addressing specific local concerns, such as increased traffic due to construction projects. These initiatives often utilize low-cost sensors that allow communities to gather real-time data on air pollutants, making the monitoring process more accessible and participatory. Morawska et al. (2018) highlight the transformative potential of low-cost sensing technologies in democratizing air quality monitoring, enabling communities to engage in discussions about environmental health and advocate for necessary changes. By fostering community involvement, these projects improve air quality data, strengthen community ties, and empower residents to take action.

Water Quality Monitoring by Local Communities

Community-led initiatives to monitor and improve water sources are vital for ensuring safe drinking water and promoting public health. In many regions, local communities have taken the initiative to assess water quality, identify contaminants, and advocate for cleaner water sources. Wang et al. (2021a, b) illustrate how community engagement in water quality monitoring can lead to significant improvements in public health outcomes, particularly in underserved populations. These initiatives often involve training community members to use testing kits and interpret results, fostering a sense of ownership and responsibility for local water resources. Bernhard et al. (2013) emphasize the importance of understanding community definitions of environmental issues, as these insights can guide effective interventions and policy changes. Communities can better address their unique challenges and advocate for improved water management practices by actively involving residents in water quality monitoring.

Urban Green Space Development Through Participatory Planning

Developing urban green spaces through participatory planning is another practical approach to community engagement in environmental health. Engaging residents in the co-design of parks and green spaces enhances the aesthetic value of neighborhoods and promotes physical and mental well-being. Participatory planning allows community members to express their preferences and priorities regarding green space design, ensuring that these areas meet local needs. Nieuwenhuijsen et al. (2017) discuss how participatory quantitative health impact assessments can inform urban planning decisions, leading to healthier environments. By involving residents in planning, cities can create green spaces that foster community interaction, improve air quality, and provide recreational opportunities. This collaborative approach not only enhances the livability of urban areas but also empowers residents to take an active role in shaping their environments.

Waste Management and Recycling Programs

Community-driven waste reduction and resource recovery solutions are essential for promoting sustainable practices and reducing environmental impacts. Engaging residents in waste management initiatives fosters a sense of community responsibility and encourages sustainable behaviors. Local organizations often facilitate workshops and educational campaigns to raise awareness about recycling and waste reduction strategies. Agnew et al. (2012) highlight the effectiveness of community organizations in promoting health initiatives, demonstrating how these groups can serve as catalysts for change. By involving residents in designing and implementing waste management programs, communities can tailor solutions to their specific needs and preferences, ultimately leading to more effective outcomes. Furthermore, citizen science initiatives can enhance community engagement in waste monitoring, allowing residents to track waste generation and recycling rates, fostering accountability, and encouraging participation (Farooqui et al. 2023).

11.7 Challenges in Implementing Participatory Environmental Health Management Projects

Implementing participatory environmental health projects presents several challenges that can hinder their effectiveness and sustainability. One of the primary challenges in implementing community-led participatory environmental health projects is the limited resources and funding. Many community initiatives struggle to secure adequate financial support, which can impede their ability to engage residents and sustain activities over time effectively. Kretchy et al. (2022) highlight that poverty and lack of resources are significant barriers to the successful implementation of participatory health projects in Ghana, where financial constraints often limit the scope and impact of interventions. Similarly, Gelius et al. (2021) discuss the economic evaluation of participatory approaches, noting that while these interventions can yield significant benefits, the additional costs associated with participatory components can complicate cost-effectiveness analyses. To address these financial constraints, community organizations must explore diverse funding sources, including grants, partnerships with local businesses, and collaborations with governmental and non-governmental organizations. Moreover, leveraging community assets and resources can help to mitigate funding challenges and enhance project sustainability.

Another significant challenge is sustaining interest and participation in participatory environmental health projects over the long term. Initial enthusiasm may wane as community members become preoccupied with other commitments or as project activities evolve. Warren et al. (2017) emphasize the importance of ongoing engagement strategies to maintain community interest in health initiatives. Effective strategies may include:

- Regular communication with community members.
- Providing updates on project progress.
- Celebrating successes reinforces the value of participation.

Additionally, incorporating community feedback into project planning and implementation can help to ensure that initiatives remain relevant and responsive to local needs. Arnold et al. (2022) suggest that fostering strong relationships between researchers and community members can enhance trust and commitment, ultimately leading to sustained engagement. By prioritizing long-term relationships and ongoing communication, participatory projects can better navigate the challenges of maintaining community involvement.

Power dynamics can significantly impact the effectiveness of participatory environmental health projects. Balancing community influence with input from external stakeholders, such as researchers and policymakers, can be challenging. Egid et al. (2021) discuss the complex role of gatekeepers in participatory research, noting that individuals or organizations that control access to communities can inadvertently marginalize certain voices. To address these power imbalances, it is crucial to establish clear communication channels and decision-making processes that prioritize community input. This may involve creating advisory boards with diverse community representatives, ensuring all voices are heard and valued. Additionally, Sørensen (2016) emphasizes the need for participatory interventions to be embedded within the community context, which can help to mitigate power imbalances and foster a more equitable participatory process. Participatory projects can enhance their legitimacy and effectiveness by actively addressing power dynamics.

Ensuring inclusivity and representation within participatory initiatives is vital for their success. Marginalization of certain groups can lead to incomplete or biased outcomes, undermining the overall effectiveness of the project. Horsham et al. (2014) highlight that participatory approaches must be designed to engage diverse populations, particularly those often overlooked in traditional health initiatives. Strategies to promote inclusivity may include targeted outreach efforts to engage underrepresented groups, providing language support, and creating accessible meeting formats. Nawrath (2024) discusses the use of participatory video to enhance representation in environmental research, emphasizing the importance of employing techniques that promote equitable participation.

11.8 Strategies for Effective Participatory Environmental Health Programs

Effective participatory environmental health programs are essential for addressing community-specific health concerns and fostering sustainable practices (Table 11.4). Collaborating with non-governmental organizations (NGOs), government agencies, and the private sector is crucial for the success of participatory environmental health programs. Strong partnerships can provide the necessary resources, expertise, and networks to enhance program implementation. Lichtveld et al. (2016) highlight the benefits of engaging community partners throughout the research process, from design to dissemination, which can improve the relevance of findings and foster trust among stakeholders. By leveraging the strengths of various partners, participatory programs can achieve greater reach and effectiveness. Also, setting objectives that reflect community priorities is essential for ensuring that participatory environmental health programs are relevant and practical. Engaging community members in the goal-setting process can help identify pressing health concerns and ensure interventions are tailored to local needs. Johnston et al. (2019) discuss how participatory air monitoring initiatives can empower communities by involving them in defining their environmental health priorities. This approach fosters ownership and enhances the likelihood of successful implementation. Furthermore, Grieger and Cummings (2022) emphasize the importance of aligning program goals with community values and needs, which can lead to more meaningful outcomes and sustained engagement.

Providing skills and knowledge for ongoing project management is vital for the sustainability of participatory environmental health programs. Training community members in relevant skills, such as data collection, analysis, and project management, can empower them to participate actively in program implementation. Addison et al. (2021) highlight the significance of capacity-building initiatives in fostering community engagement and improving health outcomes. Additionally, Oni et al. (2023) emphasize the importance of tailored training programs that address the specific needs of community members, which can enhance their ability to manage and sustain health initiatives. By investing in capacity building, programs can ensure that communities are equipped to address their environmental health challenges effectively. Leveraging digital communication and data collection tools can significantly enhance participation in environmental health programs. Technology can facilitate real-time data sharing, improve stakeholder communication, and streamline project management. Carson et al. discuss the role of digital platforms in promoting community engagement and enhancing the effectiveness of environmental health initiatives (Carson et al. 2022).

Furthermore, using mobile applications for data collection can empower community members to participate actively in monitoring their environments. This approach increases engagement and fosters a sense of ownership and responsibility among participants. Developing metrics to assess project impact and sustainability is crucial for ensuring the effectiveness of participatory environmental health programs. Continuous monitoring and evaluation can provide valuable insights into program performance and inform necessary adjustments. Haynes et al. (2016) emphasize the importance of community engagement in the evaluation process, noting that involving community members in assessing program outcomes can enhance the relevance and applicability of findings. Additionally, establishing clear evaluation metrics can help to measure changes in health outcomes, community engagement levels, and overall program effectiveness. By prioritizing monitoring and evaluation, participatory programs can ensure accountability and demonstrate their impact on community health.

Table 11.4 Techniques for effective participatory environmental health programs

Strategy	Approaches involved	Examples
Building strong partnerships	Establishing collaborations with local organizations, NGOs, universities, and government agencies to enhance resources and expertise	Partnering with local health departments and environmental NGOs to run community health screenings and pollution monitoring programs
Developing clear and achievable goals	Setting specific, measurable objectives that reflect the community's priorities and needs	Working with community members to identify primary environmental health concerns, such as air quality, and establishing pollution reduction targets
Training and capacity building	Providing community members with the skills and knowledge needed for active participation in environmental health projects	Hosting workshops on environmental data collection, such as water quality testing, empowering residents to monitor local waterways
Using technology to enhance participation	Leveraging digital tools, mobile apps, and social media to increase accessibility and engagement in environmental health initiatives	Implementing mobile apps for reporting pollution incidents or using social media platforms to share updates on environmental health projects
Ensuring inclusivity and representation	Actively involving diverse community groups, especially marginalized populations, in all stages of the project	Engaging local youth, elderly, and underrepresented groups in project planning meetings to ensure all voices are heard and considered
Ongoing communication and feedback	Establishing regular updates, open forums, and feedback loops to keep the community informed and involved	Sending out monthly newsletters on project progress and organizing quarterly community meetings for updates and feedback collection
Resource allocation and accessibility	Ensuring that participants have access to necessary resources, such as materials, funding, and information, to support their involvement	Providing monitoring equipment like air quality sensors and small grants for community-led environmental initiatives
Establishing monitoring and evaluation metrics	Developing metrics to assess the success and impact of projects, with community involvement in the evaluation process	Creating a dashboard to track reductions in local air pollution levels and inviting community members to participate in project evaluations

(continued)

Table 11.4 (continued)

Strategy	Approaches involved	Examples
Fostering community empowerment and ownership	Encouraging community members to take leadership roles and share responsibilities in environmental health projects	Forming community committees to oversee project implementation, enabling residents to lead waste reduction or water conservation initiatives
Encouraging policy advocacy and civic engagement	Supporting communities in advocating for environmental health policies that reflect their needs and priorities	Assisting residents in organizing campaigns for stricter air quality regulations or attending city council meetings to address environmental concerns

11.9 Future Directions for Community-Engaged Environmental Health Solutions

Future directions for community-engaged environmental health solutions are critical for enhancing the effectiveness and sustainability of initiatives to improve public health outcomes. Innovative participatory approaches are essential for fostering community engagement and ensuring that environmental health initiatives are responsive to local needs. Eslit (2023) emphasizes the importance of empowering grassroots initiatives by increasing awareness and education on environmental issues, which can lead to more effective community-led projects. Additionally, as discussed by Douglas et al. (2020), participatory mapping techniques can diagnose upstream determinants of health and inform policy-based interventions. These methods enhance community involvement and provide valuable data that can guide decision-making processes. By adopting innovative participatory strategies, stakeholders can create more inclusive and effective environmental health programs. Integrating technology and citizen science into community-engaged environmental health initiatives can significantly enhance participation and data collection. Digital tools facilitate real-time communication and enable community members to contribute to data gathering and analysis. For instance, as highlighted by Jelks et al. (2018), citizen science initiatives allow residents to collect spatial data on environmental stressors, empowering them to address local health concerns. Furthermore, as discussed by Fotopoulou et al., participatory data science can help prioritize stakeholders and ensure that community voices are heard in health-related projects (Fotopoulou et al. 2021). By leveraging technology and citizen science, community-engaged environmental health solutions can become more robust and data-driven, ultimately leading to improved health outcomes.

Scaling community-led initiatives is crucial for maximizing their impact on environmental health. Eslit (2023) suggests that strengthening collaboration between community members and relevant stakeholders can enhance the effectiveness and sustainability of these initiatives. Successful community-led projects can serve as models for replication in other contexts, particularly when they demonstrate tangible health benefits. For instance, the participatory approach to educating for sustainable development discussed by Sousa (2021) illustrates how community engagement can yield social learning indicators contributing to broader environmental health goals. By sharing best practices and lessons learned, communities can inspire and support one another in scaling their initiatives to achieve more significant public health impact. To effectively support participatory environmental health initiatives, policymakers must create an enabling environment that fosters community engagement. Recommendations include actively engaging policymakers in developing supportive policies and increasing funding for community-led projects. Lichtveld et al. (2019) emphasize the need for validated scales of environmental health literacy to inform policy decisions and enhance community-academic partnerships. Furthermore, integrating participatory approaches into public health policy can ensure that community voices are prioritized in decision-making processes. By advocating for policies that support participatory environmental health initiatives, stakeholders can create a more equitable and responsive public health landscape.

11.10 Conclusion

Participatory approaches to environmental health are essential for engaging communities in sustainable solutions. By fostering collaboration among stakeholders and actively empowering residents to address environmental health challenges, we can promote healthier urban environments and improve public health outcomes. Integrating Nature-Based Solutions into urban planning further enhances these efforts, providing multiple benefits for air quality, heat resilience, mental health, and community cohesion. As cities face environmental pressures, prioritizing participatory strategies and nature-based solutions will be crucial for creating resilient and sustainable urban environments. Effective community engagement in environmental health requires a comprehensive approach incorporating stakeholder identification, participatory planning, education, and community-led evaluation. By prioritizing these components, environmental health initiatives can be more inclusive, effective, and sustainable, ultimately leading to healthier communities. The approaches and techniques for community engagement in environmental health—workshops and focus groups, surveys and community mapping, participatory action research, and citizen science—are essential for fostering collaboration and ensuring community voices are integrated into decision-making processes. By employing these methodologies, environmental health initiatives can become more inclusive, effective, and responsive to the needs of the communities they serve. Communities can address local environmental challenges and promote healthier living conditions by actively involving residents in air quality monitoring, water quality assessments, urban green space development, and waste management programs. These initiatives empower residents and foster collaboration between community members, researchers, and policymakers, ultimately leading to more sustainable and effective environmental health solutions. While participatory environmental health projects offer significant community engagement and empowerment potential, they also face several challenges. Addressing resource and funding limitations, maintaining long-term engagement, navigating power dynamics, and ensuring inclusivity and representation are critical for the success of these initiatives. Implementing effective participatory environmental health programs requires a multifaceted approach that includes building strong partnerships, developing clear goals, providing training and capacity building, leveraging technology, and prioritizing monitoring and evaluation. By employing these strategies, stakeholders can enhance community engagement, improve health outcomes, and foster sustainable practices in environmental health initiatives.

References

Addison C, Jenkins B, White M, Odom D, Fortenberry M, Wilson G et al (2021) Twenty years of leading the way among cohort studies in community-driven outreach and engagement: Jackson State University/Jackson heart study. Int J Environ Res Public Health 18(2):696. https://doi.org/10.3390/ijerph18020696

Adger WN (2000) Social and ecological resilience: are they related? Prog Hum Geogr 24(3):347–364. https://doi.org/10.1191/030913200701540465

Adhikari B, James N, Newby G, Seidlein L, White N, Day N et al (2016) Community engagement and population coverage in mass anti-malarial administrations: a systematic literature review. Malar J 15(1). https://doi.org/10.1186/s12936-016-1593-y

Agnew R, Heuvel E, Tannenbaum C (2012) Efficiency of using community organisations as catalysts for recruitment to continence promotion trials. Clin Trials 10(1):151–159. https://doi.org/10.1177/1740774512460144

Agyeman J (2005) Sustainable communities and the challenge of environmental justice. NYU Press

Allen S, Fanucchi M, McCormick L, Zierold K (2019) The search for environmental justice: the story of north Birmingham. Int J Environ Res Public Health 16(12):2117. https://doi.org/10.3390/ijerph16122117

Anwar A (2023) Blockchain-based clinical trials: a meta-model framework for enhancing security and transparency with a novel algorithm. Int J Technol 14(6):1380. https://doi.org/10.14716/ijtech.v14i6.6703

Arnold D, Becker H, Böttger T, Sarma N, Bethmann A, Narimani P (2022) "What do you need? What are you experiencing?" relationship building and power dynamics in participatory research projects: critical self-reflections of researchers. Int J Environ Res Public Health 19(15):9336. https://doi.org/10.3390/ijerph19159336

Berkes F, Colding J, Folke C (2000) Rediscovery of traditional ecological knowledge as adaptive management. Ecol Appl 10(5):1251–1262. https://doi.org/10.1890/1051-0761(2000)010[1251:ROTEKA]2.0.CO;2

Bernhard M, Evans M, Kent S, Johnson E, Threadgill S, Tyson S et al (2013) Identifying environmental health priorities in underserved populations: a study of rural versus urban communities. Public Health 127(11):994–1004. https://doi.org/10.1016/j.puhe.2013.08.005

Brulle RJ, Pellow DN (2006) Environmental justice: human health and environmental inequalities. Annu Rev Public Health 27:103–124. https://doi.org/10.1146/annurev.publhealth.27.021405.102124

Bullard RD (2000) Dumping in dixie: race, class, and environmental quality. Westview Press

Cantu R, Graham M, Millard A, Flores I, Bsn R, Reyes I et al (2015) Environmental justice and community-based research in Texas borderland colonies. Public Health Nurs 33(1):65–72. https://doi.org/10.1111/phn.12187

Carr J, Lalara J, Lalara G, Lalara G, Daniels B, Clough A et al (2021) Staying strong toolbox: co-design of a physical activity and lifestyle program for aboriginal families with Machado-Joseph disease in the top end of Australia. PLoS ONE 16(2):e0244311. https://doi.org/10.1371/journal.pone.0244311

Carson M, Doberneck D, Hart Z, Kelsey H, Pierce J, Porter D et al (2022) A strategic framework for community engagement in oceans and human health. Community Sci 1(1). https://doi.org/10.1029/2022csj000001

Casson N, Cameron L, Mauro I, Friesen-Hughes K, Rocque R (2023) Perceptions of the health impacts of climate change among Canadians. BMC Public Health 23(1). https://doi.org/10.1186/s12889-023-15105-z

Cláudio L, Gilmore J, Roy M, Brenner B (2018) Communicating environmental exposure results and health information in a community-based participatory research study. BMC Public Health 18(1). https://doi.org/10.1186/s12889-018-5721-1

Cuerrier A, Brunet N, Gérin-Lajoie J, Downing A, Lévesque E (2015) The study of Inuit knowledge of climate change in Nunavik, Quebec: a mixed methods approach. Hum Ecol 43(3):379–394. https://doi.org/10.1007/s10745-015-9750-4

Douglas J, Subica A, Franks L, Johnson G, Leon C, Villanueva S et al (2020) Using participatory mapping to diagnose upstream determinants of health and prescribe downstream policy-based interventions. Prev Chronic Dis 17. https://doi.org/10.5888/pcd17.200123

Egid B, Roura M, Aktar B, Quach J, Chumo I, Dias S et al (2021) 'You want to deal with power while riding on power': global perspectives on power in participatory health research and co-production approaches. BMJ Glob Health 6(11):e006978. https://doi.org/10.1136/bmjgh-2021-006978

Ellingsen K, Yoccoz N, Tveraa T, Hewitt J, Thrush S (2017) Long-term environmental monitoring for assessment of change: measurement inconsistencies over time and potential solutions. Environ Monit Assess 189(11). https://doi.org/10.1007/s10661-017-6317-4

English P, Olmedo L, Bejarano E, Lugo H, Murillo E, Seto E et al (2017) The imperial county community air monitoring network: a model for community-based environmental monitoring for public health action. Environ Health Perspect 125(7). https://doi.org/10.1289/ehp1772

Eslit E (2023) Empowering change at the grassroots: community-led initiatives for local environmental protection. https://doi.org/10.20944/preprints202307.1055.v1

Farooqui Z, Biswas J, Saha J (2023) Long-term assessment of purpleair low-cost sensor for PM2.5 in California, USA. https://doi.org/10.20944/preprints202307.0135.v1

Folke C, Carpenter S, Walker B, Scheffer M, Elmqvist T, Gunderson L, Holling CS (2005) Regime shifts, resilience, and biodiversity in ecosystem management. Annu Rev Ecol Evol Syst 35(1):557–581. https://doi.org/10.1146/annurev.ecolsys.35.021103.105711

Fotopoulou A, Barratt H, Marandet E (2021) A data-based participatory approach for health equity and digital inclusion: prioritizing stakeholders. Health Promot Int 38(4). https://doi.org/10.1093/heapro/daab166

Franz B, Cronin C, Yeager V, Burns A, Singh S (2022) Overview of the most commonly identified public health needs and strategies in a nationally representative sample of nonprofit hospitals. Med Care Res Rev 80(3):342–351. https://doi.org/10.1177/10775587221135365

Frumkin H (2010) Environmental health: from global to local. John Wiley & Sons

Galway L (2016) Boiling over: a descriptive analysis of drinking water advisories in first nations communities in Ontario, Canada. Int J Environ Res Public Health 13(5):505. https://doi.org/10.3390/ijerph13050505

Gelius P, Sommer R, Abu-Omar K, Schätzlein V, Suhrcke M (2021) Toward the economic evaluation of participatory approaches in health promotion: lessons from four German physical activity promotion projects. Health Promot Int 36(Suppl 2):ii79–ii92. https://doi.org/10.1093/heapro/daab158

Gray K (2018) From content knowledge to community change: a review of representations of environmental health literacy. Int J Environ Res Public Health 15(3):466. https://doi.org/10.3390/ijerph15030466

Grieger K, Cummings C (2022) Informing environmental health and risk priorities through local outreach and extension. Environ Syst Decis 42(3):388–401. https://doi.org/10.1007/s10669-022-09864-0

Haklay M, Eleta I (2018) On the front line of community-led air quality monitoring, pp 563–580. https://doi.org/10.1007/978-3-319-74983-9_27

Haynes E, Elam S, Burns R, Spencer A, Yancey E, Kuhnell P et al (2016) Community engagement and data disclosure in environmental health research. Environ Health Perspect 124(2). https://doi.org/10.1289/ehp.1510411

Hoover E (2016) "We're not going to be guinea pigs;" citizen science and environmental health in a Native American Community. J Sci Commun 15(01):A05. https://doi.org/10.22323/2.15010205

Horowitz CR, Robinson M, Seifer S (2009) Community-based participatory research from the margin to the mainstream: are researchers prepared? Circulation 119(19):2633–2642. https://doi.org/10.1161/CIRCULATIONAHA.107.729863

Horsham C, Auster J, Sendall M, Stoneham M, Youl P, Crane P et al (2014). Interventions to decrease skin cancer risk in outdoor workers: update to a 2007 systematic review. BMC Res Notes 7(1). https://doi.org/10.1186/1756-0500-7-10

Ishom M, Raharjo K, Haidar M (2023) Facilitative learning strategies in community empowerment programs originated on local literature skills in realizing community-centralized sustainable development, pp 373–379. https://doi.org/10.2991/978-2-494069-95-4_44

Israel BA, Eng E, Schulz AJ, Parker EA (eds) (2010) Methods for community-based participatory research for health. John Wiley & Sons

Jelks N, Hawthorne T, Dai D, Fuller C, Stauber C (2018) Mapping the hidden hazards: community-led spatial data collection of street-level environmental stressors in a degraded, urban watershed. Int J Environ Res Public Health 15(4):825. https://doi.org/10.3390/ijerph15040825

Johnston J, Juarez Z, Navarro S, Hernandez A, Gutschow W (2019) Youth engaged participatory air monitoring: a 'day in the life' in urban environmental justice communities. Int J Environ Res Public Health 17(1):93. https://doi.org/10.3390/ijerph17010093

Kretchy I, Okoibhole L, Sanuade O, Jennings H, Strachan D, Blandford A et al (2022) Scoping review of community health participatory research projects in Ghana. Glob Health Action 15(1). https://doi.org/10.1080/16549716.2022.2122304

Lee D, Park M, Baek Y, Bae B, Heo J (2022) In-sensor image memorization and encoding via optical neurons for bio-stimulus domain reduction toward visual cognitive processing. Nat Commun 13(1). https://doi.org/10.1038/s41467-022-32790-3

Lichtveld M, Covert H, Sherman M, Shankar A, Wickliffe J, Alcala C (2019) Advancing environmental health literacy: validated scales of general environmental health and environmental media-specific knowledge, attitudes and behaviors. Int J Environ Res Public Health 16(21):4157. https://doi.org/10.3390/ijerph16214157

Lichtveld M, Goldstein B, Grattan L, Mundorf C (2016) Then and now: lessons learned from community—academic partnerships in environmental health research. Environ Health 15(1). https://doi.org/10.1186/s12940-016-0201-5

Madrigal D, Minkler M, Parra K, Mundo C, Gonzalez J, Jiménez R et al (2016) Improving Latino youths' environmental health literacy and leadership skills through participatory research on chemical exposures in cosmetics. Int Q Community Health Educ 36(4):231–240. https://doi.org/10.1177/0272684x16657734

Marsili D (2023). Environmental public health communication to engage stakeholders and foster social capacity in poorly involved communities living in industrial contaminated sites: the case study of Porto Torres (Italy). Front Commun 8. https://doi.org/10.3389/fcomm.2023.1217427

Martínez L, Rubin C, Russell B, Leslie L, Brugge D (2011) Community conceptualizations of health: implications for transdisciplinary team science. Clin Transl Sci 4(3):163–167. https://doi.org/10.1111/j.1752-8062.2011.00289.x
Minkler M, Wallerstein N (2011) Community-based participatory research for health: from process to outcomes. John Wiley & Sons
Morawska L, Thai P, Liu X, Asumadu-Sakyi A, Ayoko G, Bartoňová A et al (2018) Applications of low-cost sensing technologies for air quality monitoring and exposure assessment: how far have they gone? Environ Int 116:286–299. https://doi.org/10.1016/j.envint.2018.04.018
Nathan S, Braithwaite J, Stephenson N (2013) Facilitating the action of community representatives in a health service: the role of a community participation coordinator. BMC Health Serv Res 13(1). https://doi.org/10.1186/1472-6963-13-154
Nawrath M (2024) Using participatory video in environmental research. People Nat 6(4):1382–1393. https://doi.org/10.1002/pan3.10646
Nieuwenhuijsen M, Khreis H, Verlinghieri E, Mueller N, Rojas-Rueda D (2017) Participatory quantitative health impact assessment of urban and transport planning in cities: a review and research needs. Environ Int 103:61–72. https://doi.org/10.1016/j.envint.2017.03.022
Oetzel J, Villegas M, Zenone H, Wallerstein N, Duran B (2015) Enhancing stewardship of community-engaged research through governance. Am J Public Health 105(6):1161–1167. https://doi.org/10.2105/ajph.2014.302457
Oni B, Martin D, Bonnycastle M, Wood N, Thompson S (2023) Mino bimaadiziwin homebuilder program's impact on sustainable livelihoods among youth in Garden Hill and Wasagamack first nations: an evaluative study. Can J Nonprofit Soc Econ Res 14(S1). https://doi.org/10.29173/cjnser554
Ostrom E (2009) A general framework for analyzing sustainability of social-ecological systems. Science 325(5939):419–422. https://doi.org/10.1126/science.1172133
Pasetto R, Mattioli B, Marsili D (2019) Environmental justice in industrially contaminated sites. A review of scientific evidence in the who European region. Int J Environ Res Public Health 16(6):998. https://doi.org/10.3390/ijerph16060998
Pellow DN (2018) What is critical environmental justice? John Wiley & Sons
Qasim R, Farooqui W, Rahman A, Haroon R, Saleem M, Rafique M et al (2023) Community centred co-design methodology for designing and implementing socio-behavioural interventions to counter covid-19 related misinformation among marginalized population living in the squatter settlements of Karachi, Pakistan: a methodology paper. BMC Proc 17(S7). https://doi.org/10.1186/s12919-023-00265-y
Ramírez A, Ramondt S, Bogart K, Perez-Zuniga R (2019) Public awareness of air pollution and health threats: challenges and opportunities for communication strategies to improve environmental health literacy. J Health Commun 24(1):75–83. https://doi.org/10.1080/10810730.2019.1574320
Ramírez-Andreotta M, Brody J, Lothrop N, Loh M, Beamer P, Brown P (2016) Improving environmental health literacy and justice through environmental exposure results communication. Int J Environ Res Public Health 13(7):690. https://doi.org/10.3390/ijerph13070690
Ravenscroft J, Schell L, Cole T (2014) Applying the community partnership approach to human biology research. Am J Hum Biol 27(1):6–15. https://doi.org/10.1002/ajhb.22652
Sahan K, Pell C, Smithuis F, Phyo A, Maung S, Indrasuta C et al (2017) Community engagement and the social context of targeted malaria treatment: a qualitative study in Kayin (Karen) state, Myanmar. Malar J 16(1). https://doi.org/10.1186/s12936-017-1718-y
Schulz AJ, Israel BA, Lantz PM (2018) Assessing and strengthening characteristics of effective groups in community-based participatory research partnerships. In: Community-based participatory research for health: advancing social and health equity. John Wiley & Sons
Schure M, Kile M, Harding A, Harper B, Harris S, Uesugi S et al (2013) Perceptions of the environment and health among members of the confederated tribes of the Umatilla Indian reservation. Environ Justice 6(3):115–120. https://doi.org/10.1089/env.2013.0022

Setiawan A, Ramdhon A, Utami L (2022) Establishing a co-design framework for disaster mitigation agenda in the urban context. A case study: Sibat Solo. IOP Conf Ser Earth Environ Sci 1114(1):012025. https://doi.org/10.1088/1755-1315/1114/1/012025

Sloot D, Jans L, Steg L (2018) Can community energy initiatives motivate sustainable energy behaviours? The role of initiative involvement and personal pro-environmental motivation. J Environ Psychol 57:99–106. https://doi.org/10.1016/j.jenvp.2018.06.007

Sørensen K, Broucke S, Fullam J, Doyle G, Pelikan J, Słońska Z et al (2012) Health literacy and public health: a systematic review and integration of definitions and models. BMC Public Health 12(1). https://doi.org/10.1186/1471-2458-12-80

Sørensen O (2016) Improving the primary task. J Organ Eff People Perform 3(4):343–359. https://doi.org/10.1108/joepp-02-2016-0017

Sousa L (2021) Learning experiences of a participatory approach to educating for sustainable development in a South African higher education institution yielding social learning indicators. Sustainability 13(6):3210. https://doi.org/10.3390/su13063210

Souza R, Aguilar G, Castro A (2013) Creating meaningful partnerships between communities and environmental health researchers: the role of a direct action community organizing agency. Work Health Saf 61(8):347–352. https://doi.org/10.3928/21650799-20130716-05

Sprague N (2023) Environmental health perceptions of urban youth from low-income communities: a qualitative photovoice study and framework. Health Expect 26(5):1832–1842. https://doi.org/10.1111/hex.13776

Taylor DE (2014) Toxic communities: environmental racism, industrial pollution, and residential mobility. NYU Press

Wang A, Lin W, Liu B, Wang H, Xu H (2021a) Does smart city construction improve the green utilization efficiency of urban land? Land 10(6):657. https://doi.org/10.3390/land10060657

Wang S, Richardson M, Evans M, Johnson E, Threadgill-Matthews S, Tyson S et al (2021b) A community-engaged approach to understanding environmental health concerns and solutions in urban and rural communities. BMC Public Health 21(1). https://doi.org/10.1186/s12889-021-11799-1

Warren C, Ndwiga C, Sripad P, Medich M, Njeru A, Maranga A et al (2017) Sowing the seeds of transformative practice to actualize women's rights to respectful maternity care: reflections from Kenya using the consolidated framework for implementation research. BMC Women S Health 17(1). https://doi.org/10.1186/s12905-017-0425-8

Watkins J (2023) Human-centered participatory co-design with children and adults for a prototype lifestyle intervention and implementation strategy in a rural middle school. https://doi.org/10.21203/rs.3.rs-3711510/v1

Yadav U, Lloyd J, Baral K, Bhatta N, Mehata S, Harris M (2021) Evaluating the feasibility and acceptability of a co-design approach to developing an integrated model of care for people with multi-morbid COPD in rural Nepal: a qualitative study. BMJ Open 11(1):e045175. https://doi.org/10.1136/bmjopen-2020-045175

Zulu J, Mwamba T, Rosen A, Matenga T, Mulanda J, Kaimba M et al (2021) Community engagement for the voluntary medical male circumcision (VMMC) program: an analysis of key stakeholder roles to promote a sustainable program in Zambia. https://doi.org/10.21203/rs.3.rs-295480/v1

Chapter 12
Risk Assessment and Health Impact Studies: Strategic Tools for Managing Environmental Health

Sylvester Chibueze Izah and Matthew Chidozie Ogwu

Abstract Risk assessment and health impact studies serve as strategic tools for managing environmental health by offering structured approaches to identify, evaluate, and mitigate health risks associated with environmental exposures. These methodologies, including hazard identification, exposure assessment, dose–response relationships, and risk characterization, provide essential insights for decision-making in public health, environmental management, and occupational safety. Health Impact Assessments (HIAs), in particular, examine both direct and indirect health effects of policies and projects, delivering evidence-based recommendations that promote health equity and community well-being. The flexibility of HIAs, alongside active stakeholder participation, ensures comprehensive consideration of diverse health outcomes. Technology and data analytics advances further enhance the accuracy and responsiveness of risk assessment processes, enabling real-time analysis and predictive evaluations of emerging health risks. This evolution of risk assessment tools is critical for shaping future environmental health policies that are both scientifically grounded and capable of addressing complex environmental challenges. Nevertheless, these assessments face data uncertainty, public perception issues, and resource limitations. Overcoming these challenges demands improved data collection, enhanced risk communication, and better resource allocation for health professionals. Ultimately, integrating technological innovation, interdisciplinary collaboration, and proactive risk management is transforming the landscape of environmental

S. Chibueze Izah
Department of Community Medicine, Faculty of Clinical Sciences, Bayelsa Medical University, Yenagoa, Bayelsa State, Nigeria

Department of Microbiology, Faculty of Science, Bayelsa Medical University, Yenagoa, Bayelsa State, Nigeria

S. Chibueze Izah
e-mail: sylvester.izah@bmu.edu.ng

M. C. Ogwu (✉)
Goodnight Family Department of Sustainable Development, Appalachian State University, 212 Living Learning Center, 305 Bodenheimer Drive, Boone, NC 28608, USA
e-mail: ogwumc@appstate.edu

S. Chibueze Izah and M. Chidozie Ogwu (eds.), *Innovative Approaches in Environmental Health Management*, Environmental Science and Engineering,
https://doi.org/10.1007/978-3-031-81966-7_12

health management. As the field evolves, these strategic tools will play a vital role in advancing public health, promoting sustainable practices, and addressing health inequities in the face of increasing environmental complexities.

Keywords Risk assessment · Health impact · Environmental health · Assessments · Public health

12.1 Introduction

Risk assessment and health impact studies are critical components in understanding and managing the potential adverse effects of various environmental and health-related factors (Izah et al. 2021a, b, 2022; Uzoekwe et al. 2021; Ogamba et al. 2021; Aigberua et al. 2023, 2021). Risk assessment is defined as a systematic process for evaluating the potential risks that may be involved in a projected activity or undertaking. This process is vital for identifying hazards, assessing exposure, and determining the likelihood of adverse health outcomes, enabling informed decision-making in public health and environmental management (Krewski et al. 2014, 2022). The importance of risk assessment lies in its ability to provide a structured approach to understanding complex interactions between environmental exposures and health outcomes, thereby facilitating the development of strategies to mitigate risks and protect public health (Romero-Franco et al. 2017; Gouin et al. 2019). Risk assessment plays a crucial role in environmental health management in identifying and evaluating the potential health impacts of environmental hazards, such as pollutants, chemicals, and biological agents. Through systematic analysis of the risks associated with these hazards, health authorities can prioritize interventions and allocate resources effectively (Kaya et al. 2018; Antwi et al. 2023). Furthermore, risk assessment frameworks can incorporate various methodologies, including quantitative and qualitative approaches, to evaluate the potential health impacts of environmental exposures (Spencer et al. 2016). This comprehensive evaluation is essential for developing evidence-based policies and regulations to reduce health risks associated with environmental factors (Conlon et al. 2021).

Integrating health impact studies within the risk assessment framework enhances the understanding of how environmental factors influence health outcomes. Health impact assessments (HIAs) are used to evaluate the potential health effects of a project, policy, or program before it is implemented. This proactive approach allows stakeholders to identify potential health risks and benefits, facilitating informed decision-making (Krewski et al. 2014; Khan et al. 2023). HIAs can also engage communities in the assessment process, ensuring that the perspectives and concerns of affected populations are considered (Spencer et al. 2016; Card et al. 2013). The collaborative nature of HIAs fosters transparency and accountability in environmental health management, ultimately leading to more effective public health interventions.

Moreover, the role of risk assessment in environmental health management extends to the evaluation of cumulative risks associated with multiple exposures. Cumulative risk assessment frameworks enable the consideration of various environmental stressors and their combined effects on health outcomes (Antwi et al. 2023; Grosssteiner et al. 2023). This holistic approach is critical in addressing complex environmental issues, such as climate change and pollution, which often involve multiple interacting factors that can exacerbate health risks (Raymond et al. 2020). By assessing cumulative risks, health authorities can develop more comprehensive strategies to protect vulnerable populations and promote health equity (Kaya et al. 2018; Tang et al. 2018). Risk assessment frameworks are not limited to environmental health; they are also utilized in various sectors, including healthcare, agriculture, and industrial processes. For instance, in healthcare settings, risk assessment frameworks help identify potential hazards associated with clinical practices, thereby enhancing patient safety and quality of care (Conlon et al. 2021; Simsekler et al. 2018). Similarly, in agricultural contexts, risk assessment frameworks are employed to evaluate the health risks associated with wastewater use in irrigation, ensuring that practices are safe and sustainable (Scheierling et al. 2011). These diverse applications underscore the versatility and importance of risk assessment in safeguarding public health across different domains.

Furthermore, advancements in technology and data analytics have significantly enhanced the capabilities of risk assessment frameworks. Integrating big data, machine learning, and predictive modeling allows for more accurate risk evaluations and identifying emerging health threats (Krewski et al. 2022; Khan et al. 2023). These technological innovations enable health authorities to respond more effectively to public health challenges, such as infectious disease outbreaks and environmental disasters (Dankittipong et al. 2022). By leveraging these tools, risk assessment can become more dynamic and responsive, adapting to the evolving landscape of public health threats.

This chapter examines the strategic use of risk assessment and health impact studies to effectively manage environmental health by identifying hazards, assessing exposures, and characterizing risks. It explores the methodologies involved, their integration into policy, and the application of epidemiological data for understanding community health. Additionally, it addresses challenges such as data uncertainty and effective risk communication while examining future trends in technology, predictive modeling, and collaborative approaches to risk mitigation.

12.2 Key Concepts in Risk Assessment

Risk assessment is a systematic process crucial in identifying, evaluating, and managing potential hazards that could adversely affect human health and the environment. The key concepts in risk assessment include hazard identification, exposure assessment, dose–response relationship, and risk characterization. Each contributes to a comprehensive understanding of risks and informs decision-making processes in various fields, including public health, environmental management, and occupational safety (Table 12.1).

Although these processes outlined in Table 12.1 are linked, they are often carried out sequentially, with risk characterization central to risk management decisions (Fig. 12.1). Hazard identification is the first step in the risk assessment process, involving systematically identifying potential sources of harm or adverse effects. This process requires thoroughly examining various factors, including physical, chemical, biological, and psychosocial hazards (Saud et al. 2013; Thepaksorn et al. 2017). Effective hazard identification can utilize various tools and methodologies, such as the bow-tie diagram, which visually represents the relationship between hazards, their potential consequences, and the controls to mitigate risks (Saud et al. 2013). Furthermore, hazard identification must be continuous and adaptive, as new hazards may emerge over time due to technological changes, practices, or environmental conditions (Skripnik et al. 2023). Following hazard identification, exposure assessment evaluates the extent to which individuals or populations are exposed to identified hazards. This assessment involves quantifying the frequency, duration, and

Table 12.1 Key concepts in risk assessment

Concept	Activities	Importance in risk assessment	References
Hazard identification	The process of determining whether exposure to a chemical, substance, or environmental factor can cause harm	Essential for understanding potential health effects and environmental impacts associated with a particular hazard	Ikhajiagbe et al. (2020, 2021)
Exposure assessment	Estimating the magnitude, duration, and frequency of human or environmental exposure to the hazard	Helps quantify the risk by assessing how much, how often, and for how long populations or environments are exposed to the hazard	Izah et al. (2022), Ikhajiagbe and Ogwu (2020)
Dose–Response relationship	Describes the relationship between the amount of exposure to a hazard and the likelihood of an adverse effect	Critical for determining safe exposure levels and identifying thresholds where exposure leads to harm	Chiu and Slob (2015)
Risk characterization	Integrates hazard identification, exposure assessment, and dose–response relationship to estimate overall risk	Provides a comprehensive evaluation of the risk, including uncertainties, and supports decision-making and regulatory action	Osawaru et al. (2013a, b), Georgiadis and Younes (2021)

intensity of exposure to a hazard and identifying the pathways through which exposure occurs (Rico et al. 2016). Exposure assessment is critical for understanding the potential impact of hazards on health and the environment, as it provides the necessary data to inform subsequent steps in the risk assessment process. Geographic Information Systems (GIS) and other modeling tools can enhance exposure assessments by allowing for the visualization and analysis of spatial data related to hazard exposure (Rico et al. 2016). The dose–response relationship is a fundamental concept in risk assessment that describes the relationship between the dose of a hazard and the magnitude of the response or effect observed in exposed individuals or populations. Understanding this relationship is essential for determining the potential health effects of exposure to a hazard, as it helps to establish thresholds for safe exposure levels (Li and Wang 2023). Dose–response relationships can vary significantly among different populations and individual characteristics, necessitating the consideration of factors such as age, sex, genetic predisposition, and pre-existing health conditions when evaluating risks (Li and Wang 2023). Risk characterization integrates the information obtained from hazard identification, exposure assessment, and dose–response relationships to comprehensively assess the overall risk associated with a particular hazard. This step involves synthesizing data to estimate the likelihood and severity of adverse effects in a specified population under defined exposure scenarios (Nastos et al. 2021). Risk characterization can be qualitative or quantitative, depending on the available data and the specific context of the assessment. Clearly and understandably, it is essential to communicate risks to stakeholders, including policymakers, public health officials, and the public (Nastos et al. 2021).

The importance of these critical concepts in risk assessment cannot be overstated, as they form the foundation for effective risk management strategies. Stakeholders can make informed decisions prioritizing public health and safety by systematically identifying hazards, assessing exposure, understanding dose–response relationships,

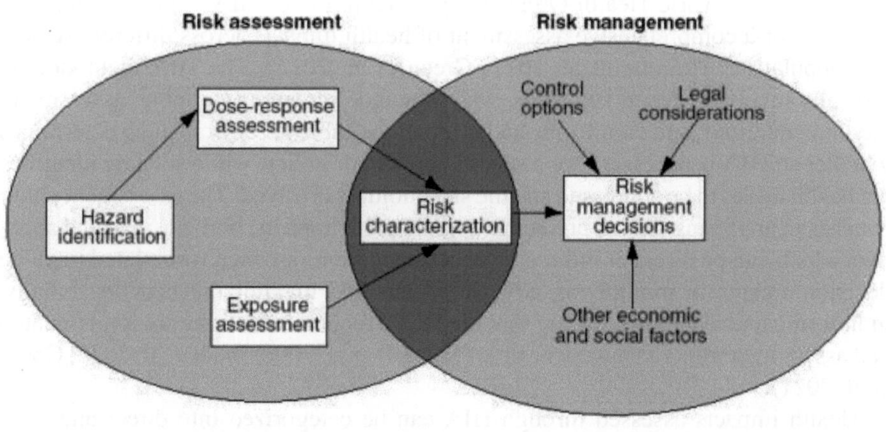

Fig. 12.1 Strategic framework of the United State's National Research Council risk assessment and risk management paradigm. *Source* Lebret (2016)

and characterizing risks (Rico et al. 2016). Moreover, integrating these concepts into regulatory frameworks and policies can help ensure appropriate measures are taken to minimize risks and protect vulnerable populations (Skripnik et al. 2023). Applying these key concepts is particularly critical from an occupational health and safety perspective. For instance, hazard identification and risk assessment methods, such as Hazard Identification Risk Assessment and Control (HIRADC), are widely used to evaluate potential hazards in various work environments (Kustono et al. 2021; Sjarifudin et al. 2023). By implementing systematic approaches to risk assessment, organizations can enhance workplace safety, reduce the incidence of occupational injuries, and improve overall employee well-being (Thepaksorn et al. 2017).

Furthermore, applying risk assessment concepts extends beyond occupational health to environmental management and public health. For example, in pesticide use, risk assessments must consider the potential ecological impacts of chemical exposure, necessitating a thorough understanding of hazard identification, exposure pathways, and dose–response relationships (Rico et al. 2016). This comprehensive approach enables the development of effective risk mitigation strategies that protect human health and the environment.

12.3 Health Impact Assessment (HIA) Framework

HIA is a systematic process that evaluates the potential health effects of a proposed policy, program, or project on a population. The methodology of HIA is designed to inform decision-makers by providing evidence-based recommendations to mitigate negative health impacts and enhance positive ones. The HIA framework integrates various health determinants, including social, environmental, and economic factors, thereby promoting a holistic approach to health in policy-making. This approach aligns with the World Health Organization's definition of HIA, which emphasizes the need for a comprehensive assessment of health impacts across different sectors and populations (Rajotte et al. 2011; Green et al. 2022). The HIA methodology typically involves several key steps, which include screening, scoping, assessment, recommendations, and monitoring and evaluation (Table 12.2). Screening determines whether an HIA is necessary for a specific project or policy, while scoping identifies the health issues to be addressed and the stakeholders involved. The assessment phase consists in analyzing data to predict potential health impacts, both positive and negative, which can be direct or indirect. Recommendations are then formulated to guide decision-makers in minimizing adverse effects and maximizing benefits. Finally, monitoring and evaluation ensure that the HIA's recommendations are implemented and assess their effectiveness over time (Mindell et al. 2008; Jabot et al. 2020; Green et al. 2021).

Health impacts assessed through HIA can be categorized into direct and indirect impacts. Direct health impacts result from the immediate effects of a policy or project, such as changes in access to healthcare services or alterations in environmental conditions that affect health. Indirect impacts, on the other hand, may arise

Table 12.2 Steps in health impact assessment

Step	Focal activities
Screening	Determines if an HIA is necessary for the proposed decision
Scoping	Establishes the framework, including which health impacts will be assessed and how
Assessment	Evaluates potential health impacts using available data, research, and stakeholder input
Recommendations	Provides strategies to promote positive health outcomes and mitigate negative impacts
Reporting	Communicates findings and recommendations to decision-makers and stakeholders
Monitoring and evaluation	Tracks the implementation of recommendations and evaluates their effectiveness

from broader social determinants of health, such as economic changes, community cohesion, or shifts in public behavior. Understanding both types of impacts is crucial for a comprehensive HIA, as they collectively influence the overall health outcomes of a population (Morteruel et al. 2020; Haigh et al. 2013; Serrano et al. 2016). The effectiveness of HIA is influenced by various factors, including the quality of stakeholder engagement, the robustness of the evidence used, and the institutional support for health considerations in policy-making. Successful HIAs often involve active participation from community members and stakeholders, enhancing the findings' relevance and applicability. Furthermore, integrating HIA into existing decision-making frameworks can facilitate a more systematic approach to health considerations in policy development (Haigh et al. 2015; Cave et al. 2021; Furber et al. 2011).

In recent years, the application of HIA has expanded globally, with increasing recognition of its value in promoting health equity and addressing health disparities. For instance, HIAs have been employed in urban planning to assess the health implications of infrastructure projects, housing developments, and transportation policies. These assessments not only inform decision-makers but also foster intersectoral collaboration, as they require input from various stakeholders, including public health officials, urban planners, and community organizations (Luo et al. 2022; Mathias and Harris-Roxas 2009; Hargrove et al. 2015). The role of HIA in promoting "Health in All Policies" (HiAP) is particularly noteworthy. HiAP is a multidisciplinary approach that seeks to improve health outcomes by integrating health considerations into all areas of policy-making. HIA serves as a critical tool within this framework, enabling policymakers to evaluate the health implications of their decisions across sectors such as transportation, housing, and education. By doing so, HIA helps to ensure that health is prioritized in all policy discussions, ultimately leading to healthier communities (Rogerson et al. 2020).

Moreover, the adaptability of HIA methodologies allows for their application in various contexts, from local to national levels. Different types of HIA, such as rapid HIA, intermediate HIA, and comprehensive HIA, cater to varying resource availability and time constraints. This flexibility makes HIA a valuable tool for diverse stakeholders, including government agencies, non-governmental organizations, and community groups, who may have different capacities and needs (Kang et al. 2011; Fakhri et al. 2015). The importance of public participation in the HIA process cannot be overstated. Engaging communities in the assessment process enriches the data collected and empowers individuals by giving them a voice in decisions that affect their health. This participatory approach has been shown to enhance the effectiveness of HIAs by ensuring that the health needs and concerns of the population are adequately addressed (Morteruel et al. 2020; Haigh et al. 2013). Furthermore, integrating HIA into environmental impact assessments (EIA) has been recognized as a best practice for ensuring that health considerations are not overlooked in environmental decision-making. By combining these two assessment processes, stakeholders can better understand the potential health impacts associated with environmental changes, facilitating more informed and health-conscious policy decisions (Dannenberg 2016; Abah 2012).

The challenges faced in implementing HIA are multifaceted, including adequate training and resources, establishing clear legal frameworks, and integrating health data into decision-making processes. Addressing these challenges is essential for maximizing the impact of HIA and ensuring that health considerations are systematically incorporated into policy development (Harris et al. 2012). One way to holistically address pertinent challenges in HIA is to integrate it into an "Environment-Health System," as suggested in the framework developed by Lebret (2016). This framework highlights the interconnected processes of risk assessment and health impact assessment within the broader environmental health context (Fig. 12.2). Figure 12.2 shows how environmental capital benefits populations by granting access to resources that influence public health.

12.4 Strategic Role of Risk Assessment in Policy and Decision-Making

The strategic role of risk assessment in policy and decision-making is increasingly recognized as a critical component in formulating effective environmental health policies (Table 12.3). Risk assessment is a systematic process for evaluating the potential adverse effects of environmental hazards on human health and the ecosystem, thereby guiding policymakers in their decisions. Integrating risk assessment into environmental health policies is essential for ensuring that decisions are based on sound scientific evidence and address the complexities and uncertainties inherent in environmental issues. This integration is not merely a technical exercise; it requires a comprehensive understanding of the social, economic, and environmental contexts in which these policies operate (Stassen et al. 2013; Ádám et al. 2014; Lund et al. 2011).

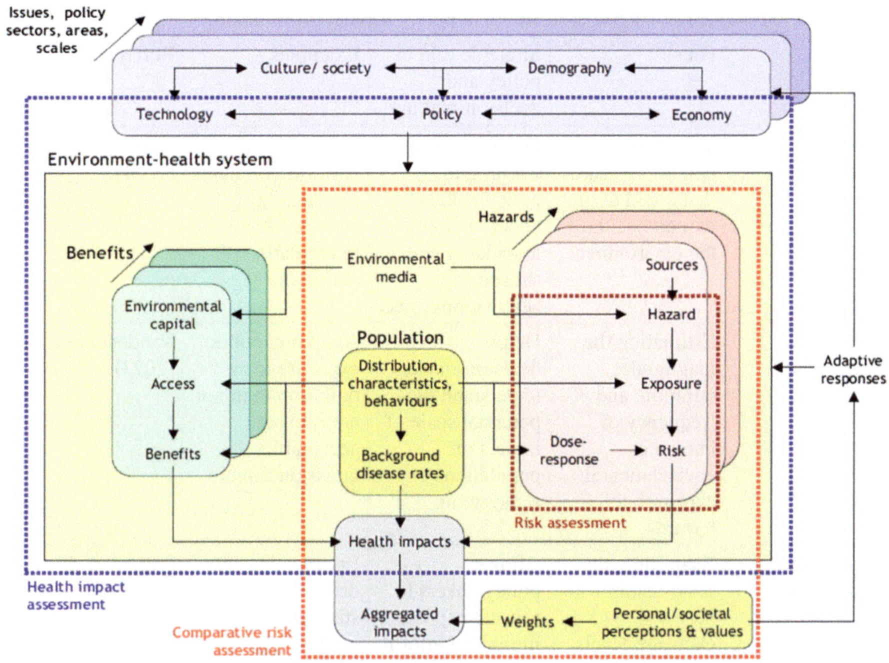

Fig. 12.2 Framework for integrated environmental health impact assessment. *Source* Lebret (2016)

At the epistemological and methodological levels, integrating risk assessment into environmental health policies involves several vital elements. These include interdisciplinary collaboration among researchers, the involvement of social scientists to understand the social dimensions of environmental health problems and the use of integrated environmental health impact assessment tools. Such tools are designed to account for various factors, including the health effects on vulnerable populations and the uncertainties associated with environmental risks (Stassen et al. 2013; Ádám et al. 2014). The Post-Normal Science epistemology, which emphasizes the need for inclusive and participatory approaches in addressing complex societal issues, is particularly relevant in this context (Stassen et al. 2013). Moreover, developing frameworks for integrated quantitative assessments of health risks has significantly advanced the field. These frameworks facilitate a comprehensive evaluation of health risks associated with environmental exposures, allowing for more informed decision-making by policymakers (Ádám et al. 2014; Neo et al. 2022). Integrating risk assessment into policymaking processes enhances the effectiveness of policies and increases their legitimacy by ensuring that diverse stakeholder perspectives are considered (Lund et al. 2011; Neo et al. 2022). This participatory approach fosters public trust and engagement in environmental health initiatives.

Table 12.3 Strategic role of risk assessment in policy and decision-making

Aspect	Focus	Strategic role in policy and decision-making	Examples	References
Hazard identification	Identifying potential hazards that could harm human health or the environment	Provides early warning to policymakers about the risks associated with a substance, activity, or project	Identification of harmful pollutants in industrial activities leading to regulation of emissions	Hassan et al. (2021)
Exposure assessment	Estimating the magnitude, duration, and frequency of human or environmental exposure to hazards	Helps decision-makers understand the potential scale of impact on populations or ecosystems	Assessing public exposure to air pollution in urban areas to set acceptable emission limits	Vandenberg et al. (2023)
Dose–response relationship	Examining the relationship between exposure levels and the likelihood of adverse health effects	Allows policymakers to set safe exposure limits and inform public health standards	Establishing safe drinking water standards based on the dose–response analysis of contaminants	Fenton et al. (2021)
Risk characterization	Integrating hazard identification, exposure assessment, and dose–response data to estimate overall risk	Provides a comprehensive evaluation of risks, aiding in risk communication and regulatory decision-making	Developing environmental safety regulations for chemical exposure in manufacturing industries	Ngwenya et al. (2023)
Uncertainty analysis	Assessing uncertainties in data and risk estimates	Helps policymakers make informed decisions in the face of uncertainty by identifying knowledge gaps	Incorporating uncertainty in climate change models to guide long-term environmental policy decisions	Lempert et al.

(continued)

Table 12.3 (continued)

Aspect	Focus	Strategic role in policy and decision-making	Examples	References
Cost–benefit analysis	Comparing the costs of mitigating risks against the potential benefits	Ensures resources are allocated efficiently, prioritizing actions that offer the most significant health or environmental benefits	Weighing the economic costs of reducing greenhouse gas emissions against the benefits of mitigating climate impacts	Scovronick et al. (2020)
Risk communication	Effectively communicating risks to the public, stakeholders, and decision-makers	Builds trust, informs public opinion, and encourages compliance with regulatory measures	Public communication regarding risks of pesticide use in agriculture, leading to better regulation and safe use	Baba and Esfandiari (2023)
Precautionary principle	Acting to prevent harm even when some cause-and-effect relationships are not fully established	Guides policymakers to take preventive action in the face of potential risks, even with incomplete data	Banning or restricting chemicals based on suspected risks before complete evidence is available (e.g., asbestos)	Hayes (2005)
Monitoring and evaluation	Continuously monitoring risks and evaluating the effectiveness of risk mitigation strategies	Ensures that policies are adaptable and can be updated based on new risk information or changing conditions	Monitoring air quality after implementing vehicle emission standards to assess policy effectiveness	Saadeh et al. (2022)

The tools and models employed in risk analysis play a vital role in shaping environmental health policies. Various methodologies, including HIA and Strategic Environmental Assessments (SEA), evaluate the potential health impacts of proposed policies or interventions (Loureiro et al. 2022; Munns et al. 2016). These tools provide a structured framework for assessing the implications of environmental changes on public health, informing policymakers about potential risks and benefits (Loureiro et al. 2022; Munns et al. 2016). Furthermore, integrating ecosystem services into risk assessment frameworks enhances the comprehensiveness of evaluations by considering the broader ecological context in which human health is situated (Munns et al. 2016, 2015).

Studies in environmental risk assessment illustrate the practical application of these tools and models in real-world scenarios. For instance, the INTARESE project in Europe employed a full-chain approach to assess the health effects of environmental stressors, demonstrating the utility of integrated assessments in informing policy decisions (Cesaroni et al. 2011). Similarly, assessing heavy metal pollution in the Yangtze River Delta highlighted the importance of health risk assessments in identifying exposure pathways and informing regulatory actions (Hu et al. 2017). These case studies highlight the value of empirical evidence in shaping effective environmental health policies and the necessity of adapting risk assessment methodologies to specific contexts. The challenges and opportunities associated with urban environmental health and sustainability further emphasize the strategic role of risk assessment in policymaking. Urban areas face many environmental health challenges, including air and water pollution, traffic congestion, and inadequate housing conditions (Vardoulakis et al. 2016). Integrated assessments considering these multifaceted risks can lead to holistic solutions promoting public health and environmental sustainability (Vardoulakis et al. 2016). The HEALTHY-POLIS initiative exemplifies this approach by focusing on interdisciplinary research to address urban health challenges through policy-relevant strategies (Vardoulakis et al. 2016).

In addition to urban contexts, integrating risk assessment into environmental health policies is vital in addressing cumulative risks associated with multiple environmental stressors. The concept of One health, which recognizes the interconnectedness of human, animal, and environmental health, highlights the need for integrated risk assessments that account for these interdependencies (Ortiz et al. 2022; Izah et al. 2023, 2024a, b). By adopting a One Health approach, policymakers can better understand the complex interactions between various environmental factors and their impacts on public health, thereby enhancing the effectiveness of interventions (Ortiz et al. 2022). Furthermore, the role of community engagement in the risk assessment process cannot be overstated. Programs like the EPA's Community Action for a Renewed Environment (CARE) initiative exemplify how community involvement can enhance the identification of local environmental hazards and inform risk reduction strategies (Hammond et al. 2010). By empowering communities to participate in the risk assessment process, policymakers can ensure that the concerns and priorities of local populations are adequately addressed in environmental health policies (Hammond et al. 2010). The evolution of risk governance, which involves integrating risk assessment and management strategies, reflects a shift towards more adaptive and participatory approaches in environmental policymaking. This evolution recognizes the complexity and uncertainty of different risk problems and advocates for tailored risk management strategies (Lebret 2015). By differentiating between types of risks and their associated uncertainties, policymakers can develop more effective and context-specific interventions that address the unique challenges posed by environmental health issues (Lebret 2015).

Moreover, integrating economic analysis into risk assessment frameworks enhances understanding environmental policies' potential costs and benefits. This integration allows for a more comprehensive evaluation of the trade-offs involved in decision-making, ultimately leading to more informed and effective policy outcomes

(Hoffmann 2011). By considering economic factors alongside health and environmental risks, policymakers can develop strategies that are effective in mitigating risks and economically viable (Hoffmann 2011).

The importance of consistent and reliable epidemiological data in risk assessment further underscores the need for robust methodologies in evaluating health risks associated with environmental exposures. Variability in the interpretation of epidemiological findings can lead to uncertainties in risk assessments, highlighting the necessity for standardized approaches to data evaluation (White et al. 2013; Morrison et al. 2014). By establishing clear criteria for assessing the consistency of epidemiological data, policymakers can enhance the credibility and reliability of risk assessments, ultimately leading to more practical regulatory frameworks (White et al. 2013).

12.5 Health Impact Studies in Practice

Health impact studies are crucial in understanding the multifaceted relationship between health and various environmental, social, and economic determinants. Epidemiological studies and data collection form the backbone of these assessments, providing the necessary evidence to inform public health interventions and policies (Table 12.4). The systematic collection of health data, particularly in community settings, allows researchers to identify health trends, risk factors, and the effectiveness of health programs. For instance, Muflih et al. (2023) highlighted the significance of community outreach initiatives in managing non-communicable diseases among the elderly, emphasizing the role of personalized health assessments in improving health outcomes. This approach enhances individual health and contributes to broader epidemiological data that can inform future health strategies.

The methods for assessing community health impacts have evolved significantly, incorporating qualitative and quantitative approaches. Shojaei et al. (2016) provide a comprehensive overview of health impact assessments (HIA) in urban development projects, illustrating the effectiveness of mixed methods approaches that include interviews and focus group discussions. Such methodologies enable a deeper understanding of community perceptions and the potential health impacts of proposed projects. Furthermore, Liu (2023) discussed the increasing recognition of health impact assessments by stakeholders, driven by a growing public concern for health and well-being. This trend underscores the importance of integrating community input into health assessments, ensuring that the voices of those affected are heard and considered in decision-making processes. Successful health impact studies often rely on community participation, which has been shown to enhance the relevance and effectiveness of health interventions. Haldane et al. (2019) conducted a systematic review emphasizing the positive outcomes of community involvement in health services development. Their findings suggest that when communities are actively

Table 12.4 Health impact studies in practice with focus on epidemiological studies and data collection

Aspect	Type of study	Strategic role in health impact studies	Examples	References
Epidemiological studies	Research that studies the patterns, causes, and effects of health and disease conditions in defined populations	Provides critical insights into the relationships between environmental exposures and health outcomes, guiding public health policies	Investigating the link between air pollution and respiratory diseases in urban areas	Rod et al. (2023)
Cohort Studies	A type of epidemiological study that follows a group of individuals over time to observe how exposures affect health	Allows for the identification of risk factors for diseases by comparing exposed and unexposed groups over time	Long-term study of the effects of air quality on lung function in populations living near industrial zones	Song and Chung (2010)
Case–control studies	A retrospective study comparing individuals with a specific health condition (cases) to those without the condition (controls)	Helps identify potential causes or risk factors by comparing exposures between affected and unaffected individuals	Study comparing cancer patients with non-patients to assess the impact of prolonged exposure to industrial chemicals	Lewallen and Courtright (1998)
Cross-sectional studies	Observational studies that analyze data from a population at a specific point in time	Useful for assessing the prevalence of health outcomes and their association with environmental exposures	Evaluating the prevalence of asthma in children in areas with high vehicular emissions	Wang and Cheng (2020)
Data collection methods	The techniques used to gather quantitative and qualitative data related to health outcomes and environmental exposure	Ensures accurate and reliable data, which is essential for identifying health impacts and informing risk assessments	Air quality monitoring, health surveys, biological sampling (e.g., blood tests), interviews, and questionnaires	Scammell (2010), Rigaud et al. (2024)

(continued)

Table 12.4 (continued)

Aspect	Type of study	Strategic role in health impact studies	Examples	References
Biomonitoring	The measurement of chemicals or their metabolites in human biological samples, such as blood or urine	Helps establish direct links between environmental exposures and health effects, providing valuable data for risk assessment	Measuring lead levels in children's blood to evaluate exposure from contaminated water or paint	Paustenbach and Galbraith (2006)
Geospatial data collection	The use of geographical data to assess environmental exposure and map health outcomes	Allows for the visualization of exposure patterns, identifying areas with higher risks of disease or pollution hotspots	Mapping cancer clusters in regions with industrial waste disposal sites	Nuckols et al. (2004), Jerrett et al. (2010)
Health surveys and questionnaires	Collects self-reported data from individuals regarding their health status and exposure to environmental factors	Gathers information on personal habits, occupation, and perceived exposure, which complements biological and observational data	National health surveys on the impact of second-hand smoke exposure in households	Margozzini et al. (2023)

engaged in the planning and evaluation of health services, there is a marked improvement in service coverage and stakeholder satisfaction. This participatory approach empowers communities and leads to more tailored and effective health interventions.

Similarly, Harris-Roxas et al. (2014) explored the impacts of equity-focused health impact assessments (EFHIAs) on health service planning, demonstrating that these assessments can significantly enhance the consideration of health equity in policymaking. Health impact studies can address disparities and promote better health outcomes for marginalized populations by prioritizing equity. Cumulative risk assessments represent another methodological advancement in health impact studies, particularly in evaluating the combined effects of multiple environmental stressors on community health. Sexton (2012) discussed the complexities involved in cumulative risk assessments, which aim to address the challenges posed by traditional single-chemical assessments. This approach is particularly relevant in communities exposed to various environmental hazards, where understanding the cumulative impact is essential for effective public health interventions. August et al. (2012) further illustrate this point by highlighting the importance of identifying high-risk communities and tailoring interventions accordingly. Researchers can better understand different populations' unique health challenges using comprehensive assessment methods. Integrating social media and digital health campaigns into health impact studies has emerged as a novel approach to engaging communities and disseminating health information. Theiss et al. (2016) examined the effectiveness of breast cancer-related digital campaigns, noting the potential of social media to enhance health communication and outreach. As digital platforms become increasingly prevalent, leveraging these tools can significantly impact public health messaging and community engagement. This trend aligns with the findings of Cameron et al. (2011), who emphasized the importance of community-designed health impact assessment tools that incorporate local knowledge and preferences. Such tools can facilitate more effective health interventions by ensuring they are culturally and contextually relevant.

In addition to these methodological advancements, the evaluation of health impact assessments has gained traction in assessing their effectiveness and influence on health policy. Bourcier et al. (2015) comprehensively evaluated HIAs in the United States, revealing that these assessments can directly impact decisions in non-health-related sectors. This finding highlights the potential of health impact assessments to inform a wide range of policy areas, ultimately leading to improved health outcomes. Moreover, Grant et al. (2015) highlighted the necessity of community health needs assessments (CHNAs) in identifying community health concerns and resources. Via systematically assessing community health needs, stakeholders can develop targeted interventions that address specific health challenges. The role of environmental assessments in health impact studies cannot be overstated, particularly in the context of tourism and urban development. Ruiz et al. explore the environmental evaluation of health tourism in Ecuador, emphasizing the importance of understanding the social perceptions of residents regarding tourism impacts (Ruiz et al. 2017). This perspective is crucial for ensuring health interventions align with community values and priorities. Furthermore, Barata-Cavalcanti et al. (2019) introduced a new community-focused methodology for assessing the collective impact of health programs funded by food and beverage companies, highlighting the need for innovative approaches to evaluate health interventions. This methodology reflects a growing recognition of the importance of industry partnerships in addressing public health challenges.

12.5.1 Adopting Lalonde Framework for Environmental Health Management

The Lalonde Framework, introduced by Canadian Health Minister Marc Lalonde in the 1974 report *A New Perspective on the Health of Canadians*, provides a comprehensive model for understanding health determinants beyond medical care alone (Lalonde 1974). This model identifies four key determinants—Biology, Environment, Lifestyle, and Healthcare Organization—that collectively shape health outcomes. Adopting the Lalonde Framework for environmental health management allows for a holistic approach to mitigating environmental health risks, recognizing

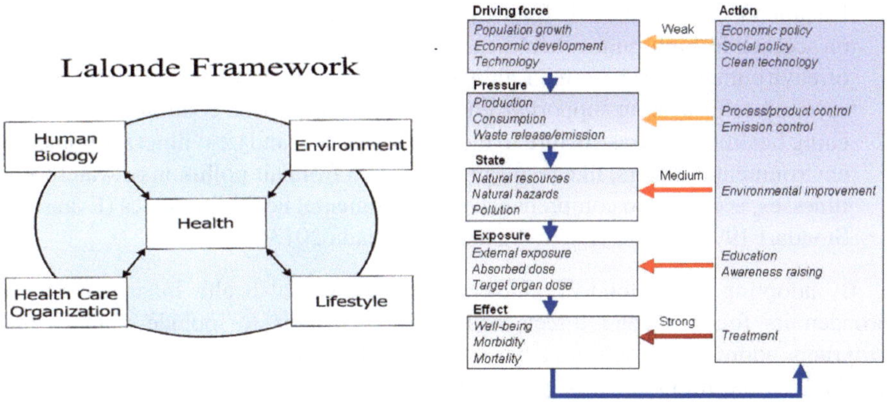

Fig. 12.3 Adopting the Lalonde framework for environmental health management. *Source* Lebret (2016)

that health is influenced by access to healthcare and a range of interconnected factors (Fig. 12.3; Lebret 2016).

1. Biology (Human Biology): Within the context of environmental health, biological factors such as genetic predispositions and physiological responses play a critical role in how individuals respond to environmental hazards. For example, genetic differences can affect susceptibility to pollutants or toxins, which supports personalized approaches to health interventions in high-risk populations (Frankish et al. 1996). While biological factors are largely beyond individual control, understanding them can improve strategies for protecting vulnerable groups from environmental exposures.
2. Environment: Lalonde's framework emphasizes the role of the environment in health, aligning closely with the goals of environmental health management. This determinant includes physical, social, and economic factors significantly impacting health, such as air and water quality, workplace conditions, and socioeconomic status. Policies targeting pollution reduction, improved sanitation, and safe housing are crucial in managing environmental health risks. Research demonstrates that populations in disadvantaged or polluted environments are at greater risk for adverse health outcomes, underscoring the need for targeted interventions in these areas (Health Canada 2001; Hancock and Perkins 1985).
3. Lifestyle: The Lalonde Framework highlights lifestyle choices, such as diet, physical activity, and substance use, which are essential to environmental health management. Lifestyle modifications can reduce exposure to environmental risks, such as avoiding areas with poor air quality or adopting safe food-handling practices to prevent contamination. Public health strategies encouraging lifestyle changes can help individuals make informed decisions that mitigate environmental risks and improve their overall health (World Health Organization [WHO] 2002; Katz and Ali 2009).

4. Healthcare Organization: Effective environmental health management also relies on accessible, high-quality healthcare services that can address the health impacts of environmental exposures. Lalonde pointed out that healthcare alone cannot sustain health without supportive environmental and social conditions. Strengthening healthcare infrastructure to monitor, diagnose, and treat illnesses related to environmental factors, like respiratory diseases from air pollution or waterborne illnesses, is critical to comprehensive environmental health strategies (Evans and Stoddart 1990; Public Health Agency of Canada 2013).

By adopting the Lalonde Framework, environmental health management can broaden its focus beyond direct medical interventions to include policies and programs addressing environmental and lifestyle factors. This holistic approach emphasizes the importance of preventive measures, health promotion, and multisectoral collaboration to create healthier environments. It aligns with modern public health models that recognize social and environmental determinants of health, thereby supporting strategies that reduce healthcare costs and improve population health outcomes. Embracing this framework encourages sustainable development and community resilience by addressing the root causes of environmental health risks, such as pollution, inadequate infrastructure, and social inequality (Raphael 2004; Lalonde 1974).

12.5.2 Challenges in Conducting Risk and Health Impact Assessments

Conducting risk and health impact assessments is fraught with challenges that can significantly hinder the effectiveness and accuracy of the evaluations. The primary challenges include uncertainty and variability in data, public perception and risk communication, and resource limitations (Table 12.5). Each of these challenges is interrelated and can compound the difficulties faced by health professionals and policymakers in assessing risks accurately and communicating them effectively to the public.

Uncertainty and variability in data are central issues in risk and health impact assessments. The inherent complexity of health systems and environmental interactions introduces significant uncertainty, which can arise from various sources, including measurement errors, model parameters, and the aggregation of data across different spatial and temporal scales (Johnson and Garcia–Menendez 2022). For instance, Johnson and Garcia–Menendez (2022) highlighted that uncertainties in modeling wildfire smoke's health impacts stem from multiple factors, including exposure–response relationships and sampling errors. Similarly, Bouwknegt et al. (2013) emphasized the importance of effectively identifying and characterizing sources of uncertainty to prioritize future research. These uncertainties can lead to a state of "uncertainty ignorance," where decision-makers are unaware of the risks associated with their choices, as noted by Grimm et al. (2019).

Table 12.5 Challenges in conducting risk and health impact assessments

Challenge	Impact	Impact on risk and Health impact assessments	Possible solutions	References
Data availability and quality	Inadequate or poor-quality data can hinder accurate assessments of risks and health impacts	Limits the ability to make evidence-based decisions and accurate predictions	Improve data collection methods, promote open data sharing, and use advanced technology for better monitoring	Bernardi et al. (2023)
Uncertainty in modeling and predictions	Uncertainty in risk and impact models due to lack of knowledge or variability in data inputs	Can lead to over- or underestimation of health impacts, affecting decision-making and policy development	Conduct uncertainty analysis and use probabilistic models to account for variability in data	Burns et al. (2014), Sadatsafavi et al., (2022)
Resource and time constraints	Limited financial and human resources, and time to complete thorough assessments	May result in incomplete assessments or rushed decisions that fail to capture the full scope of potential impacts	Allocate sufficient resources and time for comprehensive assessments; collaborate with multidisciplinary teams	Addis et al. (2023)
Lack of public and stakeholder engagement	Insufficient involvement of the public and stakeholders in the assessment process	Can result in resistance to outcomes or policy recommendations due to lack of transparency and participation	Engage stakeholders early and frequently throughout the process to ensure buy-in and incorporate diverse perspectives	Maurer et al (2022)
Complexity of health and environmental interactions	The complex interplay between various environmental factors and health outcomes is difficult to model and predict	Makes it challenging to assess cumulative risks and account for all potential health impacts	Use integrated assessment models and systems thinking to capture the multifaceted nature of health-environment interactions	Reuben et al. (2022)
Regulatory and policy limitations	Inconsistent or inadequate regulatory frameworks that do not mandate comprehensive risk or health impact assessments	Leads to gaps in risk management and insufficient protection of public health and the environment	Advocate for stronger regulations that mandate risk assessments and prioritize public health in policy decisions	Levy (2021)

(continued)

Table 12.5 (continued)

Challenge	Impact	Impact on risk and Health impact assessments	Possible solutions	References
Ethical and social concerns	Ethical concerns regarding the prioritization of certain health impacts or population groups in assessments	Can result in biased decisions or inequitable health outcomes, particularly for vulnerable populations	Ensure ethical frameworks are in place, and conduct assessments with equity considerations for vulnerable populations	Pauly et al. (2021)

Moreover, the variability in data can complicate the assessment process. For example, Yousefi et al. (2020) discussed how mapping processes in risk assessments are sensitive to expert judgments and the weighting of input variables, which can lead to inconsistent results. This variability can undermine the credibility of evaluations and lead to misinformed policy decisions. The need for a systematic approach to quantify and communicate uncertainty is critical, as highlighted by Mesa-Frias et al. (2013), who argue that understanding uncertainty is essential for supporting health policy decisions. Without addressing these uncertainties, health impact assessments may fail to provide a reliable basis for action, ultimately jeopardizing public health outcomes.

Public perception and risk communication represent another significant challenge in conducting risk and health impact assessments. Effective risk communication is essential for ensuring that the public understands the risks they face and can make informed decisions. However, Dow and Tuler (2021) noted that the complexity of risk communication strategies can lead to confusion and misinterpretation among the public. The emotional reactions of individuals to risk messages can further complicate communication efforts, as people may respond differently based on their personal experiences and beliefs (Rakow et al. 2015). This variability in public perception necessitates a nuanced approach to risk communication that considers the diverse backgrounds and perspectives of the audience.

Furthermore, the role of government agencies in risk communication is critical, as they are often responsible for disseminating information to the public. Boholm (2019) emphasizes the importance of transparency in risk communication but also points out the potential conflicts and tensions that can arise when disclosing uncertain information. For instance, revealing high levels of uncertainty may induce fear and erode public trust in government agencies, which can hinder practical communication efforts. Additionally, the need for coordinated communication across different agencies and jurisdictions can create challenges, as Chatfield et al. (2014) highlighted the complexities of interagency coordination in disaster risk communication. These challenges underscore the importance of developing comprehensive communication strategies that address both risk assessment's technical aspects and public perception's emotional dimensions.

Resource limitations pose yet another challenge in conducting risk and health impact assessments. Health systems often operate under tight budgets and limited personnel, limiting the capacity to conduct thorough inspections and implement effective communication strategies. Grutters et al. (2014) argued that moving towards uncertainty acceptance in healthcare policy is essential, but this requires resources and training that may not always be available. Additionally, developing risk communication competencies among health professionals is crucial, as identified by Dickmann et al. (2016), who emphasized the need for training in risk communication as a core public health competence. Health systems may struggle to build practical risk assessment and communication capacity without adequate resources.

Moreover, integrating risk communication into existing health frameworks can be resource-intensive. For instance, the systematic review by Jha et al. (2018) highlighted the importance of incorporating emergency risk communication into public health responses, but this requires significant investment in training and infrastructure. The challenge of resource allocation becomes even more pronounced in low-resource settings, where the burden of health risks may be disproportionately high. Adekola et al. (2016) illustrated how environmental degradation in the Niger Delta poses significant health risks, yet resource constraints often limit the capacity to assess and communicate these risks. This disparity in resource availability can lead to inequitable health outcomes and exacerbate existing health disparities.

12.6 Future Trends and Innovations in Environmental Health Management

The future of environmental health management is increasingly intertwined with advances in technology and data analytics, predictive models, and collaborative approaches for risk mitigation (Table 12.6). These trends are reshaping how to understand and respond to environmental health challenges, leading to innovative solutions that promise to enhance public health outcomes. Technology and data analytics advancements are revolutionizing environmental health management by enabling more precise monitoring and assessment of environmental factors that impact health.

The integration of big data analytics allows for the analysis of vast datasets, which can reveal patterns and correlations that were previously undetectable. For instance, mobile health applications have gained traction, facilitating self-care and health monitoring through biometric data collection and analysis (Anderson et al. 2016). These applications empower individuals to manage their health and provide valuable data for public health officials to identify trends and potential health risks in real time. Moreover, developing sophisticated environmental management systems, such as ISO 14001, has enhanced a firm's technological innovation performance by integrating environmental management practices into its operational frameworks (Yang et al. 2015). This integration fosters a culture of sustainability within organizations, encouraging proactive approaches to environmental challenges. The emphasis

Table 12.6 Innovations in environmental health management

Innovation	Tools	Strategic role in environmental health management	Examples	References
Advances in technology and data analytics	Use of modern technology such as IoT (Internet of Things), AI (Artificial Intelligence), and big data analytics	Enhances the ability to monitor environmental conditions in real time and process large datasets to detect patterns and trends	Smart sensors used in urban air quality monitoring to track pollutants and inform public health interventions	Khan and Alotaibi (2020)
Predictive models and early warning systems	Development of algorithms and models that predict environmental hazards and their potential health impacts	Provides actionable insights by forecasting potential health risks, allowing for early interventions	Predictive modeling of disease outbreaks (e.g., malaria, dengue) based on environmental factors such as temperature and precipitation	Meckawy et al. (2022)
Collaborative approaches for risk mitigation	Partnerships between governments, private sectors, and communities for collective environmental health management	Encourages shared responsibility, resource pooling, and integrated responses to mitigate environmental health risks	Multi-sector partnerships in climate adaptation strategies involving local communities, industries, and governments to reduce heat-related health risks	Gooding et al. (2022)
Remote sensing and geospatial technologies	Use of satellite imagery and geospatial tools to monitor environmental changes and assess their health impacts	Enables large-scale monitoring of land use, deforestation, water quality, and pollution hotspots	Satellite-based tracking of deforestation and its link to respiratory health problems in affected populations	Dahu et al. (2023), Li et al. (2023)
Health and environmental data integration	Integration of health data with environmental datasets to create holistic assessments of risk	Supports more comprehensive assessments of how environmental factors contribute to public health outcomes	Integrating hospital admissions data with air quality indices to understand the impact of pollution on asthma cases	Bassil et al. (2015)

(continued)

Table 12.6 (continued)

Innovation	Tools	Strategic role in environmental health management	Examples	References
Crowdsourcing and citizen science	Use of public participation to collect environmental data and report health impacts in real time	Increases data coverage and encourages community engagement in environmental health monitoring	Citizen-led water quality monitoring and reporting systems for detecting contamination in local water bodies	Christine and Thinyane (2021)

on eco-innovation is further supported by research indicating that top management's environmental awareness significantly influences the implementation of green innovations (Huang et al. 2020). This highlights the critical role of leadership in driving technological advancements that align with environmental health objectives.

Predictive models and early warning systems are essential to modern environmental health management. These systems leverage data analytics to forecast potential health risks associated with ecological changes, such as heatwaves or infectious disease outbreaks. For example, establishing heat-health warning systems in Europe has demonstrated the effectiveness of predictive modeling in reducing mortality rates during extreme heat events (Casanueva et al. 2019). By analyzing meteorological variables and their correlation with health outcomes, these systems can trigger alerts that prompt timely interventions from health authorities, thereby mitigating the adverse effects of heat stress on vulnerable populations. The importance of early warning systems is underscored by studies emphasizing the need for long-term planning and establishing clear links between warning triggers and health impacts (Casanueva et al. 2019). Such systems enhance public health preparedness and foster a culture of proactive risk management. In addition, integrating climate and weather-driven early warning systems has been identified as a crucial strategy for reducing the burden of infectious diseases, as these systems can help anticipate outbreaks and inform public health responses (Morin et al. 2018). This proactive approach is vital in the context of global environmental changes that exacerbate health risks. Collaborative approaches for risk mitigation are increasingly recognized as essential for effective environmental health management. The concept of "One Health," which emphasizes the interconnectedness of human, animal, and environmental health, has gained prominence in recent years. This approach encourages multidisciplinary collaboration among various sectors, including public health, environmental agencies, and the food industry, to address complex health challenges (Veras-Estévez and Chapman 2023). By fostering partnerships and sharing knowledge, stakeholders can develop innovative solutions that enhance resilience against emerging health threats. Research has shown that collaborative governance can lead to innovative environmental solutions involving multiple stakeholders working together to create sustainable practices (Avoyan 2022). This cooperative framework is particularly relevant in environmental health management, where diverse perspectives and expertise are necessary to tackle multifaceted issues. For instance, integrating environmental management practices within supply chains has improved technological innovation performance, highlighting the benefits of collaborative approaches in achieving sustainability goals (Yang et al. 2015).

The role of knowledge transfer in enhancing environmental performance cannot be overstated. Studies indicate that effective knowledge transfer between organizations can significantly impact environmental performance through innovation (Sulistyowatie and Pahlevi 2019). This underscores the importance of fostering a culture of continuous learning and collaboration among stakeholders in environmental health management. By sharing best practices and lessons learned, organizations can collectively enhance their capacity to address environmental health challenges. Furthermore, incorporating innovative strategies in environmental health management is crucial for adapting to the evolving landscape of public health threats. For example, developing strategic management methodologies that integrate investment and innovation activities in environmentally safe land management can provide a framework for sustainable development (Borys et al. 2021). These methodologies can guide organizations in making informed decisions that align with environmental health objectives, ultimately leading to improved public health outcomes. In addition to these collaborative efforts, the role of technology in facilitating communication and information sharing among stakeholders is paramount. The proliferation of digital platforms and mobile applications has transformed how health information is disseminated and accessed. This shift enhances public awareness of environmental health issues and empowers individuals to take proactive measures to protect their health (Deady et al. 2020). The ability to monitor health conditions and receive timely alerts through digital means represents a significant advancement in public health management. As environmental health management continues to evolve, integrating innovative technologies and collaborative approaches will be critical in addressing emerging health threats. The ongoing development of predictive models and early warning systems will enhance the ability to anticipate and respond to environmental health risks, ultimately leading to more effective public health interventions. Moreover, fostering a culture of collaboration and knowledge sharing among stakeholders will drive sustainable practices and improve health outcomes.

12.7 Conclusion

Risk assessment and health impact studies are vital for managing environmental health risks. By systematically identifying, evaluating, and mitigating the health risks associated with environmental exposures, these methodologies provide a structured, evidence-based approach that supports informed decision-making. As ecological threats become increasingly complex, these assessments play a critical role in safeguarding public health, promoting sustainability, and shaping effective policy. Risk assessment offers a comprehensive process involving hazard identification, exposure assessment, dose–response analysis, and risk characterization. This structured framework allows decision-makers to anticipate and address potential health risks, contributing to healthier environments. Health Impact Assessment (HIA) complements this by evaluating the broader health consequences of policies, projects, or programs, ensuring that health considerations are integrated into decision-making processes. HIA's emphasis on stakeholder engagement and community involvement makes it an inclusive and transparent tool, fostering public trust and equity. Both methodologies underscore the importance of scientific evidence in shaping environmental health strategies. As public health challenges—such as climate change, pollution, and urbanization—continue to rise, risk assessments and HIAs offer proactive approaches to protect communities from adverse health outcomes.

Moreover, advancements in technology and data analytics, including predictive modeling and real-time monitoring, enhance the effectiveness of these tools, enabling timely and responsive interventions. For sustainable environmental health management, these assessments help prevent adverse health effects, incorporate cumulative risks, and ensure long-term public health protection. Continued investment in refining risk assessment frameworks and integrating them with emerging technologies will be essential in navigating future environmental challenges, fostering public health, and promoting sustainability.

References

Abah S (2012) Hia practices in Nigeria. Impact Assess Project Appraisal 30(3):207–213. https://doi.org/10.1080/14615517.2012.705064

Ádám B, Molnár Á, Ádány R, Bianchi F, Bitenc K, Chereches R, Cori L, Fehr R, Kobza J, Kollarova J et al (2014) Assessment of health risks of policies. Environ Impact Assess Rev 48:47–52. https://doi.org/10.1016/j.eiar.2014.05.002

Addis BA, Gelaw YM, Eyowas FA, Bogale TW, Aynalem ZB, Guadie HA (2023) Time wasted by health professionals is time not invested in patients": time management practice and associated factors among health professionals at public hospitals in Bahir Dar, Ethiopia: a multicenter mixed method study. Front Public Health 11:1159275. https://doi.org/10.3389/fpubh.2023.1159275

Adekola J, Fischbacher-Smith M, Fischbacher-Smith D, Adekola O (2016) Health risks from environmental degradation in the niger delta, nigeria. Environ Plann C Polit Space 35(2):334–354. https://doi.org/10.1177/0263774x16661720

Aigberua AO, Izah SC, Aigberua AA (2023) Occurrence, source delineation, and health hazard of polycyclic aromatic hydrocarbons in tissues of Sarotherodon melanotheron and Chrysichthys nigrodigitatus from Okulu River, Nigeria. Environ Monitor Assess 195(3):364

Aigberua AO, Izah SC, Richard G (2021) Hazard Analysis of Trace Metals in Muscle of Sarotherodon melanotheron and Chrysichthys nigrodigitatus from Okulu River, Rivers State, Nigeria. J Environ Health Sustain Dev 6(3):1340–1356

Anderson K, Burford O, Emmerton L (2016) Mobile health apps to facilitate self-care: a qualitative study of user experiences. PLoS ONE 11(5):e0156164. https://doi.org/10.1371/journal.pone.0156164

Antwi EK, Boakye-Danquah J, Owusu-Banahene W, Dabros A, Eddy IM, Silver DA, Abolina E, Eddy B, Winder RS (2023) Risk assessment framework for cumulative effects (RAFCE). Front Environ Sci 10. https://doi.org/10.3389/fenvs.2022.1055159

August LM, Faust JB, Cushing L, Zeise L, Alexeeff GV (2012) Methodological considerations in screening for cumulative environmental health impacts: lessons learned from a pilot study in california. Int J Environ Res Public Health 9(9):3069–3084. https://doi.org/10.3390/ijerph9093069

Avoyan E (2022) Collaborative governance for innovative environmental solutions: qualitative comparative analysis of cases from around the world. Environ Manag 71(3):670–684. https://doi.org/10.1007/s00267-022-01642-7

Baba FV, Esfandiari Z (2023) Theoretical and practical aspects of risk communication in food safety: a review study. Heliyon 9(7):e18141. https://doi.org/10.1016/j.heliyon.2023.e18141

Barata-Cavalcanti O, Leung MM, Costa S, Mateo KF, Guillermin M, Palmedo PC, Crossley R, Huang TTK (2019) Assessing the collective impact of community health programs funded by food and beverage companies: a new community-focused methodology. Int Q Community Health Educ 40(2):75–89. https://doi.org/10.1177/0272684x19862359

Bassil KL, Sanborn M, Lopez R, Orris P (2015) Integrating environmental and human health databases in the great lakes basin: themes, challenges and future directions. Int J Environ Res Public Health 12(4):3600–3614. https://doi.org/10.3390/ijerph120403600

Bernardi FA, Alves D, Crepaldi N, Yamada DB, Lima VC, Rijo R (2023) Data quality in health research: integrative literature review. J Med Internet Res 25:e41446. https://doi.org/10.2196/41446

Boholm Å (2019) Risk communication as government agency organizational practice. Risk Anal 39(8):1695–1707. https://doi.org/10.1111/risa.13302

Borys B, Natalya A, Hanna T (2021) Strategic management of investment and innovation activities in the field of environmentally safe land management. Ekonomia I Środowisko 4(79):24–57. https://doi.org/10.34659/2021/4/27

Bourcier E, Charbonneau D, Cahill C, Dannenberg AL (2015) An evaluation of health impact assessments in the United States, 2011–2014. Prev Chronic Dis 12. https://doi.org/10.5888/pcd12.140376

Bouwknegt M, Knol AB, van der Sluijs JP, Evers EG (2013) Uncertainty of population risk estimates for pathogens based on qmra or epidemiology: a case study of campylobacter in the Netherlands. Risk Anal 34(5):847–864. https://doi.org/10.1111/risa.12153

Burns CJ, Wright JM, Pierson JB, Bateson TF, Burstyn I, Goldstein DA, Klaunig JE, Luben TJ, Mihlan G, Ritter L, Schnatter AR, Symons JM, Yi KD (2014) Evaluating uncertainty to strengthen epidemiologic data for use in human health risk assessments. Environ Health Perspect 122(11):1160–1165. https://doi.org/10.1289/ehp.1308062

Cameron C, Ghosh S, Eaton SL (2011) Facilitating communities in designing and using their own community health impact assessment tool. Environ Impact Assess Rev 31(4):433–437. https://doi.org/10.1016/j.eiar.2010.03.001

Card AJ, Ward JR, Clarkson PJ (2013) Trust-level risk evaluation and risk control guidance in the NHS east of England. Risk Anal 34(8):1469–1481. https://doi.org/10.1111/risa.12159

Casanueva A, Burgstall A, Kotlarski S, Messeri A, Morabito M, Flouris AD, Nybo L, Spirig C, Schwierz C (2019) Overview of existing heat-health warning systems in europe. Int J Environ Res Public Health 16(15):2657. https://doi.org/10.3390/ijerph16152657

Cave B, Pyper R, Fischer-Bonde B, Humboldt-Dachroeden S, Martin-Olmedo P (2021) Lessons from an international initiative to set and share good practice on human health in environmental impact assessment. Int J Environ Res Public Health 18(4):1392. https://doi.org/10.3390/ijerph18041392

Cesaroni G, Boogaard H, Jonkers S, Porta D, Badaloni C, Cattani G, Forastiere F, Hoek G (2011) Health benefits of traffic-related air pollution reduction in different socioeconomic groups: the effect of low-emission zoning in rome. Occup Environ Med 69(2):133–139. https://doi.org/10.1136/oem.2010.063750

Chatfield AT, Reddick CG, Inan DI, Brajawidagda U (2014, June) E-government, social media, and risk perception communication at the edge of disaster: findings from the Mt. Sinabung eruption in Indonesia. In: Proceedings of the 15th annual international conference on digital government research, pp 153–162. https://doi.org/10.1145/2612733.2612752

Chiu WA, Slob W (2015) A unified probabilistic framework for dose-response assessment of human health effects. Environ Health Perspect 123(12):1241–1254. https://doi.org/10.1289/ehp.1409385

Christine DI, Thinyane M (2021) Citizen science as a data-based practice: a consideration of data justice. Patterns (New York, N.Y.) 2(4):100224. https://doi.org/10.1016/j.patter.2021.100224

Conlon TW, Yousef N, Mayordomo-Colunga J, Tissot C, Fraga MV, Bhombal S, Suryawanshi P, Villanueva AM, Siassi B, Singh Y (2021) Establishing a risk assessment framework for point-of-care ultrasound. Eur J Pediatr 181(4):1449–1457. https://doi.org/10.1007/s00431-021-04324-4

Dahu BM, Alaboud K, Nowbuth AA, Puckett HM, Scott GJ, Sheets LR (2023) The role of remote sensing and geospatial analysis for understanding COVID-19 population severity: a systematic review. Int J Environ Res Public Health 20(5):4298. https://doi.org/10.3390/ijerph20054298

Dankittipong N, Fischer EA, Swanenburg M, Wagenaar JA, Stegeman AJ, De Vos CJ (2022) Quantitative risk assessment for the introduction of carbapenem-resistant enterobacteriaceae (CPE) into dutch livestock farms. Antibiotics 11(2):281. https://doi.org/10.3390/antibiotics11020281

Dannenberg AL (2016) Effectiveness of health impact assessments: a synthesis of data from five impact evaluation reports. Prev Chronic Dis 13. https://doi.org/10.5888/pcd13.150559

Deady M, Glozier N, Calvo R, Johnston D, Mackinnon A, Milne D, Choi I, Gayed A, Peters D, Bryant R et al (2020) Preventing depression using a smartphone app: a randomized controlled trial. Psychol Med 52(3):457–466. https://doi.org/10.1017/s0033291720002081

Dickmann P, Abraham T, Sarkar S, Wysocki P, Cecconi S, Apfel F, Nurm ÜK (2016) Risk communication as a core public health competence in infectious disease management: development of the ecdc training curriculum and programme. Eurosurveillance 21(14). https://doi.org/10.2807/1560-7917.es.2016.21.14.30188

Dow K, Tuler S (2021) Risk amplification and attenuation as communication strategies in climate adaptation in urban areas. Risk Anal 42(7):1440–1454. https://doi.org/10.1111/risa.13819

Evans RG, Stoddart GL (1990) Producing health, consuming health care. Soc Sci Med 31(12):1347–1363

Fakhri A, Harris P, Maleki M (2015) Proposing a framework for health impact assessment in iran. BMC Public Health 15(1). https://doi.org/10.1186/s12889-015-1698-1

Fenton SE, Ducatman A, Boobis A, DeWitt JC, Lau C, Ng C, Smith JS, Roberts SM (2021) Per- and polyfluoroalkyl substance toxicity and human health review: current state of knowledge and strategies for informing future research. Environ Toxicol Chem 40(3):606–630. https://doi.org/10.1002/etc.4890

Frankish CJ, Milligan CD, Reid C (1996) The determinants of health: an overview of the Canadian Institute for Advanced Research (CIAR) population health model. Am J Health Promot 10(6):382–393

Furber S, Tranter D, Harris-Roxas B, Dews C, Gray E, Goldie A, Wallace C, Mayne D, Thackway S (2011) The use of health impact assessment to determine the potential impact of an australian urban development proposal on health and well-being. Urban Policy Res 29(2):125–139. https://doi.org/10.1080/08111146.2010.546004

Georgiadis N, Younes M (2021) Risk characterization in regulatory toxicology. In: Reichl FX, Schwenk M (eds) Regulatory toxicology. Springer, Cham. https://doi.org/10.1007/978-3-030-57499-4_74

Gooding K, Bertone MP, Loffreda G, Witter S (2022) How can we strengthen partnership and coordination for health system emergency preparedness and response? Findings from a synthesis of experience across countries facing shocks. BMC Health Serv Res 22(1):1441. https://doi.org/10.1186/s12913-022-08859-6

Gouin T, Becker RA, Collot AG, Davis JW, Howard B, Inawaka K, Lampi M, Ramon BS, Shi J, Hopp PW (2019) Toward the development and application of an environmental risk assessment

framework for microplastic. Environ Toxicol Chem 38(10):2087–2100. https://doi.org/10.1002/etc.4529
Grant CG, Ramos R, Davis JL, Green BL (2015) Community health needs assessment. Health News 34(2):147–156. https://doi.org/10.1097/hcm.0000000000000057
Green L, Ashton K, Azam S, Dyakova M, Clemens T, Bellis M (2021) Using health impact assessment (hia) to understand the wider health and well-being implications of policy decisions: the covid-19 'staying at home and social distancing policy' in wales. BMC Public Health 21(1). https://doi.org/10.1186/s12889-021-11480-7
Green L, Ashton K, Bellis M, Clemens T, Douglas M (2022) Predicted and observed impacts of covid-19 lockdowns: two health impact assessments in Scotland and Wales. Health Promot Int 37(6). https://doi.org/10.1093/heapro/daac134
Grimm SE, Pouwels X, Ramaekers BL, Wijnen B, Knies S, Grutters J, Joore MA (2019) Development and validation of the transparent uncertainty assessment (trust) tool for assessing uncertainties in health economic decision models. Pharmacoeconomics 38(2):205–216. https://doi.org/10.1007/s40273-019-00855-9
Grosssteiner I, Mienne A, Lucas L, L-Yvonnet P, Trenteseaux C, Fontaine K, Sarda X (2023) Cumulative risk assessment with pesticides in the framework of MRL setting. EFSA J 21. https://doi.org/10.2903/j.efsa.2023.e211009
Grutters JP, van Asselt MB, Chalkidou K, Joore MA (2014) Healthy decisions: towards uncertainty tolerance in healthcare policy. Pharmacoeconomics 33(1):1–4. https://doi.org/10.1007/s40273-014-0201-7
Haigh F, Baum F, Dannenberg AL, Harris MF, Harris-Roxas B, Keleher H, Kemp L, Morgan R, Chok HN, Spickett J, Harris E (2013) The effectiveness of health impact assessment in influencing decision-making in australia and new zealand 2005–2009. BMC Public Health 13(1). https://doi.org/10.1186/1471-2458-13-1188
Haigh F, Harris E, Harris-Roxas B, Baum F, Dannenberg AL, Harris MF, Keleher H, Kemp L, Morgan R, Ng Chok H, Spickett J (2015) What makes health impact assessments successful? factors contributing to effectiveness in australia and new zealand. BMC Public Health 15(1). https://doi.org/10.1186/s12889-015-2319-8
Haldane V, Chuah FL, Srivastava A, Singh SR, Koh GC, Seng CK, Legido-Quigley H (2019) Community participation in health services development, implementation, and evaluation: a systematic review of empowerment, health, community, and process outcomes. PLoS ONE 14(5):e0216112. https://doi.org/10.1371/journal.pone.0216112
Hammond D, Conlon K, Barzyk T, Chahine T, Zartarian V, Schultz B(2010) Assessment and application of national environmental databases and mapping tools at the local level to two community case studies. Risk Anal 31(3):475–487. https://doi.org/10.1111/j.1539-6924.2010.01527.x
Hancock T, Perkins F (1985) The mandala of health: a conceptual model and teaching tool. Health Educ 24(1):8–10
Hargrove WL, Juárez-Carillo PM, Korc M (2015) Healthy vinton: a health impact assessment focused on water and sanitation in a small rural town on the U.S.-Mexico border. Int J Environ Res Public Health 12(4):3864–3888. https://doi.org/10.3390/ijerph120403864
Harris PJ, Kemp LA, Sainsbury P (2012) The essential elements of health impact assessment and healthy public policy: a qualitative study of practitioner perspectives. BMJ Open 2(6):e001245. https://doi.org/10.1136/bmjopen-2012-001245
Harris-Roxas B, Haigh F, Travaglia J, Kemp L (2014) Evaluating the impact of equity focused health impact assessment on health service planning: three case studies. BMC Health Serv Res 14(1). https://doi.org/10.1186/1472-6963-14-371
Hassan BT, Jiawen G, Farzaneh H (2021) Air Pollution Health Risk Assessment (AP-HRA), principles and applications. Int J Environ Res Public Health 18(4):1935. https://doi.org/10.3390/ijerph18041935
Hayes AW (2005) The precautionary principle. Arh Hig Rada Toksikol 56(2):161–166

Health Canada (2001) The population health template: Key elements and actions that define a population health approach. Health Canada, Ottawa

Hoffmann S (2011) Overcoming barriers to integrating economic analysis into risk assessment†. Risk Anal 31(9):1345–1355. https://doi.org/10.1111/j.1539-6924.2011.01674.x

Hu B, Jia X, Hu J, Xu D, Xia F, Li Y (2017) Assessment of heavy metal pollution and health risks in the soil-plant-human system in the yangtze river delta, china. Int J Environ Res Public Health 14(9):1042. https://doi.org/10.3390/ijerph14091042

Huang SZ, Lu JY, Chau KY, Zeng HL (2020). Influence of ambidextrous learning on eco-innovation performance of startups: moderating effect of top management's environmental awareness. Front Psychol 11. https://doi.org/10.3389/fpsyg.2020.01976

Ikhajiagbe B, Ogwu MC (2020) Hazard quotient, microbial diversity and plant composition of spent crude oil polluted-soil. Beni-Suef Univ J Basic Appl Sci 9(26). https://doi.org/10.1186/s43088-020-00052-0

Ikhajiagbe B, Ogwu MC, Lawrence AE (2020) Single-tree influence of *Tectonia grandis* Linn. f. on plant distribution and soil characteristics in a planted forest. Bull Natl Res Centre 44: Article number 29. https://doi.org/10.1186/s42269-020-00285-0

Ikhajiagbe B, Ogwu MC, Lato NF (2021) Growth and yield response of soybean (*Glycine max* [L.] Merr.) accessions after exposure to cadmium. Vegetos Int J Plant Res Biotechnol. https://doi.org/10.1007/s42535-021-00189-y

Izah SC, Richard G, Stanley HO, Sawyer WE, Ogwu MC, Uwaeme OR (2024a) Prospects and application of multivariate and reliability analyses to one health risk assessments of toxic elements. Toxicol Environ Heal Sci 16(2):127–134

Izah SC, Amabie T, Aigberua AO (2024b) Technological advancements for sustainable clean air. In: Ogwu MC, Izah SC (eds) Sustainable Strategies for Air Pollution Mitigation: Development, Economics and Technologies. Part of the Bok Series: The Handbook in Environmental Chemsitry, Springer, Berlin, Heidelberg. https://link.springer.com/chapter/10.1007/698_2024_1120

Izah SC, Richard G, Stanley HO, Sawyer WE, Ogwu MC, Uwaeme OR (2023) Integrating the one health approach and statistical analysis for sustainable aquatic ecosystem management and trace metal contamination mitigation. ES Food Agrofor 14:1012. https://doi.org/10.30919/esfaf1012

Izah SC, Aigberua AO, Richard G (2022) Concentration, source, and health risk of trace metals in some liquid herbal medicine sold in Nigeria. Biol Trace Elem Res 200:3009–3302

Izah SC, Richard G, Aigberua AO, Ekakitie O (2021a) Variations in reference values utilized for the evaluation of complex pollution indices of potentially toxic elements: a critical review. Environ Chall 5:100322. https://doi.org/10.1016/j.envc.2021.100322

Izah SC, Uzoekwe SA, Aigberua AO (2021b) Source, geochemical spreading and risks of trace metals in particulate matter 2.5 within a gas flaring area in Bayelsa State, Nigeria. Adv Environ Technol 7(2):101–118

Jabot F, Tremblay E, Rivadeneyra A, Diallo TA, Lapointe G (2020) A comparative analysis of health impact assessment implementation models in the regions of montérégie (québec, canada) and nouvelle-aquitaine (france). Int J Environ Res Public Health 17(18):6558. https://doi.org/10.3390/ijerph17186558

Jerrett M, Gale S, Kontgis C (2010) Spatial modeling in environmental and public health research. Int J Environ Res Public Health 7(4):1302–1329. https://doi.org/10.3390/ijerph7041302

Jha A, Lin L, Short S, Argentini G, Gamhewage G, Savoia E (2018) Integrating emergency risk communication (ERC) into the public health system response: systematic review of literature to aid formulation of the 2017 who guideline for erc policy and practice. PLoS ONE 13(10):e0205555. https://doi.org/10.1371/journal.pone.0205555

Johnson MM, Garcia-Menendez F (2022) Uncertainty in health impact assessments of smoke from a wildfire event. Geohealth 6(1). https://doi.org/10.1029/2021gh000526

Kang E, Park HJ, Kim JE (2011) Health impact assessment as a strategy for intersectoral collaboration. J Prev Med Public Health 44(5):201–209. https://doi.org/10.3961/jpmph.2011.44.5.201

Katz DL, Ali A (2009) Preventive medicine, integrative medicine and the health of the public. Am J Prev Med 37(6):473–477
Kaya GK, Ward JR, Clarkson PJ (2018) A framework to support risk assessment in hospitals. Int J Qual Health Care 31(5):393–401. https://doi.org/10.1093/intqhc/mzy194
Khan S, Haleem A, Khan MI (2023) A risk assessment framework using neutrosophic theory for the halal supply chain under an uncertain environment. Arab Gulf J Sci Res. https://doi.org/10.1108/agjsr-12-2022-0312
Khan ZF, Alotaibi SR (2020) Applications of artificial intelligence and big data analytics in m-Health: a healthcare system perspective. J Healthc Eng 2020:8894694. https://doi.org/10.1155/2020/8894694
Krewski D, Saunders-Hastings P, Baan RA, Barton-Maclaren TS, Browne P, Chiu WA, Gwinn M, Hartung T, Kraft AD, Lam J et al (2022) Development of an evidence-based risk assessment framework. Altex 39(4):667. https://doi.org/10.14573/altex.2004041
Krewski D, Westphal M, Andersen ME, Paoli GM, Chiu WA, Al-Zoughool M, Croteau MC, Burgoon LD, Cote I (2014) A framework for the next generation of risk science. Environ Health Perspect 122(8):796–805. https://doi.org/10.1289/ehp.1307260
Kustono D, Puspitasari P, Al Irsyad M, Nursabrina A, Adesta EYT (2021) Hazard identification and risk assessment (hira) analysis of nanotechnology laboratory in universities in indonesia. Int J Mech Eng Technol Appl 2(1):41. https://doi.org/10.21776/mechta.2021.002.01.6
Lalonde M (1974) A new perspective on the health of Canadians: a working document. Government of Canada, Ottawa
Lebret E (2016) Integrated environmental health impact assessment for risk governance purposes; across what do we integrate? Int J Environ Res Public Health 13(1):71. https://doi.org/10.3390/ijerph13010071
Lebret E (2015) Integrated environmental health impact assessment for risk governance purposes; across what do we integrate? Int J Environ Res Public Health 13(1):71. https://doi.org/10.3390/ijerph13010071
Levy JI (2021) Accounting for health risk inequality in regulatory impact analysis: barriers and opportunities. Risk Anal: Offic Public Soc Risk Anal 41(4):610–618. https://doi.org/10.1111/risa.13714
Lewallen S, Courtright P (1998) Epidemiology in practice: case-control studies. Community Eye Health 11(28):57–58
Li L, Wang C (2023) Microbiological risk assessment and its use in food industry. Int J Biol Life Sci 2(1):14–16. https://doi.org/10.54097/ijbls.v2i1.5163
Li L, Xin X, Zhao J, Yang A, Wu S, Zhang H, Yu S (2023) Remote sensing monitoring and assessment of global vegetation status and changes during 2016–2020. Sensors (Basel, Switzerland) 23(20):8452. https://doi.org/10.3390/s23208452
Liu X (2023) Current status and influencing factors of policy identification in health impact assessment: a case study of Zhejiang province. Health Res Policy Syst 21(1). https://doi.org/10.1186/s12961-023-01064-9
Loureiro A, Partidário MDR, Santana P (2022) Strategic assessment of neighbourhood environmental impacts on mental health in the Lisbon region (Portugal): a strategic focus and assessment framework at the local level. Sustainability 14(3):1547. https://doi.org/10.3390/su14031547
Lund K, Dinse K, Callewaert J, Scavia D (2011) Benefits of using integrated assessment to address sustainability challenges. J Environ Stud Sci 1(4):289–295. https://doi.org/10.1007/s13412-011-0047-7
Luo W, Deng Z, Zhong S, Deng M (2022) Trends, issues and future directions of urban health impact assessment research: a systematic review and bibliometric analysis. Int J Environ Res Public Health 19(10):5957. https://doi.org/10.3390/ijerph19105957
Margozzini P, Tolonen H, Bernabe-Ortiz A, Cuschieri S, Donfrancesco C, Palmieri L, Sanchez Romero LM, Mindell JS, Oyebode O (2023) National Health Examination Surveys: an essential piece of the health planning puzzle. medRxiv: the preprint server for health sciences, 2023.07.11.23292221. https://doi.org/10.1101/2023.07.11.23292221

Mathias KR, Harris-Roxas B (2009) Process and impact evaluation of the greater christchurch urban development strategy health impact assessment. BMC Public Health 9(1). https://doi.org/10.1186/1471-2458-9-97

Maurer M, Mangrum R, Hilliard-Boone T, Amolegbe A, Carman KL, Forsythe L, Mosbacher R, Lesch JK, Woodward K (2022) Understanding the influence and impact of stakeholder engagement in patient-centered outcomes research: a qualitative study. J Gen Intern Med 37(Suppl 1):6–13. https://doi.org/10.1007/s11606-021-07104-w

Meckawy R, Stuckler D, Mehta A, Al-Ahdal T, Doebbeling BN (2022) Effectiveness of early warning systems in the detection of infectious diseases outbreaks: a systematic review. BMC Public Health 22(1):2216. https://doi.org/10.1186/s12889-022-14625-4

Mesa-Frias M, Chalabi Z, Vanni T, Foss AM (2013) Uncertainty in environmental health impact assessment: quantitative methods and perspectives. Int J Environ Health Res 23(1):16–30. https://doi.org/10.1080/09603123.2012.678002

Mindell JS, Boltong A, Forde I (2008) A review of health impact assessment frameworks. Public Health 122(11):1177–1187. https://doi.org/10.1016/j.puhe.2008.03.014

Morin CW, Semenza JC, Trtanj JM, Glass GE, Boyer C, Ebi KL (2018) Unexplored opportunities: use of climate- and weather-driven early warning systems to reduce the burden of infectious diseases. Curr Environ Health Reports 5(4):430–438. https://doi.org/10.1007/s40572-018-0221-0

Morrison J, Pons-Vigués M, Bécares L, Burström B, Gandarillas A, Domínguez-Berjón F, Diez È, Costa G, Ruiz M, Pikhart H et al (2014) Health inequalities in european cities: perceptions and beliefs among local policymakers. BMJ Open 4(5):e004454. https://doi.org/10.1136/bmjopen-2013-004454

Morteruel M, Bacigalupe A, Aldasoro E, Larrañaga I, Serrano E (2020) Health impact assessments in Spain: have they been effective? Int J Environ Res Public Health 17(8):2959. https://doi.org/10.3390/ijerph17082959

Muflih M, Suwarsi S, Asmarani FL, Erwanto R, Amigo TAE(2023) Enhancing non-communicable disease management in elderly families: a community outreach initiative. https://doi.org/10.21203/rs.3.rs-3202285/v1

Munns WR Jr, Poulsen V, Gala WR, Marshall SJ, Rea AW, Sorensen MT, von Stackelberg K (2016) Ecosystem services in risk assessment and management. Integr Environ Assess Manag 13(1):62–73. https://doi.org/10.1002/ieam.1835

Munns WR Jr, Rea AW, Suter GW, Martin L, Blake-Hedges L, Crk T, Davis C, Ferreira G, Jordan S, Mahoney M, Barron MG (2015) Ecosystem services as assessment endpoints for ecological risk assessment. Integr Environ Assess Manag 12(3):522–528. https://doi.org/10.1002/ieam.1707

Nastos PT, Dalezios NR, Faraslis IN, Mitrakopoulos K, Blanta A, Spiliotopoulos M, Sakellariou S, Sidiropoulos P, Tarquis AM (2021) Risk management framework of environmental hazards and extremes in Mediterranean ecosystems. Nat Hazard 21(6):1935–1954. https://doi.org/10.5194/nhess-2020-155

Neo EX, Hasikin K, Mokhtar MI, Lai KW, Azizan MM, Razak SA, Hizaddin HF (2022) Towards integrated air pollution monitoring and health impact assessment using federated learning: a systematic review. Front Public Health. 10. https://doi.org/10.3389/fpubh.2022.851553

Ngwenya S, Mashau NS, Mhlongo SE, Traoré AN (2023) A systematic review of the risk management frameworks for potentially toxic chemical elements. Toxicol Ind Health 39(12):679–686. https://doi.org/10.1177/07482337231209092

Nuckols JR, Ward MH, Jarup L (2004) Using geographic information systems for exposure assessment in environmental epidemiology studies. Environ Health Perspect 112(9):1007–1015. https://doi.org/10.1289/ehp.6738

Ogamba EN, Charles EE, Izah SC (2021) Distributions, pollution evaluation and health risk of selected heavy metal in surface water of Taylor creek, Bayelsa State, Nigeria. Toxicol Environ Health Sci 13(2):109–121. https://doi.org/10.1007/s13530-020-00076-0

Ortiz P, Torres-Sánchez A, López-Moreno A, Cerk K, Ruiz-Moreno Á, Monteoliva-Sánchez M, Ampatzoglou A, Aguilera M, Gruszecka-Kosowska A (2022) Impact of cumulative environmental and dietary xenobiotics on human microbiota: risk assessment for one health. J Xenobiot 12(1):56–63. https://doi.org/10.3390/jox12010006

Osawaru ME, Ogwu MC, Braimah L (2013a) Growth responses of two cultivated Okra species (*Abelmoschus caillei* (A. Chev.) Stevels and *Abelmoschus esculentus* (Linn.) Moench) in crude oil contaminated soil. Nig J Basic Appl Sci 21(3):215–226

Osawaru ME, Ogwu MC, Chime AO (2013b) Assessment of growth performance of two Okra species (*Abelmoschus esculentus* [L.] Moench and *Abelmoschus caillei* [A. Chev.] Stevels) exposed to crude oil contaminated soil. Nig J Biotechnol 26:11–20

Pauly B, Revai T, Marcellus L, Martin W, Easton K, MacDonald M (2021) The health equity curse": ethical tensions in promoting health equity. BMC Public Health 21(1):1567. https://doi.org/10.1186/s12889-021-11594-y

Paustenbach D, Galbraith D (2006) Biomonitoring and biomarkers: exposure assessment will never be the same. Environ Health Perspect 114(8):1143–1149. https://doi.org/10.1289/ehp.8755

Public Health Agency of Canada (2013) What makes Canadians healthy or unhealthy? http://www.phac-aspc.gc.ca/ph-sp/determinants/determinants-eng.php

Rajotte BR, Ross CL, Ekechi CO, Cadet VN (2011) Health in all policies: addressing the legal and policy foundations of health impact assessment. J Law Med Ethics 39(S1):27–29. https://doi.org/10.1111/j.1748-720x.2011.00560.x

Rakow T, Heard CL, Newell BR (2015) Meeting three challenges in risk communication. Policy Insights Behav Brain Sci 2(1):147–156. https://doi.org/10.1177/2372732215601442

Raphael D (2004) Social determinants of health: Canadian perspectives. Canadian Scholars' Press

Raymond C, Matthews T, Horton RM (2020) The emergence of heat and humidity too severe for human tolerance. Sci Adv 6(19). https://doi.org/10.1126/sciadv.aaw1838

Reuben A, Manczak EM, Cabrera LY, Alegria M, Bucher ML, Freeman EC, Miller GW, Solomon GM, Perry MJ (2022) The interplay of environmental exposures and mental health: setting an agenda. Environ Health Perspect 130(2):25001. https://doi.org/10.1289/EHP9889

Rico A, Van den Brink PJ, Gylstra R, Focks A, Brock TC (2016) Developing ecological scenarios for the prospective aquatic risk assessment of pesticides. Integr Environ Assess Manag 12(3):510–521. https://doi.org/10.1002/ieam.1718

Rigaud M, Buekers J, Bessems J, Basagaña X, Mathy S, Nieuwenhuijsen M, Slama R (2024) The methodology of quantitative risk assessment studies. Environ Health: A Global Access Sci Source 23(1):13. https://doi.org/10.1186/s12940-023-01039-x

Rod NH, Broadbent A, Rod MH, Russo F, Arah OA, Stronks K (2023) Complexity in epidemiology and public health. Addressing complex health problems through a mix of epidemiologic methods and data. Epidemiology (Cambridge, Mass.) 34(4):505–514. https://doi.org/10.1097/EDE.0000000000001612

Rogerson B, Lindberg R, Baum F, Dora C, Haigh F, Simoncelli AM, Parry Williams L, Peralta G, Pollack Porter KM, Solar O (2020) Recent advances in health impact assessment and health in all policies implementation: lessons from an international convening in barcelona. Int J Environ Res Public Health 17(21):7714. https://doi.org/10.3390/ijerph17217714

Romero-Franco M, Godwin HA, Bilal M, Cohen Y (2017) Needs and challenges for assessing the environmental impacts of engineered nanomaterials (enms). Beilstein J Nanotechnol 8:989–1014. https://doi.org/10.3762/bjnano.8.101

Ruiz GL, García CD, Covarrubias RR (2017) Environmental assessment for health tourism in Baños de Agua Santa, Ecuador. J Environ Protect 8(7):799–809. https://doi.org/10.4236/jep.2017.87050

Saadeh R, Khader Y, Malkawi M, Allouh MZ (2022) Communicating the risks of air pollution to the public: a perspective from Jordan and Lebanon. Environ Health Insights 16:11786302221127852. https://doi.org/10.1177/11786302221127851

Sadatsafavi M, Yoon Lee T, Gustafson P (2022) Uncertainty and the value of information in risk prediction modeling. Med Decis Making: Int J Soc Med Decis Making 42(5):661–671. https://doi.org/10.1177/0272989X221078789

Saud YE, Israni K, Goddard J (2013) Bow-tie diagrams in downstream hazard identification and risk assessment. Process Saf Prog 33(1):26–35. https://doi.org/10.1002/prs.11576

Scammell MK (2010) Qualitative environmental health research: an analysis of the literature, 1991–2008. Environ Health Perspect 118(8):1146–1154. https://doi.org/10.1289/ehp.0901762

Scheierling SM, Bartone CR, Mara DD, Drechsel P (2011) Towards an agenda for improving wastewater use in agriculture. Water Int 36(4):420–440. https://doi.org/10.1080/02508060.2011.594527

Scovronick N, Ferranna M, Dennig F, Budolfson M (2020) Valuing health impacts in climate policy: ethical issues and economic challenges. Health Affairs (Project Hope) 39(12):2105–2112. https://doi.org/10.1377/hlthaff.2020.01117

Serrano E, Larrañaga I, Morteruel M, Baixas de Ros MD, Basterrechea M, Martinez D, Aldasoro E, Bacigalupe A (2016) Urban regeneration as population health intervention: a health impact assessment in the bay of pasaia (Spain). Int J Equity Health 15(1). https://doi.org/10.1186/s12939-016-0424-7

Sexton K (2012) Cumulative risk assessment: an overview of methodological approaches for evaluating combined health effects from exposure to multiple environmental stressors. Int J Environ Res Public Health 9(2):370–390. https://doi.org/10.3390/ijerph9020370

Shojaei P, Karimlou M, Mohammadi F, Malekafzali H (2016) Health impact assessment of urban development project. Global J Health Sci 8(9):224. https://doi.org/10.5539/gjhs.v8n9p224

Simsekler ME, Ward JR, Clarkson PJ (2018) Design for patient safety: a systems-based risk identification framework. Ergonomics 61(8):1046–1064. https://doi.org/10.1080/00140139.2018.1437224

Sjarifudin D, Kurnia H, Nuryono A, Tambunan EBM (2023) Hazard identification risk assessment and determining control (HIRADC) method for shoe cutting dies production. Jurnal Sistem Teknik Industri 25(2):322–333. https://doi.org/10.32734/jsti.v25i2.12186

Skripnik I, Savelev D, Kaverzneva T, Rumyantseva N (2023) Implementation of a risk-based ohs management system at imc mining company. E3s Web Conf 376:05031. https://doi.org/10.1051/e3sconf/202337605031

Song JW, Chung KC (2010) Observational studies: cohort and case-control studies. Plast Reconstr Surg 126(6):2234–2242. https://doi.org/10.1097/PRS.0b013e3181f44abc

Spencer B, Lawler J, Lowe C, Thompson L, Hinckley T, Kim SH, Bolton S, Meschke S, Olden JD, Voss J (2016) Case studies in co-benefits approaches to climate change mitigation and adaptation. J Environ Plann Manag 60(4):647–667. https://doi.org/10.1080/09640568.2016.1168287

Stassen KR, Smolders R, Leroy P (2013) Sensitizing events as trigger for discursive renewal and institutional change in flanders' environmental health approach, 1970s–1990s. Environ Health 12(1). https://doi.org/10.1186/1476-069x-12-46

Sulistyowatie SL, Pahlevi RW (2019) Innovation of business and environmental performance in kebon indah batik association klaten: the impact of transfer of knowledge and quality management. Jurnal Manajemen Bisnis 10(2). https://doi.org/10.18196/mb.10178

Tang L, Wang L, Li Q, Zhao J (2018) A framework designation for the assessment of urban ecological risks. Int J Sust Dev World 25(5):387–395. https://doi.org/10.1080/13504509.2018.1434570

Theiss SK, Burke RM, Cory JL, Fairley TL (2016) Getting beyond impressions: an evaluation of engagement with breast cancer-related facebook content. Mhealth 2:41–41. https://doi.org/10.21037/mhealth.2016.10.02

Thepaksorn P, Thongjerm S, Incharoen S, Siriwong W, Harada K, Koizumi A (2017) Job safety analysis and hazard identification for work accident prevention in para rubber wood sawmills in southern thailand. J Occup Health 59(6):542–551. https://doi.org/10.1539/joh.16-0204-cs

Uzoekwe SA, Izah SC, Aigberua AO (2021) Environmental and human health risk of heavy metals in atmospheric particulate matter (PM10) around gas flaring vicinity in Bayelsa State, Nigeria. Toxicol Environ Health Sci 13(4):323–335. https://doi.org/10.1007/s13530-021-00085-7

Vandenberg LN, Rayasam SDG, Axelrad DA, Bennett DH, Brown P, Carignan CC, Chartres N, Diamond ML, Joglekar R, Shamasunder B, Shrader-Frechette K, Subra WA, Zarker K, Woodruff TJ (2023) Addressing systemic problems with exposure assessments to protect the public's health. Environ Health: A Global Access Sci Source 21(Suppl 1):121. https://doi.org/10.1186/s12940-022-00917-0

Vardoulakis S, Dear K, Wilkinson P (2016) Challenges and opportunities for urban environmental health and sustainability: the healthy-polis initiative. Environ Health 15(S1). https://doi.org/10.1186/s12940-016-0096-1

Veras-Estévez BA, Chapman HJ (2023) Incorporating one health in health professions curriculum in the dominican republic. Innovare Revista De Ciencia Y Tecnología 12(3):122–123. https://doi.org/10.5377/innovare.v12i3.17171

Wang X, Cheng Z (2020) Cross-sectional studies: strengths, weaknesses, and recommendations. Chest 158(1S):S65–S71. https://doi.org/10.1016/j.chest.2020.03.012

White RH, Fox MA, Cooper GS, Bateson TF, Burke TA, Samet JM (2013) Workshop report: evaluation of epidemiological data consistency for application in regulatory risk assessment. Open Epidemiol J 6(1)

World Health Organization (2002) The world health report 2002: reducing risks, promoting healthy life. World Health Organization

Yang J, Qing-lan H, Zhou J, Yuan C (2015) The influence of environmental management practices and supply chain integration on technological innovation performance—evidence from china's manufacturing industry. Sustainability 7(11):15342–15361. https://doi.org/10.3390/su71115342

Yousefi S, Pourghasemi HR, Emami SN, Pouyan S, Eskandari S, Tiefenbacher JP (2020) A machine learning framework for multi-hazards modeling and mapping in a mountainous area. Sci Reports 10(1). https://doi.org/10.1038/s41598-020-69233-2

Index

S. Chibueze Izah and M. Chidozie Ogwu (eds.), *Innovative Approaches in Environmental Health Management*, Environmental Science and Engineering,
https://doi.org/10.1007/978-3-031-81966-7

MIX
Papier aus verantwortungsvollen Quellen
Paper from responsible sources
FSC® C105338

If you have any concerns about our products,
you can contact us on
ProductSafety@springernature.com

In case Publisher is established outside the EU,
the EU authorized representative is:
Springer Nature Customer Service Center GmbH
Europaplatz 3, 69115 Heidelberg, Germany

Printed by Libri Plureos GmbH
in Hamburg, Germany